MICROCONTROLLER PROGRAMMING

AN INTRODUCTION

MICROCONTROLLER PROGRAMMING

AN INTRODUCTION

SYED R. RIZVI

CRC Press
Taylor & Francis Group
Boca Raton   London   New York

CRC Press is an imprint of the
Taylor & Francis Group, an **informa** business

CRC Press
Taylor & Francis Group
6000 Broken Sound Parkway NW, Suite 300
Boca Raton, FL 33487-2742

© 2012 by Taylor & Francis Group, LLC
CRC Press is an imprint of Taylor & Francis Group, an Informa business

No claim to original U.S. Government works

Printed in the United States of America on acid-free paper
Version Date: 2011912

International Standard Book Number: 978-1-4398-5077-0 (Hardback)

Library of Congress Cataloging-in-Publication Data

Rizvi, Syed R.
   Microcontroller programming : an introduction / author, Syed R. Rizvi.
      p. cm.
   Includes bibliographical references and index.
   ISBN 978-1-4398-5077-0 (hardcover : alk. paper)
      1. Microcontrollers--Programming. 2. Programmable controllers. I. Title.

TJ223.P76R59 2012
629.8'95--dc23   2011035866

Visit the Taylor & Francis Web site at
http://www.taylorandfrancis.com

and the CRC Press Web site at
http://www.crcpress.com

Dedication

To my beloved wife Susan, for providing encouragement and understanding throughout the preparation of this text; and to my daughters, Zahra and Sophia.

Contents

Preface

A few decades ago there were only mainframes, each shared by many people. In the last decade or so, we witnessed the personal computing era, that is, person and machine interaction through desktops. With the popularity of mobile, handheld, and smart devices, we have already entered into the third wave of computing, which can be called Ubiquitous Computing. We rely on these equipments and devices in our lives to such an extent that one cannot imagine the world without these multifunctional gadgets.

Microcontrollers are inevitably used in equipments and devices ranging from smart cards and GPSs to power plants and space shuttles. A microcontroller is a complete computer control system on a single chip. For example, Motorola's 68HC11 family of microcontrollers contain a central processing unit, three kinds of memory (ROM, RAM, and EEPROM), analog-to-digital converter, both synchronous and asynchronous serial interfaces, an on-board clock and pulse accumulator subsystem, plus a range of input and output ports.

With the rapidly advancing and affordable technology, and growing demand for one device to perform multiple tasks (e.g., mobile phone acting as a phone but also as a games console, music player, camera, GPS, etc.), the programming domain for these creative devices is faced with several daunting challenges. In order to meet these challenges microcontroller programmers must have systematic familiarity of application development, systems programming, and I/O operation, as well as system timing and memory management. Best problem-solving and programming techniques are essential in order to achieve microcontroller product realization.

The present book is based on my experience, extending over a period of about 15 years of teaching and working in the field of electrical engineering and computer science. During this period, the unavailability of a book that contained practical microcontroller programming projects emphasizing efficient and reliable programming through problem decomposition and traceability techniques made me conscious of the need for one on the subject from which the student could solve a few problems and could decide from himself or herself whether he or she proceeded rightly or not. Additionally, one of the most difficult hurdles to overcome that is faced by students taking an introductory course in microcontroller programming is getting started on interfacing hardware set-up and configuration. It is not uncommon to observe that getting the first assembly program to run is a painstaking experience. All of these problems could be solved if in a book the theory is supplemented by examples designed to illustrate how concepts may be applied, and help readers develop hands-on experience in both software and hardware. This book has been written to meet the genuine demand of the students. It provides basic concepts reinforced by plentiful illustrations, examples, exercises, and applications.

Unique Learning Tools

With a focus on the capabilities provided by the Motorola 68HC11 microcontroller, this book provides a hands-on introduction to the lower- and higher-level languages, tools, and techniques needed to build embedded applications. The information in this book is structured to help prepare you for the real-world challenges you may face when building embedded applications.

Special features included in this textbook to enhance the learning and comprehension process are:

1. Performance-based objectives at the beginning of each chapter outline the goals to be achieved. Each chapter begins with a set of learning objectives written in the form of specific actions that the reader should be able to perform.
2. Each chapter begins with an easy-to-understand introduction that helps put the material in that chapter in context.
3. Several annotations are given in the page margin throughout the text to highlight particular points that were made on the page.
4. Over 150 examples are worked out step by step to clarify problems that are normally stumbling blocks.
5. Over 200 detailed illustrations give readers visual explanations and serve as the basis for all discussions.
6. Timing waveforms are used to illustrate the timing analysis techniques used in industry.
7. Several photographs illustrate specific devices and circuits discussed in the text.
8. Full-length projects are covered in the final five chapters, providing hands-on embedded application development experience. These projects include requirements analysis, engineering design, program code, testing and execution, and discussion on results.
9. Section review quizzes with answers provide readers periodic self-evaluation.
10. A summary at the end of each chapter provides a quick review of the key points covered.
11. Over 450 end-of-chapter quizzes, questions, and problems help reinforce concepts. A complete range of problems, from straightforward to very challenging, are included.

12. A glossary at the end of each chapter summarizes essential terminologies introduced in the chapter. The reason this section is included at the end of the text is to make it convenient to loop-up the meaning of a new word or term.

13. An extensive appendix supplement of key topics includes a comprehensive glossary.

14. Source code CD-ROM accompanying this text provides assembly and C programs that are referenced in the text.

Level and Audience

This book is ideal for readers with little programming and/or hardware experience who want to master the basics of requirements analysis, engineering design, troubleshooting, and programming a microcontroller. Therefore this text can be used as a first course in microcontrollers at either the associate or bachelor's level of an engineering major. It can also be used by more experienced engineers who must study the specifics of the Motorola's 68HC11 microcontroller. The expected audience would include students studying courses such as microcontrollers, embedded design, embedded programming, engineering design, instrumentation for process control, electro-mechanical control systems, and measurement and control systems, as well as manufacturing, mechanical, mechatronics, production, and instrumentation engineering programs that use embedded controllers as part of their curricula. The book can also help the interested hacker or hobbyist in getting familiar with the Motorola's 68HC11 microcontroller.

The early chapters of this text cover digital theory and devices. In later chapters this information is applied to microcontrollers and, finally, you will learn about the construction and operation of embedded systems. The later chapters discuss advanced microcontroller topics, such as handshaking, interrupts, analog to digital conversion, and higher-level programming. When finished with this book, you will have a deep and solid understanding of how a microcontroller can be programmed in order to build robust embedded applications. With that kind of foundation you will find it relatively easy to branch out to other microcontroller systems.

Part 1	Basic Electronics	Chapter 1 and Chapter 2
Part 2	Microcontroller Hardware and Interfacing	Chapter 3
Part 3	Fundamental Microcontroller Programming	Chapter 4 and Chapter 7
Part 4	Advanced Microcontroller Programming	Chapter 8 to Chapter 12

Part 1 is for readers with little or no knowledge of basic electronics. Part 2 is well suited for the readers who never had hands-on experience with microcontroller hardware and circuit building. Part 3 is the core of this book and is a must for a reader who is new to assembly programming but will serve as a refresher for an experienced programmer. More experienced engineers and programmers who want to study specifics of the Motorola's 68HC11 microcontroller can start directly from Part 4. Thus, this book can also be used for a graduate-level microcontroller course in electrical or computer engineering. In general, this text is well suited for engineers, programmers, and science and engineering students who want to sharpen their skills in microcontroller application development.

An attempt has been made to present the problem decomposition approach in an easy-to-understand form, especially for those who have little or no aptitude for computer programming. It is hoped that the present book will meet the needs of everyone, student or professional engineer, who has a system to design that needs an embedded controller.

To the Instructors

The primary objective for an instructor in a course dealing with microcontrollers should be to have students learn the practical skills required to design and troubleshoot actual microcontroller-based applications that they will see on the job. This book makes a strong effort to use requirements analysis and design development techniques when creating a microcontroller-based application. The text covers the basic fundamentals of microcontroller design so that the students, knowing the basic building blocks, can teach themselves the newest technology when faced with it on the job. Also, material is provided to help reduce the anxiety that students feel when they first start a new job and are faced with schematics of large systems to analyze. When building embedded applications, instructors frequently are so focused on teaching how to use the instruction set that they forget to emphasize the general concepts that are critical for the students to understand and use the hardware and software efficiently.

This book can be used for a one- or two-semester course (depending on the number of class periods per week) in microcontroller technology and is intended for students of technology and engineering programs. Core fundamentals are presented without being intermingled with advanced or peripheral topics. Several theoretical and practical examples are provided in the text to develop student's analytical and programming skills. The examples give the students experience working with real-world large-scale schematics like those that they will see on the job. Several annotations are given in the page margin throughout the text. These are intended to highlight particular points that were made on the page. Six different type of side annotations are Common Practice, Helpful Hints, Common Misconception, Best Practice, Team Discussion, and Self-Learning. The text content is sequenced to support a laboratory with a class lecture. The full-length projects covered in the last five chapters are provided as a guide in analyzing, designing, implementing, testing, and debugging microcontroller-based projects. Most of the projects ensure that the circuit is made up of actual integrated circuits and the specifications are taken from actual manufacturer's data sheets. Please refer to the Unique Learning Tools section to learn more about the salient features of this text.

I would like to share with you some teaching strategies that I would recommend for using this text.

1. Fine tune your syllabus by referring to the performance-based objectives at the beginning of each chapter.

2. Lecture using the PowerPoint® slides provided in the supplements package to instructors adopting the text. These slides contain significant figures of the text.

3. Encourage team discussions. Take advantage of the Team Discussion margin annotation as early as possible in your course as a means to develop cooperative learning by encouraging student interaction.

4. Take the 10-minute quiz each week. Separate chapterwise quiz files are available in the supplements package to the instructors adopting the text. The quiz must be closed book, but the students should be allowed to carry a formula sheet for the quiz or test.

5. Base your homework or assignment on the questions and problems at the end of the chapter. No solution or answers to the end of the chapter quizzes, questions, and problems are provided in this text. The instructor's solution manual contains all the answers and solutions.

6. Explain in detail similar sample solutions before you start hands-on laboratory work. Enough examples of programs and projects are given in the second half of the text.

7. If the laboratory work is completed before time, ask your students to work on appropriate Self-Learning margin annotation. This would improve their research and writing abilities.

Extensive Supplements Package (Instructor Resources)

An extensive package of supplementary material is available to aid in the teaching and learning process.

- Instructor's Solution Manual: Includes worked-out solutions to the entire end-of-chapter exercises.
- CD-ROM: This CD-ROM includes a complete set of unique presentation slides. These innovative slides are coordinated with each chapter and are an excellent tool to supplement classroom presentations. The CD-ROM also contains end-of-chapter quizzes in separate files so that they can be used to test the students in a 10-minute-long quiz session every week.
- Online laboratory exercises.
- Online assembly program bank.
- Online video tutorials to getting started on the hardware interfacing.

NOTE: Instructors can contact CRC Press for a free copy of the Instructor's Solution Manual.

To the Students

You are beginning your study of microcontrollers at a good time. Technological advances made in the past 30 years have provided us with integrated circuits that can perform complex tasks with a minimum amount of abstract theory and complicated circuitry. The study of the microcontroller can provide you job skills that open doors to a multitude of highly paid jobs related to computer and microprocessor-based systems. A strong grounding in the fundamentals of microcontroller programming will prepare you for the highly skilled jobs of the future. I have featured the HC11 because it is an ideal subject to study for a fundamental microcontroller programming textbook. Once you understand the 68HC11, you pass a major hurdle and things begin to make sense in the microcontroller world. The types of instructions and programming techniques that are used with the 8-bit HC11 are similar to those used by all 8-bit microcontrollers. Thus, once you become proficient at programming the fairly sophisticated HC11, it should be relatively easy to learn how to program other 8/16/32-bit microcontrollers.

A strong effort was made to make the text easy to read and understand so that motivated students can teach themselves topics that require extra work without the constant attention of the instructor. There are ample illustrations, examples, and review questions to help students reach a point where they can reason out the end-of-chapter problems on their own. This book is specially written as a learning tool, not just as a reference. The concept and theory of each topic is presented first. Then an explanation of its operation is given. This is followed by several worked-out examples and, in some cases, full-length projects. The end of chapter quizzes, questions, and problems will force you to dig back into the reading to see that you have met the learning objectives given at the beginning of the chapter. Please refer to the Unique Learning Tools section to learn more about the salient features of this text.

The problems at the end of each chapter will provide you with practicing analytical reasoning and sharpening your programming skills. The procedures for their solutions in the majority of the cases are already given to you in the examples. One good way to prepare for homework or assignment problems and tests is to cover up the solutions to the examples and try to solve them out yourself. If you get stuck, you've got the answer and an explanation for the answer right there.

I also suggest that you take advantage of the Web site www. microcontrollerguide.com. An online assembly program bank and video tutorials demonstrate getting started on the hardware interfacing is provided for this text. The more hands-on practice you get, the easier the course will be. I wish you the best of luck in your studies and future employment.

Accompanying Student Resources

CD-ROM. Packaged with each text, this CD includes relevant assembly and C source code to practice programming skills.

Acknowledgments

This book wouldn't have been possible except for the patience and enthusiasm of Nora Konopka for trying out new ideas. What I appreciate most of all was the time she devoted to me, listening to my ideas and helping me to come up with the book concept presented here.

I would like to take a moment to recognize all the people who contributed to making this book a success. I am grateful to my undergraduate instructors Nirmal-Kumar C. Nair (University of Auckland, New Zealand) and Sukumar Brahma (New Mexico State University), who introduced me to the world of microcontrollers about two decades ago. I would like to thank my graduate advisor, Stephan Olariu (Old Dominion University) who taught me the meaning of the famous quote, "You cannot plough a field by turning it over in your mind." One of my role models is Jalaiah Unnam (President, Analytical Services & Materials, Inc.). He has been a constant source of inspiration to me over the last decade. I would also like to thank Karl Wiedemann (Chief Scientist, Analytical Services & Materials, Inc.). He has been my guru from the time I started my real-world application development for NASA Langley Research Center. Among several other things, he taught me how to efficiently use the concept of abstraction in problem solving. Even today, I look forward to our animated discussions on anything and everything in computer science. I would like to thank Gopalan Balasubramanian (Research Scientist, Analytical Services & Materials, Inc.) for providing valuable suggestions that made many improvements throughout.

I would like to thank Rasha Morsi (Norfolk State University), Cristina M. Pinotti (University of Perugia, Italy), Gongjun Yan (Indiana University–Kokomo) and Zainab Zaidi (NICTA, Australia) for the technical guidance and assistance they always gave willingly and promptly. Without their help this book would not exist. I thank Andrew Velkey (Sigma Xi, The Scientific Research Society) for letting me judge many undergraduate and graduate research works, which helped me in forming a good understanding of what areas in embedded systems are difficult for students to apply in the real world.

Reviewers of my manuscript drafts provided a variety of viewpoints on what I should include and what level of presentation I should use. Although the final decision may not reflect their views—which often differed considerably from one another—each reviewer forced me to reflect on every page of my manuscript. I thank Fanny Popo Limon Duparcmeur (NASA Langley Research Center) for her valuable comments and advice when reviewing sections of the preliminary manuscript. Helpful advice was also received from Johannes Kerimo (Texas A&M University), Aisha Hasan (Virginia Tech), Ketaki Patel (MyLife.com), Rohin Sethi (Qualcomm), and Cici Burghardt

(Christopher Newport University) during review of the manuscript. I am grateful to Sanya Rizvi (NIIT Technologies), Ketan Badgujar (LD College of Engineering), and Rajendra Shirhatti (Fuji Films) for valuable information about chapter organization. Also, I thank my students and teaching colleagues whose experience led me to craft this first edition.

I am indebted to Snober Sajjad (University of Western Sydney) and Susan Zehra (Norfolk State University) who helped me in confirming the answers in the solution manual. They also tested all the assembly and higher-level programs included in the text and the solution manual. I thank Basma Rizvi (Aligarh Muslim University) for the wonderful cover art. Thanks to staff at Freescale Semiconductor, Inc., who have answered numerous questions, both technical and copyright related.

I give my sincerest appreciation to Jill Jurgensen, CRC Press, for her timely support, excellent advice, and kind words of encouragement. The staff at CRC Press are consummate professionals, and I have the highest regard for them. Without their support, this book certainly would not have been possible.

Finally, a very special thanks to my wife Susan, who provided invaluable emotional support. Thank you for letting me spend all those hours in front of my computer on yet another after-hour project. I also thank my 5-year-old daughter Zahra, and 1-year-old daughter Sophia, who all provide much joy.

Syed Rizvi
Norfolk, Virginia
March 2011

The Author

Since 2002, **Syed Rizvi** has been working as a research scientist at Analytical Services & Materials, Inc., Hampton, Virginia, where he is developing advanced technologies, primarily for NASA Langley Research Center. Previously, he served on the faculty of three engineering universities in India for nearly 4 years. His research interests are related to embedded systems, vehicular ad-hoc networks, sensor networks, satellite networks, QoS provisioning, wireless multimedia, computer architecture, and software engineering. He has published several book chapters and articles in archival journals and conference proceedings. He received his M.S. in computer science from Old Dominion University (U.S.) and B.S. in electrical engineering. He has served on the editorial board and review committees of several journals and books. Mr. Rizvi maintains the Web site Microcontroller Guide (www.microcontrollerguide.com), which contains hands-on tutorials, discussions, and interesting microcontroller-based projects for both students and hobbyists.

1

Number Systems, Operations, and Codes

... Read Euler, because in his writings all is clear, well calculated, because they teem with beautiful examples, and because one must always study the sources.

—**Joseph L. Lagrange**[*] **(on his deathbed, 1813)**

OUTLINE

---

[*] Grattan-Guinness, I. 1985, 'A Paris curiosity, 1814: Delambre's obituary of Lagrange, and its "supplement."'

OBJECTIVES

Upon completion of this chapter, you should be able to

1. Appreciate the difference between digital and analog quantities.
2. Determine the weighting factor for each digit position in the decimal, binary, octal, and hexadecimal numbering systems.
3. Convert any number from one of the four number systems (decimal, binary, octal, or hexadecimal) to its equivalent value in any of the remaining three numbering systems.
3. Apply arithmetic operations to the binary and hexadecimal number systems.
4. Describe the format and use of binary-coded-decimal (BCD) numbers.
5. Convert a binary number to its 1's and 2's complement form.
6. Express positive and negative numbers in sign-magnitude, 1's complement, and 2's complement form.
7. Determine the ASCII code for any alphanumeric data by using the ASCII code translation table.

Key Terms: Alpha Numeric, Analog, ASCII Code, BCD, Binary, Bit, Byte, Decimal, Digital, Hexadecimal, Least Significant Bit (LSB), Least Significant Digit (LSD), Most Significant Bit (MSB), Most Significant Digit (MSD), Nibble, Octal, Octet, Parity

1.1 Introduction

There are many ways in which we can represent a numeric value. Each convention for representing numeric values is called a number system. The term "number," to most of us, immediately brings to mind the familiar decimal number system with its 10 digits: 0, 1, 2, 3, 4, 5, 6, 7, 8, and 9. It may sound bizarre to discuss number systems with only two digits, or five, or perhaps eight, but in reality, the only reason we use the decimal numbers is because we have ten fingers that led us to the ten basic symbols of the decimal number system. In this chapter, we will discuss the four number systems (decimal, binary, octal, or hexadecimal), and study their relationship to each other. These systems are essential to the study of digital computing. The binary number system and digital codes are fundamental to computing devices such as microcontrollers and to digital electronics in general. The major reason why binary digits are used in computers is the simplicity with which electrical, magnetic, and mechanical devices can represent binary digits. Almost all counting and computing circuits use binary numbers (or BCD), whereas devices for information communication to and from digital system frequently use octal and hexadecimal numbers. The primary goal of this chapter is to introduce the various number system concepts as a frame of reference for further detailed study in the succeeding chapters.

1.2 Digital versus Analog Quantities

A system that deals with continuously varying physical quantities such as voltage, temperature, pressure, or velocity is called an analog system. Most quantities in nature occur in analog, yielding an infinite number of different levels. Figure 1.1 illustrates an audio signal waveform as recorded with an oscilloscope. Note how the voltage level or amplitude of this signal changes over time. Each sound has its own outline. Signals like those shown in Figure 1.1 are referred to as analog because they are allowed to take on a continuous range of values over time.

On the other hand, a system that deals with discrete digits or quantities is called a digital system. Digital electronics deals exclusively with 1's and 0's, or ONs and OFFs. Digital codes (such as ASCII) are then used to convert the 1's and 0's to a meaningful number, letter, or symbol for some output display.

Discrete elements of information are represented in a digital system by physical quantities called signals. Electrical signals such as voltage and

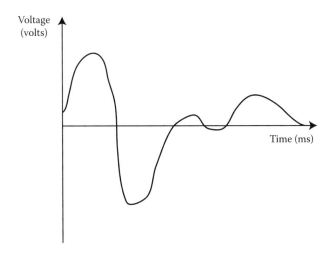

FIGURE 1.1
Oscilloscope display of a voice voltage signal.

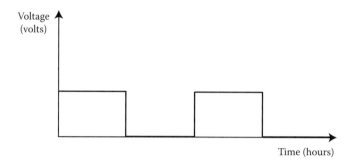

FIGURE 1.2
Oscilloscope display of a furnace on-off control signal.

currents are the most common. A discrete signal is a time series consisting of a sequence of quantities. In other words, it is a time series that is a function over a domain of discrete integers. Unlike a continuous-time signal (analog), a discrete-time signal (digital) is not a function of a continuous argument. Therefore, one can say that a discrete signal has a countable domain, like the natural numbers. In contrast to the analog signal of Figure 1.1 is the digital signal as shown in Figure 1.2. Note that the voltage level of this signal changes in "steps" or has discrete values.

Section 1.2 Review Quiz

An analog signal has a theoretically infinite resolution. (True/False)

1.3 Digital Numbering System (Base 10)

In everyday use, numbers are represented in the decimal (base 10) system that has 10 symbols (0 to 9). The position of each digit in a decimal number indicates the magnitude of the quantity represented and can be assigned a weight. The decimal system is weighted in that it utilizes a positional notation wherein the power of the base that multiplies a particular digit is determined by its position in the sequence of digits that represents a given number. The weight for whole numbers and fractional numbers are the positive and negative powers of ten, respectively. The value of a decimal number is the sum of the digits after each digit has been multiplied by its weight. Consider the base 10 number 982979. We place a subscript 10 to the number 982979 in order to make it clear to the reader that 982979_{10} represents a decimal number. The digit 9 occurs three times in the sequence, but each occurrence has a different weight because the digit occupies a different position corresponding to a power of the base.

Example 1.1

Express the decimal number 982979 as a sum of the values of each digit.

SOLUTION

The following are the column weights and the digits for the decimal number 982979.

10^5	10^4	10^3	10^2	10^1	10^0	Column weights
9	8	2	9	7	9	Digits

$982979 = 9 \times 10^5 + 8 \times 10^4 + 2 \times 10^3 + 9 \times 10^2 + 7 \times 10^1 + 9 \times 10^0$

$= 900{,}000 + 80{,}000 + 2{,}000 + 900 + 70 + 9$

The left-most 9 is weighted by 10^5, the next digit 8 by 10^4. This positional notation is easily extended by decimal fractions, in which case, negative powers of the base 10 are used. For example, $0.653 = 6 \times 10^{-1} + 5 \times 10^{-2} + 3 \times 10^{-3}$.

Common Practice: When necessary to avoid confusion, the base of a number can be appended as a subscript. For example, 100 decimal becomes 101_{10}, and 100 binary becomes 100_2.

Section 1.3 Review Quiz

The decimal number 0.831 may be expressed as $8 \times 10^{-3} + 3 \times 10^{-1} + 1 \times 10^{-0}$. (True/False)

1.4 Binary Numbering System (Base 2)

A number system is nothing more than a code representing quantity. It is possible to express a number in any base. The binary number system is another way to represent quantities. It is less complicated than the decimal system because it has only two digits. In the binary case, the base is 2, and only two symbols are needed (0 and 1). Each digit is called a "bit" and, again, a positional notation is used. The position of a 1 or 0 in a binary number indicates its weight, or value within the number, just as a position of a decimal digit determines the value of that digit. The right-most bit is the LSB (least significant bit) in a binary whole number and has a weight of $2^0 = 1$. The left-most bit is the MSB (most significant bit) and its weight depends on the size of the binary number.

Helpful Hint: The weight or value of a bit increases from right to left in a binary number.

TABLE 1.1

Number System Correspondence Table

Decimal (Base 10)	Binary (Base 2)	Octal (Base 8)	Hexadecimal (Base 16)
0	0	0	0
1	1	1	1
2	10	2	2
3	11	3	3
4	100	4	4
5	101	5	5
6	110	6	6
7	111	7	7
8	1000	10	8
9	1001	11	9
10	1010	12	A
11	1011	13	B
12	1100	14	C
13	1101	15	D
14	1110	16	E
15	1111	17	F

A binary count of 0 through 15 is shown in Table 1.1. To find the decimal equivalent of any binary number, merely write the decimal equivalent of each of the powers of 2, multiply by the appropriate binary digit, and add the results.

Example 1.2

Express the binary number 1000101.1011 as a decimal (base 10) number.

SOLUTION

Since the integer part has seven digits (bits), the most significant has a weight of 2^6 or 64. Its decimal equivalent may be easily computed as

$$1000101 = 1 \times 2^6 + 0 \times 2^5 + 0 \times 2^4 + 0 \times 2^3 + 1 \times 2^2 + 0 \times 2^1 + 1 \times 2^0$$
$$= 1 \times 64 + 0 \times 32 + 0 \times 16 + 0 \times 8 + 1 \times 4 + 0 \times 2 + 1 \times 1$$
$$= 69_{10}$$

For the decimal part,

$$.1011 \quad = 1 \times 2^{-1} + 0 \times 2^{-2} + 1 \times 2^{-3} + 1 \times 2^{-4}$$
$$= 1 \times 0.5 + 0 \times 0.25 + 1 \times 0.125 + 1 \times 0.0625$$
$$= 0.6875_{10}$$

Since the binary numbers require only two symbols, they are ideally suited for representation by electronic devices since only two easily distinguishable states, such as ON and OFF (conducting and nonconducting), are required.

Group Discussion: Discuss the mathematical joke "There are only 10 types of people in the world—those who understand binary, and those who don't."

Section 1.4 Review Quiz

What is the decimal equivalent of 1111_2?

1.5 Octal Numbering System (Base 8)

The octal numeral system (oct for short) is quite important in the computing world. Those familiar with file permissions under Unix systems would have come across this system. To begin with, the octal number system has a base of eight, meaning that it has the first eight decimal digits. In other words, the digits of the octal number system are 0 to 7. Note that there is no 8 or 9. The question arises—how does one count beyond 7 with octal numbers? As with binary and decimal numbers, after running out of basic symbols, we merely form two-digit combinations, taking the second digit followed by the first digit, then the second followed by the second, and so forth. Therefore, after reaching 7 in octal numbers, the next number is 10, then 11, and so on. We count as follows with octal numbers:

0, 1, 2, 3, 4, 5, 6, 7,
10, 11, 12, 13, 14, 15, 16, 17,
20, 21, 22, 23, 24, 25, 26, 27, …, etc.

To find the decimal equivalent of any octal number, simply write the decimal equivalent of each of the powers of 8, multiply by the appropriate octal digit, and add the results.

Example 1.3

Express the octal number 257 as a decimal (base 10) number.

SOLUTION

$$257_8 = 2 \times 8^2 + 5 \times 8^1 + 7 \times 8^0$$
$$= 2 \times 64 + 5 \times 8 + 7 \times 1$$
$$= 128 + 40 + 7$$
$$= 175_{10}$$

TABLE 1.2

Names of 3-Bit Groups

Digit Name	3-Bit Binary	Digit Symbol
Zero	000	0
One	001	1
Two	100	2
Three	011	3
Four	100	4
Five	101	5
Six	110	6
Seven	111	7

One interesting thing about various numbering systems is that they are formed from the binary code by grouping bits by threes or fours. Here the octal number system is formed from the binary code by grouping bits by threes, then reading each group by itself as a number from 000 (zero) to 111 (seven). This is illustrated in Table 1.2. Since only eight groups are possible, the octal number is base-8. As mentioned earlier, the octal numbers will never contain an 8 or 9 since these digits cannot be written in 3 bits.

Group Discussion: Discuss the mathematical joke "Why do mathematicians always confuse Halloween and Christmas? Because 31 Oct = 25 Dec."

Section 1.5 Review Quiz

The digit 8 is found in the octal numbering system. (True/False)

1.6 Hexadecimal Numbering System (Base 16)

After knowing that the octal numbering system is formed by grouping three binary bits, one can try to see what would happen if the grouping is done with four bits. Will there be any advantage with a four-bit grouping over three? The answer to this question is that by moving from three- to four-bit grouping, we will encounter what is called the hexadecimal numbering system. The hexadecimal notation extends the grouping idea to 4 bits and constitutes a base 16 number system. In Table 1.3, each 4-bit group has a name. For example, 0011 is called three, and 1000 is called eight. Notice that there are 16 possible groups of 4 bits. This makes it impossible to use a decimal symbol for every name since there are only 10 decimal symbols. Therefore, we use a letter of the alphabet for numbers from 10 up to 15, which in turn allows us to still use one symbol for each name. The hexadecimal symbols 0 to 9 are the decimal equivalents of the first ten 4-bit binary groups. To represent the last six groups, we use the first six letters of the alphabet as shown in Table 1.3.

TABLE 1.3

Names of 4-Bit Groups

Digit Name	3-Bit Binary	Digit Symbol
Zero	0000	0
One	0001	1
Two	0100	2
Three	0011	3
Four	0100	4
Five	0101	5
Six	0110	6
Seven	0111	7
Eight	1000	8
Nine	1001	9
Ten	1010	A
Eleven	1011	B
Twelve	1100	C
Thirteen	1101	D
Fourteen	1110	E
Fifteen	1111	F

Before we go further, it is important to define terms such as bit, nibble, byte, and octet. We will be using these terms frequently from now onward. A binary digit, abbreviated as bit, is the smallest unit of information in digital computing. In the binary world, a single bit can hold only one of two values: 0 or 1. More meaningful information is obtained by combining consecutive bits into larger units. For example, a nibble is composed of 4 consecutive bits. Since a nibble contains 4 bits, there are sixteen (2^4) possible values, so a nibble corresponds to a single hexadecimal digit. For example, $F_{16} = 1111_2$. A byte is also an ordered collection of bits. The size of a byte is typically hardware dependent, but the modern de facto standard is 8 bits, as this is a convenient power of 2. The term octet explicitly denotes a sequence of 8 bits because of the ambiguity associated with the term byte, and is widely used in digital computing. In short, a nibble is half a byte (octet). A byte (octet) is represented by two hexadecimal digits; therefore, it is common to display a byte of information as two nibbles. Thus, patterns like 3E and D2 are possible in a hexadecimal system, and a number like BF might not even look like a number at all. However, this system of writing binary numbers where every 8-bit byte becomes a two-digit hexadecimal (base 16) code only takes two print characters to write. This is a more compact way of writing the binary contents of a computer memory. In the hexadecimal system, 16 keys are needed for a keyboard, but the striking rate is only one-fourth of that required with a binary keyboard.

Group Discussion: Why is the striking rate with hexadecimal keyboard only one-fourth of that required with a binary keyboard?

Example 1.4

Express the hexadecimal number A85 as a decimal (base 10) number.

SOLUTION

$A85_{16}$ = A   8   5

     = 1010   1000   0101   (binary equivalent for A, B, and 5)

     = $2^{11} + 2^9 + 2^7 + 2^2 + 2^0$ (ignoring 0s and taking only the 1s)

     = 2048 + 512 + 128 + 4 + 1

     = 2693_{10}

Section 1.6 Review Quiz

The symbol H is found in the Hexadecimal Numbering System. (True/False)

1.7 Binary-Coded-Decimal System

The BCD system is used to represent each of the 10 decimal digits as a 4-bit binary code. Each 4-bit group (nibble) is a separate decimal place. This code is useful for outputting to displays that are always numeric (0 to 9), such as those found in digital clocks or digital voltmeters.

Example 1.5

Express the BCD 0111 0101 1000 as a decimal (base 10) number.

SOLUTION

0111   0101 1000_{BCD} contains 3 sets of 4-bit binary code

0111 = 7_{10}

0101 = 5_{10}

1000 = 8_{10}

Therefore, the decimal equivalent is 758.

The number 2986_{10}, for example, would be encoded in this system as

2	9	8	6
0010	1001	1000	0110

where each nibble is read separately as one decimal digit, and numbers larger than 1001, although possible, are never used. In the BCD system, numbers such as 1010, 1011, and 1111 are illegal code.

Section 1.7 Review Quiz

The term BCD stands for (a) Binary-Converted Decimal, (b) Binary-Coding Directives, (c) Binary-Containing Decimals, (d) Binary-Coded Decimal. (Choose one)

1.8 Binary Conversions

Since the binary number system is universally employed in digital computing, it is valuable to understand the general properties of number systems and the techniques of conversion from one to another. The conversion from binary to decimal is usually performed by the digital computer for ease of interpretation by the person reading the number. On the other hand, when a person enters a decimal number into a digital computer, that number must be converted to binary before it can be worked. In Example 1.2, we converted a binary number into its decimal equivalent. Let us look at some examples to understand some more binary conversions.

Example 1.6

Convert 152_{10} to binary using successive division.

SOLUTION

$152 \div 2 = 76$ remainder 0 (LSB)

$76 \div 2 = 38$ remainder 0

$38 \div 2 = 19$ remainder 0

$19 \div 2 = 9$ remainder 1

$9 \div 2 = 4$ remainder 1

$4 \div 2 = 1$ remainder 0

$2 \div 2 = 1$ remainder 0

$1 \div 2 = 0$ remainder 1 (MSB)

Common Misconception: Students often reverse the LSB and MSB when recording the binary solution. While recording the binary solution, be careful and remember not to reverse the LSB and MSB.

Arranging the bits from the above $152_{10} = 1001\ 1000_2$

Converting from binary to octal is simply a matter of grouping the binary positions in groups of three (starting at the least significant position) and writing down the octal equivalent.

Example 1.7

Convert 00 1101_2 and 1011 1000_2 to octal.

SOLUTION

In 00 1101_2, we can see that there are two sets of 3-bit groups. They are 001_2 and 101_2, which are equal to 1_8 and 5_8, respectively. Therefore, 00 $1101_2 = 15_8$.

On the other hand, in 1011 1000_2, we can observe that, moving from the right-hand side, there are two sets of 3-bit groups (i.e., 111_2 and 000_2), and the third set is a 2-bit group (i.e., 10_2). To convert this 2-bit to a 3-bit group, we add a leading zero to it. Thus 10_2 becomes 010_2. Now the groups are 010_2, 111_2, and 000_2 which are equal to 2_8, 7_8, and 0_8, respectively. Therefore, 1011 $1000_2 = 270_8$.

Best Practice: It is common practice to write binary numbers in groups of 4 bits each. This makes it easier to recognize the hexadecimal digits. Thus, 1011101000101 is more often written as 1011 1010 0010. We follow this practice as much as possible throughout the text.

To convert from binary to hexadecimal, group the binary numbers in groups of four (starting in the least significant position) and write down the equivalent hex digit.

Example 1.8

Convert 0100 1101_2 to hex.

SOLUTION

In 0100 1101_2, we can see that there are two sets of 4-bit groups. They are 0100_2 and 1101_2, which are equal to 4_{16} and D_{16}, respectively. Therefore, 0100 $1101_2 = 4D_{16}$.

Sectional 1.8 Review Quiz

What is the binary equivalent of 1111_{10}?

1.9 Binary Operations

Binary arithmetic is essential in all digital computers and in many other types of digital systems. To understand digital systems, we must know the basics of binary addition, subtraction, multiplication, and division. This section provides techniques that will be used in later chapters.

1.9.1 Binary Addition

The four basic rules for addition of binary digits (bits) are as follows:

$0 + 0 = 0$ Sum of 0 with a carry of 0
$0 + 1 = 1$ Sum of 1 with a carry of 0
$1 + 0 = 0$ Sum of 1 with a carry of 0
$1 + 1 = 10$ Sum of 0 with a carry of 1

Note the difference between the first three and the final one. The difference lies in the result where the first three are single bit but the final one is a binary two (10). When binary numbers are added, the last condition creates a sum of 0 in a given column and a carry of 1 over to the next column to the left. When there is a carry of 1, we have a situation in which three bits are being added (a bit in each of the two numbers and a carry bit). This situation is illustrated as follows:

1(carry bit) $+ 0 + 0 = 01$ Sum of 1 with a carry of 0
1(carry bit) $+ 0 + 1 = 10$ Sum of 0 with a carry of 1
1(carry bit) $+ 1 + 0 = 10$ Sum of 0 with a carry of 1
1(carry bit) $+ 1 + 1 = 11$ Sum of 1 with a carry of 1

Example 1.9 illustrates binary addition, the equivalent decimal addition is also shown.

Example 1.9

Add the following binary numbers:

(a) 11 + 11
(b) 111 + 10
(c) 111 + 11
(d) 110 + 100

SOLUTION

(a)
```
 11      3
+11     +3
───     ──
110      6
```

(b)
```
 111      7
 +10     +2
────     ──
1001      9
```

(c)
```
 111      7
 +11     +3
────     ──
1010     10
```

(d)
```
 110      6
+100     +4
────     ──
1010     10
```

1.9.2 Binary Subtraction

The four basic rules for subtracting bits are as follows:

$0 - 0 = 0$
$1 - 1 = 0$
$1 - 0 = 1$
$10 - 1 = 1$ $0 - 1$ with a borrow of 1

When subtracting numbers, we sometimes have to borrow from the next column to the left. A borrow is required in binary only when we try to subtract a 1 from a 0. In this case, when a 1 is borrowed from the next column to the left, a 10 is created in the column being subtracted.

   Example 1.10 illustrates binary subtraction; the equivalent decimal subtraction is also shown.

Example 1.10

Perform the following binary subtractions:

   (a) 10 – 00
   (b) 11 – 01
   (c) 11 – 10
   (d) 101 – 011

SOLUTION

(a)   10   2
       −00   −0
       ‾‾‾   ‾‾
        10   2

(b)   11   3
       −01   −1
       ‾‾‾   ‾‾
        10   2

(c)   11   3
       −10   −2
       ‾‾‾   ‾‾
        01   1

(d)   0101   5
      −011   −3
      ‾‾‾‾   ‾‾
       010   2

Let us examine exactly how we got $101_2 - 011_2 = 010_2$. Starting from the right-most column, the first subtraction is 1–1=0. This was quite simple. The next column is 0–1. For this we borrow 1 from the next column to the left, making a 10 in this column, and replacing 1 with 0 in the column from where we borrowed the 1 (shown as struck out above). Now the current column is 10–1 which yields a 1.

What remains is the final column that originally was 1−0 but with what happened in column two has become 0−0. The subtraction 0−0 yields a 0.

1.9.3 Binary Multiplication

The four basic rules for multiplying bits are as follows:

$0 \times 0 = 0$

$0 \times 1 = 0$

$1 \times 0 = 0$

$1 \times 1 = 1$

Multiplication is performed with binary numbers in the same manner as with decimal numbers. It involves forming partial products, shifting each successive partial product left one place, and then adding all the partial products. Example 1.11 illustrates binary multiplication; the equivalent decimal multiplication is also shown.

Example 1.11

Perform the following binary multiplications:

   (a) 01 × 10
   (b) 11 × 11
   (c) 101 × 111

SOLUTION

```
(a)      01        1
        ×10       ×2
        ────      ──
         00        2
       +01X
        ────
        010
```

```
(b)      11        3
        ×11       ×3
        ────      ──
         11        9
       +11X
        ────
        1001
```

```
(c)     111        7
       ×101       ×5
       ────      ──
        111       35
      +000X
       ──────
       100011
```

Helpful Hint: Binary multiplication of two bits is the same as multiplication of decimal digits 0 and 1.

1.9.4 Binary Division

Division in binary follows the same procedure as division in decimal, as Example 1.12 illustrates. The equivalent decimal divisions are also given.

Example 1.12

(a) 110 ÷ 11
(b) 110 ÷ 10

SOLUTION

(a)
$$
\begin{array}{r}
10 \\
11\overline{)110} \\
11 \\
\hline
000
\end{array}
$$

The equivalent decimal division is

$$
\begin{array}{r}
2 \\
3\overline{)6} \\
6 \\
\hline
0
\end{array}
$$

(b)
$$
\begin{array}{r}
11 \\
10\overline{)110} \\
10 \\
\hline
10 \\
10 \\
\hline
00
\end{array}
$$

The equivalent decimal division is

$$
\begin{array}{r}
3 \\
2\overline{)6} \\
6 \\
\hline
0
\end{array}
$$

Section 1.9 Review Quiz

Which of the following are true in binary (a) $1 \times 1 = 1$, (b) $1 - 1 = 0$, (c) $1 + 1 = 2$?

1.10 Octal Conversions

In Example 1.3, we converted an octal number into its decimal equivalent. Let us look at some examples to understand some more octal conversions.

To convert from decimal to octal, the successive-division procedure can be used.

Example 1.13

Convert the decimal number 123.456 to an equivalent octal (base 8) number.

SOLUTION

The integer conversion is performed as follows:

$123 \div 8 = 15$ remainder 3 (Least Significant Digit (LSD))

$15 \div 8 \;\; = 1$ remainder 7

$1 \div 8 \;\;\; = 0$ remainder 1 (Most Significant Digit (MSB))

From the above, read up to form 173

The fractional conversion is performed as follows:

$0.456 \times 8 = 3.648$ generated integer 3

$0.648 \times 8 = 5.184$ generated integer 5

$0.184 \times 8 = 1.472$ generated integer 1

$0.472 \times 8 = 3.776$ generated integer 3

The process has been arbitrarily terminated

$123.456_{10} = 173.3513_8$

Helpful Hint: At this point, you may think that since calculators can perform these conversions, we don't need to learn conversion from one number system to another using a pencil-and-paper method. It is important to master these conversion procedures through the pencil-and-paper method in order to help you understand the basic concepts.

Example 1.14

Convert $4\,8\,6_{10}$ to octal.

SOLUTION

$486 \div 8 = 60$ remainder 6   (LSD)

$60 \div 8 = 7$ remainder 4

$7 \div 8 = 0$ remainder 7   (MSB)

Therefore, $4\,8\,6_{10} = 746_8$

To check if the answer 746_8 is correct, we perform the same procedure as in Example 1.3.

$$746_8 = 7 \times 8^2 + 4 \times 8^1 + 6 \times 8^0$$
$$= 448 + 32 + 6$$
$$= 486_{10}$$

Example 1.15

Convert 5 2 4_8 to binary.

SOLUTION

5, 2, and 4 are equivalent to 101_2, 010_2, and 100_2 respectively.
  Therefore, 5 2 4_8 = 1 0101 0100_2.

You will probably not run into many occasions that call for the conversion of octal numbers to hex. Should the need arise, conversion is a two-step procedure. Convert the octal number to binary; then convert the binary number to hex.

Example 1.16

Convert 7526_8 to hex.

SOLUTION

The conversion of 7526_8 to hex is in two steps. First, we convert the octal number to binary. Then we regroup the binary digits into groups of four and add zeros where needed to complete groups; finally, we convert the binary number to hex.
  The numbers 7, 5, 2, and 6 are equivalent to 111_2, 101_2, 010_2, and 110_2 respectively.
  Therefore, 7526_8 = 111 101 010 110_2
  Regrouping the binary digits into groups of four, we get 1111, 0101, and 0110, which have the hex equivalents F_{16}, 5_{16}, and 6_{16}, respectively.
  Thus, 7526_8 = 1111 0101 0110_2 = $F56_{16}$.

Section 1.10 Review Quiz

Convert 1_8, 11_8, 111_8 and 1111_8 into hex.

1.11 Hexadecimal Conversions

The procedures for converting hexadecimal numbers to binary and octal are the reverse of the binary and octal conversions to hexadecimal. In Example 1.4, we converted a hexadecimal number into its decimal equivalent. Let us look at some examples to understand some more hexadecimal conversions.

Example 1.17

Convert 498_{10} to hex.

SOLUTION

$489 \div 16 = 31$ remainder 2 (LSD)

$31 \div 16 = 1$ remainder 15 (which is equal to F_{16})

$1 \div 16 = 0$ remainder 1 (MSD)

Therefore, $498_{10} = 1\ F\ 2_{16}$.

Example 1.18

Convert $A9_{16}$ to binary.

SOLUTION

$A_{16} = 1010_2$

$9_{16} = 1001_2$

Therefore, $A9_{16} = 1010\ 1001_2$.

Example 1.19

Convert $F56_{16}$ to octal.

SOLUTION

The conversion of 7526_8 to octal is in two steps. First we convert the hex number to binary. Then we regroup the binary digits into groups of three and add zeros where needed to complete groups; finally, we convert the binary number to octal.

The numbers F, 5, and 6 are equivalent to 1111_2, 0101_2, and 0110_2, respectively. Therefore, $F56_{16} = 1111\ 0101\ 0110_2$.

Regrouping the binary digits into groups of three, we get 111_2, 101_2, 010_2, and 110_2 with binary equivalents 7, 5, 2, and 6, respectively.

Thus, $F56_{16} = 1111\ 0101\ 0110_2 = 7526_8$.

Team Discussion: Which is the smallest number—132_8, 132_{10}, or 132_{16}?

Section 1.11 Review Quiz

Convert 1_{16}, 11_{16}, 111_{16}, and 1111_{16} to octal.

1.12 Hexadecimal Operations

As in the case of binary, hexadecimal arithmetic is essential to all digital computing. To understand digital systems, we must know the basics of hexadecimal addition and subtraction. This section provides techniques that will be used in later chapters.

1.12.1 Hexadecimal Addition

We are used to addition in the decimal number system. If you recall, the basic concepts in adding decimal numbers are that there are 10 digits in counting (0 to 9), and when you reach 10, you carry a "1" over to the next column. Also, the number after 9 is 10. Now, applying the same concept but changing the base from 10 to 16 will give us the method to add hexadecimal numbers. In the hexadecimal number system, there are 16 digits in counting (0 to F). When we reach 16, we carry a "1" over to the next column. The number after F (decimal 15) is 10 in hex (or 16 in decimal). We will use the following steps to perform hexadecimal addition:

1. Add one column at a time.
2. Convert to decimal (base 10) and add the numbers.
3a. If the result of step two is 16 or larger, subtract the result from 16 and carry 1 to the next column.
3b. If the result of step two is less than 16, convert the number to hexadecimal.

The example below illustrates hexadecimal addition.

Example 1.20

Add the following hexadecimal numbers:

(a) $21_{16} + 16_{16}$
(b) $DF_{16} + AD_{16}$

SOLUTION

(a) $\quad 21_{16}$ right column: $1_{16} + 6_{16} = 1_{10} + 6_{10} = 7_{10} = 7_{16}$
$\quad +16_{16}$ left column: $\;\;2_{16} + 1_{16} = 2_{10} + 1_{10} = 3_{10} = 3_{16}$
  ─────
$\quad\;\; 37_{16}$

(b) $\quad DF_{16}$ right column: $F_{16} + D_{16} = 15_{10} + 13_{10} = 28_{10}$
$\quad +AD_{16}$ $\qquad\qquad\qquad 28_{10} - 16_{10} = 12_{10} = C_{16}$ with a 1 carry
  ─────   left column: $\;\;D_{16} + A_{16} + 1_{16} = 13_{10} + 10_{10} + 1_{10} = 24_{10}$
$\quad 18C_{16}$ $\qquad\qquad\qquad 24_{10} - 16_{10} = 8_{10} = 8_{16}$ with a 1 carry

1.12.2 Hexadecimal Subtraction

There are many ways to perform hexadecimal subtraction. Since it involves using the concept of 2's complement, we will cover it in section 1.13.2.

Section 1.12 Review Quiz

Add the following hexadecimal numbers:

(a) $21_{16} + 21_{16}$
(b) $AA_{16} + 33_{16}$

1.13 1's and 2's Complements of Binary Numbers

Previously, we discussed subtraction where we made use of a borrow and produced a difference. In practice, however, subtraction is accomplished by the same hardware that is used for addition through the use of complementary arithmetic. The 1's complement and the 2's complement of a binary number are important because they permit the representation of negative numbers. The method of 2's complement arithmetic is commonly used in computers to handle negative numbers.

1.13.1 Finding the 1's Complement

The 1's complement of a binary number is found by changing all 1's to 0's and all 0's to 1's, as illustrated in the following example.

Example 1.21

Find the 1's complement for the following binary numbers:

(a) 0101 0101
(b) 0010 1011

SOLUTION

Replacing all 1's to 0's, and all 0's to 1's:

(a) 1010 1010
(b) 1101 0100

1.13.2 Finding the 2's Complement

The 2's complement of a binary number is obtained by exchanging the 1's and 0's of the original number and addition of 1 to the result. In other words, the 2's complement of a binary number is found by the addition of 1 to the LSB of the 1's complement.

$$2\text{'s complement} = (1\text{'s complement}) + 1$$

Subtracting a given number X from another binary number Y is accomplished by taking the 2's complement of X to convert it to $-X$ and adding this to Y. In this method, the left-most digit is interpreted as a sign bit (0 for positive, 1 for negative), which is treated as any other bit except that a carry-out from the addition of sign bits is neglected.

Example 1.22

Find the 2's complement of 1011 0010.

SOLUTION

```
1011 0010  Binary number
0100 1101  1's complement
+        1  Add 1
─────────
0100 1110  2's complement
```

Example 1.23

Subtract 230_{10} from 185_{10} by converting to binary and using 2's complement arithmetic.

SOLUTION

Note that the number 230 is the larger of the two given in the problem. Also, it takes 8 bits to represent 230_{10}. We add one additional bit for the sign. Therefore, $230_{10} = 011100110_2$, and its 2's complement is obtained by inverting the 1's and 0's and adding 1.

```
011100110   Binary number
+       1   Add 1
─────────
100011010   2's complement
```

Therefore, $-230_{10} = 100011010_2$

Next, we add this to the binary equivalent of 185_{10}:

```
−230₁₀  = 100011010₂
+185₁₀  = 010111001₂
+____   +  _____
−45₁₀     111010011₂
```

The left-most bit is a 1, indicating that the result is negative. To obtain the desired magnitude, we take the 2's complement of our result since $-(-X) = X$.

$11101001 1_2$ Result from above
000101100_2 1's complement
$+$ 1 Add 1

_____

000101101_2

The decimal equivalent is 45, which we have already determined to be negative.

There are many ways to get the 2's complement of a hexadecimal number. The most common is to convert the hexadecimal number to binary. Then take the 2's complement of the binary number and convert the result to the hexadecimal. Example 1.24 illustrates the subtraction performed on hexadecimal numbers.

Example 1.24

Subtract the hexadecimal number $C4_{16} - 0B_{16}$.

SOLUTION

First, converting hex to binary, we get $0B_{16} = 0000\ 1011_2$
Then, the 2's complement of $00001011_2 = 1111\ 0101_2$
Next, converting binary back to hex, we get $1111\ 0101_2 = F5_{16}$

$C4_{16}$
$+F5_{16}$ Add

_____

$B9_{16}$ Drop carry, as in 2's complement addition

The difference is $B9_{16}$

Section 1.13 Review Quiz

What is the 2's complement of 1100_2?

1.14 Signed Numbers

In digital computing, signed number representations are required to encode negative numbers in binary number systems. As in mathematics, negative numbers in any base are represented by prefixing them with a "−" sign; in computer hardware, numbers are represented in binary only without extra symbols, requiring a method of encoding the minus sign. There are three forms in which signed integer (whole) numbers can be represented in binary: sign-magnitude, 1's complement, and 2's complement. Of these, the 2's complement is the most important, and the sign-magnitude is the least used. Noninteger and very large or small numbers can be expressed in floating-point format.

1.14.1 The Sign Bit

The left-most bit in a signed binary number is the sign bit, which tells you whether the number is positive or negative. A 0 sign bit indicates a positive number, and a 1 sign bit indicates a negative number.

1.14.2 Sign-Magnitude Form

Sign-magnitude representation is possibly the simplest way to represent a signed number. The representation consists of one sign bit and other bits denoting the magnitude, or absolute value, of the number. For example, using 4 bits $+5_{10}$ and -5_{10} can be represented as 0101_2 and 1101_2, respectively. Using an 8-bit signed binary number having the sign-magnitude form $+25_{10}$ and -25_{10} can be represented as $0001\ 1001_2$, and $1001\ 1001_2$, respectively. The only difference between +25 and −25 is the sign because the magnitude bits are in true binary for both positive and negative numbers. In a sign-magnitude form, a negative number has the same magnitude bits as the corresponding positive number, but the sign bit is a 1 rather than a 0.

Example 1.25

Express the decimal number −38 as an 8-bit number in the sign-magnitude form.

SOLUTION

Converting the decimal number +38 to binary yields $0010\ 0110_2$. In the sign-magnitude form, −38 is produced by changing the sign bit to a 1 and leaving the magnitude bits as they are. We get 10100110.

1.14.3 1's Complement Form

There is no difference between the representation of positive numbers in 1's complement form and the positive sign-magnitude numbers. But the negative numbers are different. For a negative number, we compute the 1's complement of its positive number. This generates the number in negative. Let us take a simple example for better understanding. Here we will use eight bits. The decimal number −27 may be expressed as the 1's complement of +27 ($0001\ 1011_2$) as $1110\ 0100_2$.

Example 1.26

Express the decimal number −38 as an 8-bit number in the 1's complement form.

SOLUTION

Converting the decimal number +38 to binary yields $0010\ 0110_2$. In the 1's complement form, −38 is produced by taking the 1's complement of +38 ($0010\ 0110_2$). We get $1101\ 1001_2$.

1.14.4 2's Complement Form

There is no difference between the representation of positive numbers in 2's complement form and the positive sign-magnitude numbers. But the negative numbers are different. For a negative number, we compute the 2's complement of its positive number. This generates the number in negative. Let us take a simple example for better understanding. Here we will use eight bits. We just saw that the decimal number −27 may be expressed as the 1's complement of +27 ($0001\ 1011_2$) as $1110\ 0100_2$. Now we add 1 in order for it to produce the corresponding 2's complement. Therefore, $1110\ 0100_2 + 1_2 = 1110\ 0101_2$.

Example 1.27

Express the decimal number −38 as an 8-bit number in the 2's complement form.

SOLUTION

As seen before, converting the decimal number +38 to binary yields $0010\ 0110_2$. In the 2's complement form, −38 is produced by taking the 2's complement of +38 ($0010\ 0110_2$) as follows:

$0010\ 0110_2$	Binary +38
$1101\ 1001_2$	1's complement
$+\qquad 1$	
$1101\ 1010_2$	2's complement

Self-Learning: Do some Internet research and read about parity method for error detection.

Section 1.14 Review Quiz

Signed number representations are needed to encode negative numbers in binary number systems. (True/False)

1.15 The ASCII Code

Alphanumeric is a combination of alphabetic and numeric, and is used to describe the collection of Latin letters and Arabic digits. Not all data stored and processed by computer is numerical. There are letters and symbols such as punctuation marks used in day-to-day processing. Information, just as names, addresses, and item descriptions, must be input and output in a readable

TABLE 1.4

ASCII Table

Key	ASCII	Hexadecimal	Key	ASCII	Hexadecimal
A	100 0001	41	Space	010 0000	20
B	100 0010	42	(	010 1000	28
C	100 0011	43	)	010 1001	29
D	100 0100	44	+	010 1011	2B
E	100 0101	45	0	011 0000	30
F	100 0110	46	1	011 0001	31
G	100 0111	47	2	011 0010	32
H	100 1000	48	3	011 0011	33
I	100 1001	49	4	011 0100	34
J	100 1010	4A	5	011 0101	35
K	100 1011	4B	6	011 0110	36
L	100 1100	4C	7	011 0111	37
M	100 1101	4D	8	011 1000	38
N	100 1110	4E	9	011 1001	39
O	100 1111	4F			
P	101 0000	50			
Q	101 0001	51			
R	101 0010	52			
S	101 0011	53			
T	101 0100	54			
U	101 0101	55			
V	101 0110	56			
W	101 0111	57			
X	101 1000	58			
Y	101 1001	59			
Z	101 1010	5A			

format. A complete alphanumeric code would include the 26 lowercase letters, 26 uppercase letters, 10 numeric digits, 7 punctuation marks, and anywhere from 20 to 40 other characters, such as +, –, /, $, *, &, and so on. Because of the practical advantages of the binary system, other sorts of such data are stored in two-valued (binary-like) form as well. Most of the industry has settled on an input/output (I/O) alphanumeric code called the American Standard Code for Information Interchange (ASCII, pronounced as "askee"), of which some representative keyboard characters are presented in Table 1.4. This list is not intended to be exhaustive; merely illustrative. The ASCII code uses 7 bits to represent all the alphanumeric data used in computer I/O. Note that decimal digits are listed as BCD-encoded digits preceded by 011 (9 = 011 1001). Other three-digit prefixes are used for nonnumeric data. Each time a key is depressed on an ASCII keyboard, that key is converted into its ASCII code and processed by the computer. Then, before outputting the computer contents to a display terminal or printer, all information is converted from ASCII into standard English.

Example 1.28

Using Table 1.4, determine

(a) the ASCII code for the lowercase letters P, Q, and R.
(b) what $100\ 0111_{ASCII}$ represents.

SOLUTION

(a) P = 101 0000
Q = 101 0001
R = 101 0010

(b) MSB is the 3-bit group which is 100
LSB is the 4-bit group which is 0111
From the Table 1.4, $100\ 0111_{ASCII}$ = G.

Example 1.29

Use an exhaustive ASCII code set to determine the binary ASCII codes that are entered from the computer's keyboard when the following C language program statement is typed in. Also express other code in hexadecimal.

$$\text{if } (y > 8)$$

SOLUTION

From an exhaustive ASCII code set, the ASCII code for the corresponding symbol is used below.

Symbol	binary	hexadecimal
i	1101001	69_{16}
f	1100110	66_{16}
space	0100000	20_{16}
(	0101000	28_{16}
y	1111001	79_{16}
>	0111110	$3E_{16}$
8	0111000	38_{16}
)	0101001	29_{16}

Example 1.30

Interpret the following ASCII-coded sequence:

1000001 1000100 1000100 0100000 0110011 0111001 0110100

SOLUTION

Using Table 1.4 we convert each 7 bit group as

$1000001_2 = A_{ASCII}$
$1000100 = D_{ASCII}$
$1000100 = D_{ASCII}$
$0100000 = Blank_{ASCII}$
$0110011 = 3_{ASCII}$
$0111001 = 9_{ASCII}$
$0110100 = 4_{ASCII}$

Team Discussion: The basic ASCII set uses 7 bits for each character, giving it a total of 128 unique symbols. The extended ASCII character set uses 8 bits, which gives it additional characters. What are these extra characters used for?

Therefore, 1000001 1000100 1000100 0100000 0110011 0111001 0110100 = "ADD 394."

Section 1.15 Review Quiz

In the ASCII code table, which of the following are true?

(a) All uppercase come before lowercase letters, that is, "Z" before "a."
(b) Digits and many punctuation marks come before letters, that is, "3" is before "one."

1.16 Summary

1. The two basic ways of representing the numerical value of physical quantities are analog (continuous) and digital (discrete).
2. Numerical quantities occur naturally in analog form but must be converted to digital form to be used by computers or digital circuitry.
3. Digital or logic circuits operate on voltages that fall in prescribed ranges that represent either a binary 0 (OFF state) or a binary 1 (ON state).
4. Given n binary bits, there are 2^n different binary combinations. The largest number that can be represented with n bits is $2^n - 1$.
5. Any number system can be converted to decimal by multiplying each digit by its weighting factor.
6. The weighting factor of the least significant digit in any numbering system is always 1.
7. Binary numbers can be converted to octal by forming groups of 3 bits, and to hexadecimal by forming groups of 4 bits.
8. The successive-division procedure can be used to convert from decimal to either binary, octal, or hexadecimal.
9. The 1's complement of a binary number is derived by changing 1's to 0's and 0's to 1's.

10. Two's complement is a code in which the MSB identifies the binary number as negative or positive.

11. The 2's complement of a binary number can be derived by adding 1 to the 1's complement.

12. Binary subtraction can be accomplished with addition by using the 1's and 2's complement method.

13. A positive binary number is represented by a 0 sign bit.

14. A negative binary number is represented by a 1 sign bit.

15. BCD is a popular code for use with seven-segment displays. It is similar to hexadecimal but is defined only for the decimal digits 0 to 9.

16. ASCII code is a 7-bit code used by computers to represent all letters, numbers, and symbols in digital form.

Glossary

Alphanumeric: Characters that contain alphabet letters as well as numbers and symbols.

Analog: A system that deals with continuously varying physical quantities such as voltage, temperature, pressure, or velocity. Most quantities in nature occur in analog, yielding an infinite number of different levels.

ASCII Code: American Standard Code for Information Interchange. ASCIII is a 7-bit code used in digital systems to represent all letters, symbols, and numbers to be input or output to the outside world.

BCD: Binary-coded decimal. A 4-bit code used to represent the 10 decimal digits 0 to 9.

Binary: The base 2 numbering system. Binary numbers are made up of 1's and 0's, each position being equal to a different power of 2.

Bit: A single binary digit. The binary number 1101 is a 4-bit number.

Byte: A group of eight bits.

Decimal: The base 10 numbering system. The 10 decimal digits are 0 to 9. Each decimal position is a different power of 10.

Digital: A system that deals with discrete digits or quantities. Digital electronics deals exclusively with 1's and 0's, or ONs and OFFs. Digital codes (such as ASCII) are then used to convert the 1's and 0's to a meaningful number, letter, or symbol for some output display.

Hexadecimal: The base 16 numbering system. The 16 hexadecimal digits are 0 to 9 and A to F. Each hexadecimal position represents a different power of 16.

Least Significant Bit (LSB): The bit having the least significance in a binary string. The LSB will be in a position of the lowest power of 2 within the binary number.

Least Significant Digit (LSD): The digit having the least significance in a digital string.

Most Significant Bit (MSB): The bit having the most significance in a binary string. The MSB will be in a position of the highest power of 2 within the binary number.

Most Significant Digit (MSD): The digit having the most significance in a digital string.

Nibble: A group of four bits.

Octal: The base 8 numbering system. The eight octal numbers are 0 to 7. Each octal position represents a different power of 8.

Octet: A group of eight bits.

Parity: In relation to binary codes, the condition of evenness or oddness of the number of 1's in a code group.

Answers to Section Review Quiz

1.2   True
1.3   False
1.4   15
1.5   False
1.6   False
1.7   (d)
1.8   10001010111
1.9   (a) and (b)
1.10 1, 9, 49, 249
1.11 1, 21, 421, 10421
1.12 (a) 42, (b) DD
1.13 0100_2
1.14 True
1.15 (a) and (b)

True/False Quiz

1. The decimal number system is a weighted system with ten digits.
2. The octal number system is a weighted system with two digits.

3. $433_{10} = 1A7_{16}$.

4. LSD stands for lowest singular byte.

5. In binary, $1 + 1 = 2$.

6. $B2F_{16} = 5457_8$.

7. The 1's complement of the binary number 1010 is 1111.

8. The 2's complement of the binary number 0001 is 1110.

9. $137_{10} \neq 10001001_2$.

10. The right-most bit in a signed binary number is the sign bit.

11. The hexadecimal number system has 16 characters, six of which are alphabetic characters.

12. BCD stands for binary-coded decimal.

13. ASCII stands for American standard code for intelligence information.

14. A parity bit is usually an extra bit that is attached to a code group that is being transferred from one location to another.

15. A nibble is a string of nine bits.

Questions

QUESTION 1.1

What are the advantages of using the octal number system and the hexadecimal number system over the binary number system?

QUESTION 1.2

(a) Describe the method of successive division by 8 that is used in decimal-to-octal conversion.
(b) Describe the method of successive multiplication by 8 that is used in decimal-to-octal conversion.

QUESTION 1.3

What is the difference between 1's and 2's complement? Is there a significant advantage to using one over another in terms of software execution speed? Why?

QUESTION 1.4

List some applications of BCD.

QUESTION 1.5

Describe one major advantage of UNICODE over ASCII.

Problems

PROBLEM 1.1

What is the largest number that can be represented using eight bits?

PROBLEM 1.2

Convert the following decimal numbers to octal:

(a) 39   (b) 82   (c) 418

PROBLEM 1.3

Convert the following octal numbers to decimal:

(a) 23   (b) 377   (c) 47321

PROBLEM 1.4

Convert the following octal numbers to binary:

(a) 20   (b) 377   (c) 527133   (d) 217

PROBLEM 1.5

Convert the hex number $3A5_{16}$ into binary and find its 2's complement.

PROBLEM 1.6

Multiply 101_2 by 11_2.

PROBLEM 1.7

Convert the following hex numbers to decimal:

(a) 2C   (b) A09   (c) FFFF

PROBLEM 1.8

Convert the following decimal numbers to hex:

(a) 423   (b) 214   (c) 9999

PROBLEM 1.9

Convert the following hex numbers to binary:

(a) 20   (b) 3C   (c) FFFF

PROBLEM 1.10

Express the hexadecimal number B73D as an equivalent octal number.

PROBLEM 1.11

Show steps to determine the decimal values of the signed binary numbers expressed in 2's complement:

(a) 0101 0110   (b) 1010 1010

PROBLEM 1.12

Show the steps required to calculate the difference $173_8 - 173_8$.

PROBLEM 1.13

Show the steps required to calculate the difference $84_{16} - 2A_{16}$.

PROBLEM 1.14

Convert the following BCD to their decimal equivalent:

(a) 1001010001110000   (b) 001101010001

PROBLEM 1.15

Convert the BCD number 0010 0111 1001 0101 into hexadecimal.

PROBLEM 1.16

(a) How many bytes are in a 32-bit string?
(b) What is the largest decimal value that can be represented in binary using two bytes?

PROBLEM 1.17

A small microcontroller used in a wireless sensor network uses octal codes to represent its 12-bit memory address.

(a) How many octal digits are required?
(b) What is the range of address in octal?
(c) How many memory locations are there?

PROBLEM 1.18

Using 2's complement, subtract 176 from 204.

PROBLEM 1.19

How many bits are needed to represent decimal values ranging from 0 to 12,500?

PROBLEM 1.20

Generally, the keyboards that we use in our computers are ASCII type. Each keystroke produces the ASCII equivalent of the designated character. If you type

    PRINT A

what is the output of an ASCII keyboard?

2

Semiconductors and Digital Logic

Thinking is the hardest work there is, which is probably the reason why so few engage in it.

—Henry Ford (Industrialist)

OUTLINE

OBJECTIVES

Upon completion of this chapter, you should be able to

1. Understand the basic concept of the seven basic gates (NOT, AND, OR, NAND, NOR, EX-OR, and EX-NOR).
2. Describe the operation and use of all the seven gates.
3. Recognize and use the distinctive shape logic gate symbols.
4. Construct truth tables for two-, three-, and four-input gates.
5. Construct timing diagrams showing the proper time relationships of inputs and outputs for the various logic gates.
6. Understand the concept of parity.
7. Understand the working of an Odd-Parity Generator.

Key Terms: AND Gate, Bias, Boolean Algebra, Boolean Equation, CMOS, Complement, Controller Inverter, Disabled, Enabled, Even Parity, Exclusive-NOR Gate, Exclusive-OR (XOR) Gate, Gate, Inverter, Logic, Logic Circuit, NAND Gate, NOR Gate, Odd Parity, OR Gate, Timing Diagram, Truth Table, Word

2.1 Introduction

In the year 1847, English mathematician George Boole published *The Mathematical Analysis of Logic*. In this book, he demonstrated how the use of a specific set of logic can help one to move through piles of data to find the required information. His approach toward logic was the most important part of his work. By integrating logic into mathematics, Boole was able to determine what formed the base of Boolean logic or algebra. Each variable in Boolean algebra has either of two values: true or false. The original purpose of this two-state algebra was to solve logic problems. Boolean algebra had no practical application until 1937 when Claude Shannon used it to analyze telephone switching circuits. Shannon is credited with founding both digital computer and digital circuit design theory in 1937, when, as a 21-year-old master's student at MIT, he wrote a thesis demonstrating that electrical application of Boolean algebra could construct and resolve any logical, numerical relationship.

This chapter introduces the logic gates that are fundamental building blocks for forming any electronics circuitry. Gates are often called logic circuits because they can be analyzed with Boolean algebra. A logic gate has one output terminal and one or more input terminals. They are digital (two-state) circuits because the input and output signals are either low or high voltages. Its output will be HIGH (1) or LOW (0) depending on the digital levels at the input terminals. Through the use of logic gates, we can design digital systems that will evaluate digital input levels and produce a specific output response based on that particular logic circuit design. The seven logic gates are AND, OR, NAND, NOR, INVERTER, exclusive-OR, and exclusive-NOR. Microcontroller applications covered in the succeeding chapters utilize integrated circuits that contain various logic gates. A microcontroller is made up of hundreds of thousands or even millions of logic gates.

2.2 Diode Logic

In electronics, a diode is a two-terminal electronic component that conducts electric current in only one direction. The diode conducts conventional current in a direction from the p-type side (called the anode terminal) to the n-type side (called the cathode terminal), but not in the opposite direction. The switching (or ideal) diode has two states. If the bias (voltage polarity) applied to the diode makes the anode end more positive than the cathode end, current will flow through the diode. If the anode is not more positive than the cathode, no significant current will flow. As shown in Figure 2.1, the condition that permits current to flow in the diode is called forward

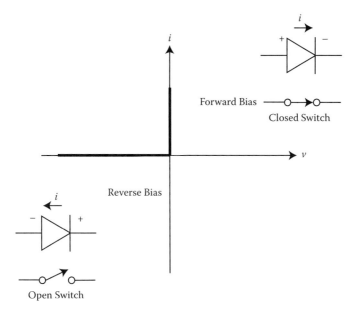

FIGURE 2.1
Ideal diode.

bias. On the other hand, the condition in which current is blocked is called reverse bias. For switching purposes, forward bias can be considered as a short-circuit condition. A reverse bias can be considered as an open-circuit condition. Because the semiconductor diode lets current pass easily in one direction but not in the other, we can think (though an approximation in reality) of a diode as an automatic switch—it is closed when current tries to pass in one direction but open when it tries to pass in the other direction. This unidirectional behavior is called rectification and is used to convert alternating current to direct current and to extract modulation from radio signals in radio receivers.

Section 2.2 Review Quiz

A diode is a two-terminal electronic component that conducts electric current in only one direction. (True/False)

2.3 The Inverter

An inverter is a gate with only one input signal and one output signal where the output state is always the opposite of the input state. An inverter is also called a NOT gate due to the fact that it implements logical negation. It represents perfect switching behavior, which is the defining assumption in digital electronics.

2.3.1 Inverter Truth Table

A transistor inverter is shown in Figure 2.2. The basic concept that we want to focus on here is that this common-emitter amplifier switches between cutoff and saturation. When V_{IN} is low (approximately 0 V), the transistor cuts off and V_{OUT} becomes high. Furthermore, a high V_{IN} saturates the transistor, forcing V_{OUT} to become low. This operation is summarized in Table 2.1 (A) where a low input produces a high output, and a high input produces a low output. Table 2.1 (B) conveys the same message in binary form where 0

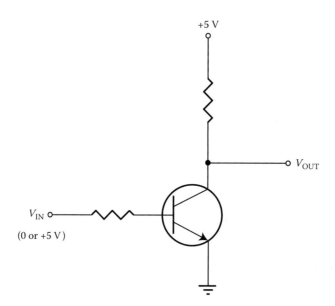

FIGURE 2.2
An example of a transistor inverter.

TABLE 2.1

Illustration of Truth
Table for an Inverter

	V_{IN}	V_{OUT}
(A)	Low	High
	High	Low
(B)	0	1
	1	0

stands for low voltage and binary 1 for high voltage. Since Table 2.1 (B) shows the output for each possible input in terms of levels and corresponding bits, it is called a truth table. As mentioned earlier, an inverter is also called a NOT gate because the output is not the same as the input. The output is sometimes called the complement of the input due to its opposite nature to the input.

2.3.2 Inverter Symbol

An inverter symbol is shown in Figure 2.3 (a). The negation indicator is a "bubble" that indicates inversion or complementation when it appears on the input and output of any logic element. Generally, inputs are on the left of a logic symbol, and the output is on the right. When appearing on the input (Figure 2.3 (b)), the bubble means that a 0 is the active or asserted input state, and the input is called an active-LOW input. When appearing on the output as in Figure 2.3 (a), the bubble means that a 0 is the active and asserted output state, and the output is called an active-LOW output. The absence of a bubble on the input or output means that a 1 is the active or asserted state, and, in this case, the input or output is called active-HIGH. Note that a change in the placement of the negation or polarity indicator does not imply a change in the way an inverter operates.

If two inverters are cascaded as in Figure 2.3 (c), it forms a noninverting amplifier. Figure 2.3 (d) is the symbol for the noninverting amplifier. Regardless of the circuit design, the action is always the same, that is, a high input voltage produces a high output voltage, and a low input voltage results in a low output voltage. The primary use of the noninverting amplifier is buffering or isolating two other circuits.

Helpful Hint: Whenever you see the bubble symbol, remember that the output is the complement of the input.

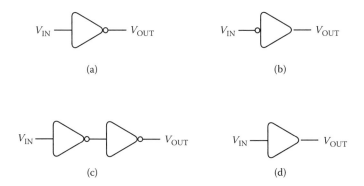

FIGURE 2.3
Logic symbols: (a) active-HIGH input inverter, (b) active-LOW input inverter, (c) double inverter, (d) buffer.

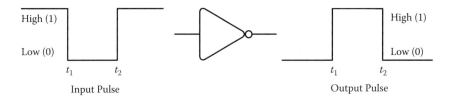

FIGURE 2.4
Inverter operation with a pulse input.

2.3.3 Operation of an Inverter

Figure 2.4 shows the output of an inverter for a pulse input, where t_1 and t_2 indicate the corresponding points on the input and output pulse waveform. When the input is LOW (0), the output is HIGH (1); when the input is HIGH (1), the output is LOW (0), thereby producing an inverted output pulse.

2.3.4 Timing Diagrams

A timing diagram is a representation of a set of signals in the time domain. It can contain many rows, usually one of them being the clock. It is a tool ubiquitous in digital electronics, which, besides providing an overall description of the timing relationships, helps find and diagnose digital logic problems. The time relationship of the output pulse to the input pulse that was illustrated in Figure 2.4 can be shown with a simple timing diagram by aligning the two pulses so that the occurrences of the pulse edges appear in the proper time relationship. The rising edge of the input pulse and the falling edge of the output pulse ideally occur at the same time. Similarly, the falling edge of the input pulse and the rising edge of the output pulse ideally occur at the same time. This timing relationship is shown in Figure 2.5. In practice, there is a very small delay from the input transition until the corresponding output transition.

Example 2.1

A waveform is applied to an inverter in Figure 2.6. Determine the output waveform corresponding to the input and show the timing diagram. According to the position of the bubble, what is the active output state? If the inverter is shown with the bubble (negative indicator) on the input instead of the output, how is the timing diagram affected?

SOLUTION

Team Discussion: Timing diagrams are especially useful for illustrating the time relationship of digital waveforms with multiple pulses. How?

The output waveform is exactly opposite or inverted to the input, as shown in the timing diagram in Figure 2.7. The active or asserted output state is 0. In the case when the bubble (negative indicator) of the inverter is on the input instead of the output, the timing diagram will not be affected.

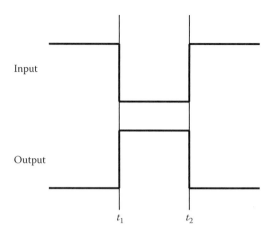

FIGURE 2.5
Timing diagram for Figure 2.4.

FIGURE 2.6
Timing diagram for Example 2.1.

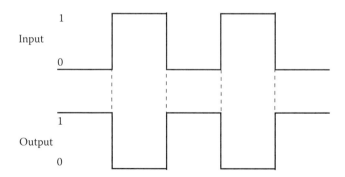

FIGURE 2.7
Timing diagram solution for Example 2.1.

2.3.5 Logic Expressions for an Inverter

In Boolean algebra, a complement of a variable is designated by a bar over the letter. A variable can take on a value of either 1 or 0. If a given variable is 1, its complement is 0, and vice versa. The operation of an inverter can be

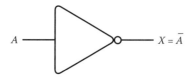

FIGURE 2.8
Logic expressions for an inverter.

expressed as follows: If the input variable is called A and the output variable is called X, then

$$X = \overline{A}$$

This expression states that the output is the complement of the input, so if $A = 0$, then $X = 1$, and if $A = 1$, then $X = 0$. Figure 2.8 illustrates the input variable A becoming $\overline{A}$ as the output. The complement variable $\overline{A}$ can be read as "A bar" or "not A."

Section 2.3 Review Quiz

An inverter is also called a ____ gate due to the fact that it implements logical negation. (NOT/ NON/ NOW)

2.4 The AND Gate

The AND gate has only one output but can have two or more input signals. It performs what is known as logical multiplication. All input must be high to get a high output. The AND gate is one of the basic gates that can be combined to form any logic function.

2.4.1 AND Gate Symbol

The AND gate is composed of two or more inputs and a single output, as indicated by the standard logic symbol shown in Figure 2.9. This figure shows the logic symbols for 2-, 3-, and 4- input gates. Inputs are on the left, and the output is on the right in each symbol.

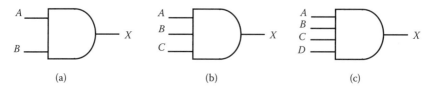

FIGURE 2.9
AND gate symbols.

2.4.2 Operation of an AND Gate

An AND gate produces a high output only when all the inputs are high. When any of the inputs is low, the output is low. Therefore, the function of an AND gate is to establish when certain conditions are simultaneously true, as indicated by high levels on all of its inputs, to produce a logical high on its output as an indication that all these conditions are true. The inputs of the 2-input AND gate in Figure 2.9 are labeled A and B, and the output is labeled X.

Figure 2.10 (a) illustrates a very simple way of building an AND gate. In this circuit the input can be either low (ground) or high (+5 V). In Figure 2.10 (b), both inputs are low, and therefore, both diodes conduct and pull the output down to a low voltage. If one of the inputs is low and the other high as shown in the Figure 2.10 (c), the diode with the low input conducts, and this pulls the output down to a low voltage. Additionally, the diode with the high input is reverse-biased or cut off, symbolized by the dark shading in Figure 2.10 (c). When both inputs are high as illustrated in Figure 2.10 (d), both diodes are cut off and act like an open switch. Since there is no current in the resistor,

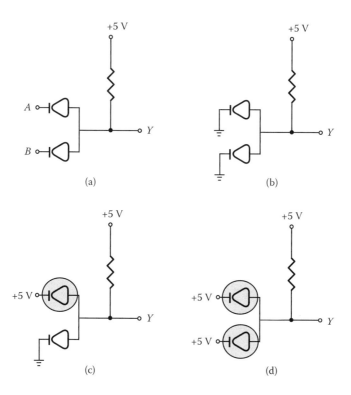

FIGURE 2.10
A simple AND gate construction with various scenarios.

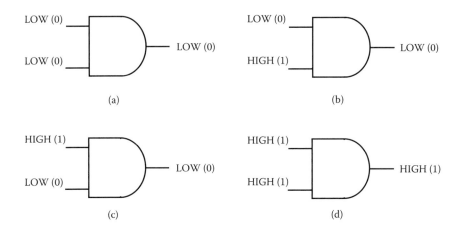

FIGURE 2.11
Operation of a 2-input AND gate.

the supply voltage pulls the output up to a high voltage (+5 V). In a nutshell, the gate operation can be stated as follows:

> For a 2-input AND gate, output X is HIGH only when input A and B are HIGH; X is LOW when either A or B is LOW, or when both A and B are LOW.

Figure 2.11 illustrates a 2-input AND gate with all four possibilities of input combinations and the resulting output for each.

2.4.3 AND Gate Truth Table

As we have discussed before in the section dealing with NOT gates (inverter), the logical operation of a gate can be expressed with a truth table that lists all input combinations with the corresponding outputs. Such a truth table for an AND gate is shown in Table 2.2. As usual, binary zero stands for low voltage, and binary 1 for high voltage. As can be observed, inputs A and B must be set to logical high to get a high output. This is the reason why the circuit is named an AND gate. Figure 2.9 (b) is a 3-input AND gate. If all inputs

TABLE 2.2

Truth Table for a
2-Input AND Gate

A	B	Output (X)
0	0	0
0	1	0
1	0	0
1	1	1

TABLE 2.3

Truth Table for a
3-Input AND Gate

A	B	C	Output
0	0	0	0
0	0	1	0
0	1	0	0
0	1	1	0
1	0	0	0
1	0	1	0
1	1	0	0
1	1	1	1

are low, all diodes conduct and pull the output down to a low voltage. Even one conducting diode will pull the output down to a low voltage, and therefore, the only way to get a logical high output is to set all the inputs to logical high. When all inputs are set to logical high, all diodes become nonconducting, and the supply voltage draws the output up to a high voltage. Table 2.3 summarizes the 3-input AND gate. The output is 0 for all input words except 111. In a nutshell, all inputs must be high in order to get a high output.

AND gates can have as many inputs as desired; just add one diode for each additional input. For instance, 8 diodes result in an 8-input AND gate, 16 diodes in a 16-input AND gate, and 32 diodes in a 32-input AND gate. The truth table can be extended to any number of inputs. The total number of possible combinations of binary inputs to a gate is determined by the following formula: $N = 2^n$, where N is the number of possible input combinations, and n is the number of input variables. Thus, the combinations will be 4, 8, and 16 when the input variables are 2, 3, and 4 in numbers. One can find the number of input bit combinations for gates with any number of inputs by using the formula $N = 2^n$.

Helpful Hint: No matter how many inputs an AND gate has, all inputs must be set to logical high in order to get a high output.

Example 2.2

Develop the truth table for a 4-input AND gate.

SOLUTION

There are 16 possible input combinations ($2^4 = 16$) for a 4-input AND gate. Table 2.4 shows the 4-input AND gate truth table.

Example 2.3

Determine the total number of possible input combinations for a 5-input AND gate.

TABLE 2.4

Truth Table for a 4-Input
AND Gate

A	B	C	D	Output
0	0	0	0	0
0	0	0	1	0
0	0	1	0	0
0	0	1	1	0
0	1	0	0	0
0	1	0	1	0
0	1	1	0	0
0	1	1	1	0
1	0	0	0	0
1	0	0	1	0
1	0	1	0	0
1	0	1	1	0
1	1	0	0	0
1	1	0	1	0
1	1	1	0	0
1	1	1	1	1

Common Misconception: When you build a truth table, you might mistakenly omit certain input combinations if you do not set the variables up as a binary counter. A systematic approach is key here.

SOLUTION

$N = 2^n$, where n is the number of inputs and N is the total number of possible combinations. Therefore, $N = 2^5 = 32$. There are 32 possible combinations of inputs for a 5-input AND gate.

2.4.4 Timing Diagrams

Recall that a timing diagram is a graphical method of showing the exact output behavior of a logic circuit for every possible set of input conditions. If we follow the truth table of an AND gate, both the input vary, but the output is always 0 except when both the inputs are 1. Let us take the example of the waveform shown in Figure 2.12. During the time interval t_1, input A is 0 and input B is 1, so the output is 0. During time interval t_2, both input A and B change their values from that of t_1 such that the new input A is 1 and input B is 0. This results in the output still remaining at 0. During the time interval t_3, input A is 0 and input B is 1, so the output is 0. During time interval t_4, input A is 0, input B is 0, and the output is therefore 0. Inputs A and B are both 1 during the time interval t_5, making output $X = 1$ during this interval. Finally, during the time interval t_6, input A is 0 and B is still at 1, and thus the output comes back to 0.

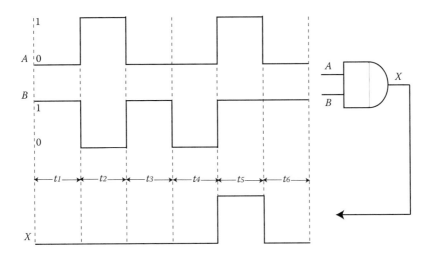

FIGURE 2.12
Example of an AND Gate operation with a timing diagram showing input and output relationships.

Example 2.4

If two waveforms, *A* and *B*, are applied to the AND gate inputs as in Figure 2.13, what is the resulting output waveform?

SOLUTION

As shown in the timing diagram in Figure 2.13, the output waveform *X* is at logical 1 only when *A* and *B* waveforms are at logical 1 simultaneously.

Helpful Hint: To solve a timing analysis problem, it is useful to look at the gate's truth table to see what the unique occurrence is for the gate. In the case of the AND gate, the odd occurrence is when the output goes high due to all high inputs.

2.4.5 Logic Expressions for an AND Gate

The Boolean equation for the AND function can more simply be written as

$$X = A.B$$

Here, by placing a dot between the two variables, as *A.B*, or by simply writing the adjacent letters without the dot, as *AB*, the inputs are expressed to be logically equivalent to the output *X*. Boolean multiplication follows the same basic rules governing binary multiplication as follows:

$0.0 = 0$
$0.1 = 0$
$1.0 = 0$
$1.1 = 1$

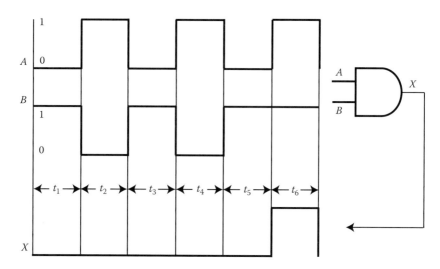

FIGURE 2.13
Timing diagram for a 2-input AND gate.

When more input variables are required, the same concept can be extended by simply using a new letter for each input variable. The function of a 3-input AND gate, for example, can be expressed as $X = ABC$, where A, B, and C are the input variables. The expression for a 4-input AND gate can be $X = ABCD$, and so on.

Helpful Hint: Boolean multiplication is the same as the AND function.

Example 2.5

The 6-bit register of Figure 2.14 stores the word FEDCBA. The *ENABLE* input is digital. Describe the behavior of the circuit.

SOLUTION

The circuit uses the *ENABLE* input as an input to the AND gates. The 6-bit registers are the input for these AND gates also. A low *ENABLE* blocks the register contents from the final output, but a high *ENABLE* transmits the register contents. For example, when *ENABLE* = 0, each AND gate has a low *ENABLE* input. This means that irrespective of the values of the 6-bit register, the output will always be LOW. In other words, the final word will be $Y_5Y_4Y_3Y_2Y_1Y_0 = 000000$. Additionally, if the *ENABLE* is HIGH, that is, *ENABLE* = 1, the output of each AND gate depends on the data inputs (F, E, D, C, B, and A). A low data input results in a low output, and a high data input in a high output. For example, if $FEDCBA = 111001$, a high *ENABLE* gives $Y_5Y_4Y_3Y_2Y_1Y_0 = 111001$. In short, a high *ENABLE* transmits the register contents to the final output to get $Y_5Y_{4A}Y_3Y_2Y_1Y_0 = FEDCBA$.

Common Practice: Computers utilize all the basic logic operations when it is necessary to selectively manipulate certain bits in one or more bytes of data. Selective bit manipulations are done with a mask. For example, to clear (make all 0s) the left four bits in a data byte but keep the right four bits, ANDing the data byte with 00001111 will do the job.

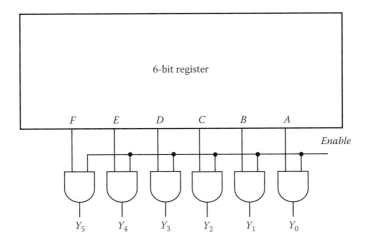

FIGURE 2.14
Using AND gates to block or transmit data.

Section 2.4 Review Quiz

The AND gate performs what is known as _____ multiplication. (logical/systemic/machine/simple)

2.5 The OR Gate

The OR gate has two or more input signals but only one output signal. A logical 1 at the output results if one or both the inputs to the gate are set to logical 1. If neither input is set to logical 1, a logical 0 output results. In another sense, the function of OR effectively finds the maximum between two binary digits, just as the complementary AND (covered in the next section) function finds the minimum. In the OR gate, if any input signal is high, the output signal is high. The OR gate performs what is known as logical addition.

2.5.1 OR Gate Symbol

The OR gate is composed of two or more inputs and a single output, as indicated by the standard logic symbol shown in Figure 2.15. This figure shows the logic symbols for 2-, 3-, and 4-input gates. Inputs are on the left, and the output is on the right in each symbol.

2.5.2 Operation of an OR Gate

An OR gate produces a logical 1 on the output when any of its inputs is logical 1. The output is logical 0 only when all of the inputs are set to logical 0.

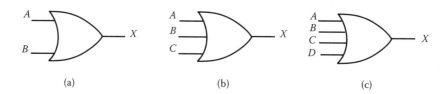

FIGURE 2.15
OR gate symbols.

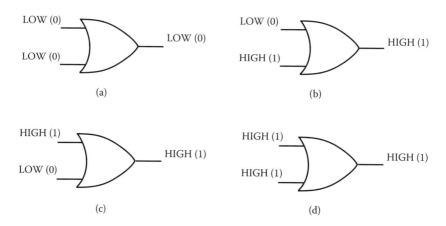

FIGURE 2.16
Operation of a 2-input OR gate.

Therefore, an OR gate determines when one or more of its inputs are 1 and produces a 1 on its output to indicate this condition. The inputs of the 2-input OR gate in Figure 2.16 are labeled A and B, and the output is labeled X. The HIGH level is the active or asserted output level for the OR gate. Figure 2.16 illustrates the operation for a 2-input OR gate for all four possible input combinations. The operation of the gate can be stated as follows:

> *For a 2-input gate, output X is HIGH when either input A or input B is HIGH, or when both A and B are HIGH; X is LOW only when both A and B are LOW.*

Figure 2.17 shows a simple method of building an OR gate. If both inputs are low, the output is low. If either input is high, the diode with the high input conducts, and the output is high. Because of the two inputs, we call this circuit a 2-input OR gate.

Helpful Hint: For an OR gate, at least one HIGH input produces a HIGH output.

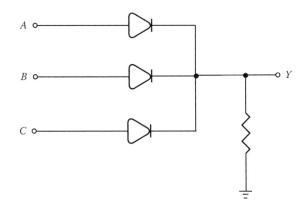

FIGURE 2.17
A 3-input diode OR gate.

2.5.3 OR Gate Truth Table

Table 2.5 summarizes the action of an OR gate. Binary 0 stands for low voltage, and binary 1 for high voltage. Notice that one or more high inputs produces a high output, and that is the reason why this circuit is called an OR gate. Figure 2.16 (b) shows a 3-input OR gate. If all inputs are low, all diodes are off and the output is 0. If one or more inputs are high, the output is high. Table 2.5 illustrates the OR gate truth table. An OR gate can have as many inputs as desired. Just add one diode for each additional input. Seven diodes result in a 7-input OR gate, 10 diodes in a 10-input OR gate. No matter how many inputs, the action of any OR gate is to produce HIGH output if one or more inputs are high. Bipolar transistors and MOSFETs can also be used to build OR gates. No matter what devices are used, OR gates always produce a high output when one or more inputs are high.

Best Practice: When constructing a truth table, always list the input words in a binary progression like 000, 001, 010, ... 111. This guarantees that all input possibilities will be accounted for. It also improves the readability of the truth table.

TABLE 2.5

Truth Table for a
2-Input OR Gate

A	B	Output (X)
0	0	0
0	1	1
1	0	1
1	1	1

TABLE 2.6

Truth Table for a 4-Input
OR Gate

A	B	C	D	Output
0	0	0	0	0
0	0	0	1	1
0	0	1	0	1
0	0	1	1	1
0	1	0	0	1
0	1	0	1	1
0	1	1	0	1
0	1	1	1	1
1	0	0	0	1
1	0	0	1	1
1	0	1	0	1
1	0	1	1	1
1	1	0	0	1
1	1	0	1	1
1	1	1	0	1
1	1	1	1	1

Example 2.6

Show a truth table of a 4-input OR gate.

SOLUTION

Let A, B, C, D stand for input bits, and Z for the output bit. Recall that the number of input words in a truth table always equals 2^n, where n is the number of input bits. A 2-input OR gate has a truth table with 2^2- or 4-input words, a 2-input OR gate has a truth table with 2^3- or 8-input words, and a 4-input OR gate has a truth table with 2^4- or 16-input words. Table 2.6 shows the truth table of a 4-input OR gate such that the output is LOW (0) only when all the inputs are LOW (0), that is, $A = 0$, $B = 0$, $C = 0$, and $D = 0$.

2.5.4 Timing Diagram

Let us now explore the operation of an OR gate with pulse waveform inputs. Again, the important thing to remember in the analysis of gate operation with pulse waveforms is the time relationship of all the waveforms involved. For example, in Figure 2.18, let us first take the time interval t_1. Here the input A is 0 and input B is 1, so the output is 1. During time interval t_2, both input A and B change their values from that of t_1 such that the new input A is 1 and input B is 0. This results in the output still remaining at 1. During the time interval t_3, input A is 0 and input B is 1, so the output is 1. During time interval t_4, input A is 0, input B is 0, and the output is therefore 0. Inputs A

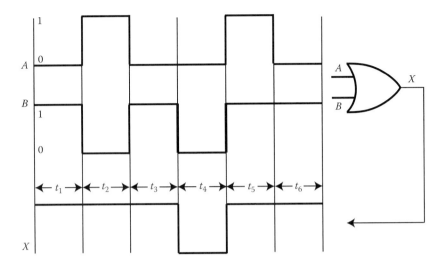

FIGURE 2.18
Example of OR gate operation with a timing diagram showing input and output relationships.

and B are both 1 during the time interval, t_5, making output $X = 1$ during this interval. Finally, during the time interval t_6, input A is 0 and B is still at 1, and thus the output remains at 1. In this illustration, we have applied the truth table operation of the OR gate to each of the time intervals during which the levels are nonchanging.

Example 2.7

If the 2-input waveforms, A and B, in Figure 2.19 are applied to the OR gate, what is the resulting output waveform?

SOLUTION

The output waveform X of a 2-input OR gate is HIGH when either or both input waveforms are HIGH as shown in the timing diagram. In this case, both input waveforms are never HIGH at the same time.

Example 2.8

For the 3-input waveforms, A, B, and C, in Figure 2.20, show the output waveform with its proper relation to the inputs.

SOLUTION

When either or both input waveforms are HIGH, the output is HIGH as shown by the output waveform X in the timing diagram.

Helpful Hint: To solve a timing analysis problem, it is useful to look at the gate's truth table to see what the unique occurrence is for the gate. In the case of the OR gate, the odd occurrence is when the output goes low due to all low inputs.

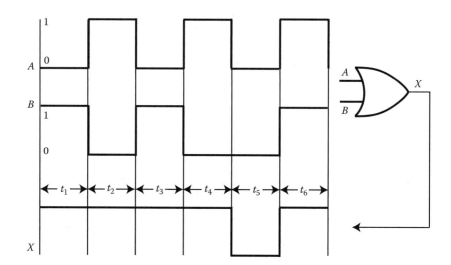

FIGURE 2.19
Timing diagram with a 2-input OR gate.

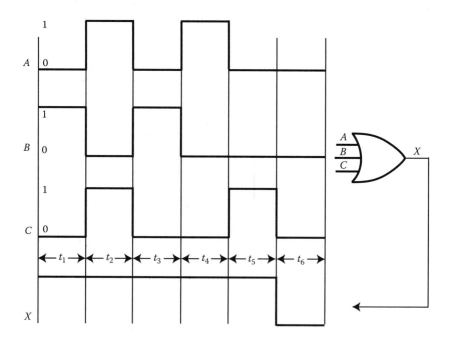

FIGURE 2.20
Timing diagram with a 3-input OR gate.

2.5.5 Logic Expressions for an OR Gate

The logical OR function of two variables is represented mathematically by a + between the two variables, for example, $A + B$. The plus sign is read as "OR." Addition in Boolean algebra involves variables whose values are either binary 1 or binary 0. The basic rules for Boolean addition are as follows:

$$0 + 0 = 0$$
$$0 + 1 = 1$$
$$1 + 0 = 1$$
$$1 + 1 = 1$$

Notice that Boolean addition differs from binary addition in the case where two 1s are added. There is no carry in Boolean addition. The operation of a 2-input OR gate can be expressed as follows: If one input variable is A, if the other input variable is B, and if the output variable is X, then the Boolean expression is $X = A + B$. Figure 2.21 shows the OR gate logic symbol with two input variables and the output variable labeled. To extend the OR expression to more than two input variables, a new letter is used for each additional variable. For instance, the function of a 3-input OR gate can be expressed as $X = A + B + C$. The expression for a 4-input OR gate can be written as $X = A + B + C + D$, and so on. The OR gate operation can be evaluated by using the Boolean expressions for the output X by substituting all possible combinations of 1 and 0 values for the input variables, as shown in Table 2.7 for a 2-input OR gate. This evaluation shows that the output X of an OR gate is a 1 (HIGH) when any one or more of the inputs are 1 *Helpful Hint:* Boolean addition (HIGH). A similar analysis can be extended to OR gates is the same as the OR function. with any number of input variables.

Figure 2.22 illustrates a decimal-to-binary encoder that has 10 inputs and provides 4 outputs. This circuit converts decimal to binary. At any one time, only one input line has a value of 1. Table 2.8 shows the truth table of a decimal-to-binary encoder. The switches are push-button switches like those of a cell phone. The bits out of the OR gates form a 4-bit word, designated $Y_3Y_2Y_1Y_0$. When push button D is pressed, the Y_1 and Y_0 OR gates have high

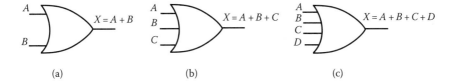

(a)   (b)   (c)

FIGURE 2.21
Boolean expressions for OR gates with two, three, and four inputs.

TABLE 2.7

Boolean Expression for the Truth
Table of a 2-Input OR Gate

A	B	Output (X)
0	0	$0 + 0 = 0$
0	1	$0 + 1 = 1$
1	0	$1 + 0 = 1$
1	1	$1 + 1 = 1$

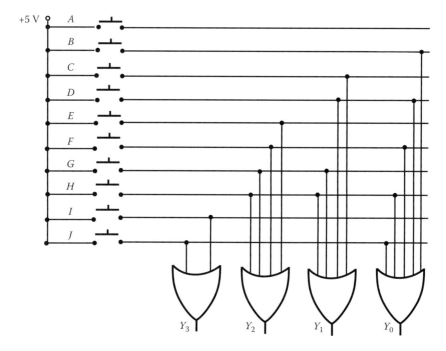

FIGURE 2.22
Decimal-to-binary encoder.

inputs, therefore, the output word is $Y_3Y_2Y_1Y_0 = 0011$. If button F is keyed, the Y_2 and Y_0 OR gates have high inputs and the output word becomes $Y_3Y_2Y_1Y_0 = 0101$. When switch J is pressed, $Y_3Y_2Y_1Y_0 = 1001$.

Section 2.5 Review Quiz

The OR gate performs what is known as logical _____.
(addition/subtraction/multiplication/division)

TABLE 2.8

Truth Table for a Decimal-to-Binary Encoder

A	B	C	D	E	F	G	H	I	J	Y_3	Y_2	Y_1	Y_0
1	0	0	0	0	0	0	0	0	0	0	0	0	0
0	1	0	0	0	0	0	0	0	0	0	0	0	1
0	0	1	0	0	0	0	0	0	0	0	0	1	0
0	0	0	1	0	0	0	0	0	0	0	0	1	1
0	0	0	0	1	0	0	0	0	0	0	1	0	0
0	0	0	0	0	1	0	0	0	0	0	1	0	1
0	0	0	0	0	0	1	0	0	0	0	1	1	0
0	0	0	0	0	0	0	1	0	0	0	1	1	1
0	0	0	0	0	0	0	0	1	0	1	0	0	0
0	0	0	0	0	0	0	0	0	1	1	0	0	1

2.6 The NAND Gate

The NAND gate is the opposite of the digital AND gate and behaves in a manner that corresponds to the opposite of the AND gate. The NAND gate is a universal gate in the sense that any Boolean function can be implemented by NAND gates. They can also be made with more than two inputs, yielding an output that is LOW if all of the inputs are HIGH, and an output that is HIGH if any of the inputs is LOW.

Common Practice: Digital systems employing certain logic circuits take advantage of NAND's functional completeness. In complicated logical expressions, normally written in terms of other logic functions such as AND, OR, and NOT, writing these in terms of NAND saves on cost because implementing such circuits using the NAND gate yields a more compact result than the alternatives.

2.6.1 NAND Gate Symbol

The NAND gate is composed of two or more inputs and a single output, as indicated by the standard logic symbol shown in Figure 2.23. This figure shows the logic symbols for 2-, 3-, and 4- input gates. Inputs are on the left, and the output is on the right in each symbol.

2.6.2 Operation of a NAND Gate

A NAND gate produces a logical 0 at the output only when all the inputs are set to logical 1. When any of the inputs is low, the output will be high. For the specific case of a 2-input NAND gate, as shown in Figure 2.23 (a) with the inputs labeled A and B and the output labeled X, the operation can be stated as follows:

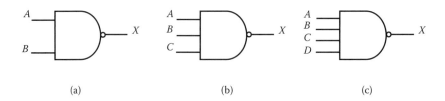

FIGURE 2.23
NAND gate symbols.

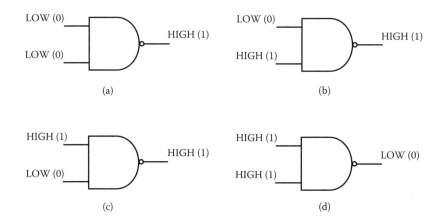

FIGURE 2.24
Operation of a 2-input NAND gate.

> *For a 2-input NAND gate, output X is LOW only when inputs A and B are HIGH; X is HIGH when either A or B is LOW, or when both A and B are LOW.*

This operation is the opposite of the AND in terms of the output level. In a NAND gate, the LOW level (0) is active or asserted output level, as indicated by the bubble on the output. Figure 2.24 illustrates the operation of a 2-input NAND gate for all four input combinations, and Table 2.9 is the truth table summarizing the logical operation of the 2-input NAND gate.

Helpful Hint: In a NAND gate, the only time a LOW output occurs is when all the inputs are HIGH.

2.6.3 Timing Diagram

Now let us explore the pulse waveform operation of a NAND gate. The next two examples illustrate the operation of a NAND gate with pulse waveform inputs. Again, as with the other types of gates, we will simply follow the truth table operation to determine the output waveforms in the proper time relationship to the inputs.

TABLE 2.9

Truth Table for a
2-Input NAND Gate

A	B	Output (X)
0	0	1
0	1	1
1	0	1
1	1	0

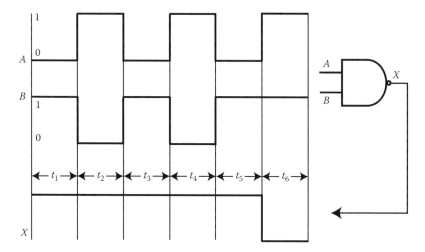

FIGURE 2.25
Timing diagram for a 2-input NAND gate.

Example 2.9

If the two waveforms A and B shown in Figure 2.25 are applied to the NAND gate inputs, determine the resulting output waveform.

SOLUTION

Output waveform X is LOW only during the four time intervals when both input waveforms A and B are HIGH as shown in the timing diagram.

Example 2.10

What would be the effect on the output of the timing diagram in Example 2.9 if an input C was added that always remained at logical 0?

SOLUTION

Following the truth table of a 3-input NAND gate, it can be easily established that the output waveform X will not have any effect if an input C that always remains at 0 is added to the timing diagram of Example 2.9.

Helpful Hint: To solve a timing analysis problem, it is useful to look at the gate's truth table to see what the unique occurrence is for the gate. In the case of the NAND gate, the odd occurrence is when the output goes low due to all high inputs.

2.6.4 Negative-OR Equivalent Operation of a NAND Gate

Inherent in a NAND gate's operation is the fact that one or more LOW inputs produce a HIGH output. Table 2.9 shows that output X is 1 when any of the inputs, A and B, is 0. From this viewpoint, a NAND gate can be used for an OR operation that requires one or more low inputs to produce a high output. This aspect of the NAND operation is referred to as negative-OR. The term negative in this context means that the inputs are defined to be in the active or asserted state when LOW. For a 2-input NAND gate performing a negative-OR operation, output X is high when either input A or input B is low, or when both A and B are low. When a NAND gate is used to detect one or more lows on its inputs rather than all highs, it is performing the negative-OR operation and is represented by the standard logic symbol shown in Figure 2.26. Although the two symbols in Figure 2.26 represent the same physical gate, they serve to define its role or mode of operation in a particular application.

Helpful Hint: Some students find it easier to analyze a NAND gate by solving it as an AND gate and then inverting the result.

Example 2.11

For a 3-input NAND gate in Figure 2.27, operating as a negative-OR, determine the output with respect to the inputs.

SOLUTION

The output waveform X is HIGH any time an input waveform is LOW as shown in the timing diagram.

NAND   Negative-OR

FIGURE 2.26
Standard symbols representing the two equivalent operations of a NAND gate.

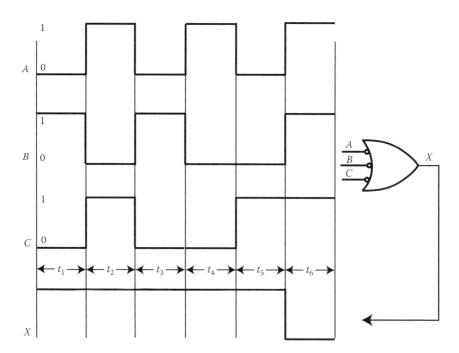

FIGURE 2.27
Timing diagram for Example 2.11.

2.6.5 Logic Expressions for a NAND Gate

The Boolean expression for the output of a 2-input NAND gate is $X = \overline{AB}$. This expression says that the two input variables, A and B, are first ANDed and then complemented, as indicated by the bar over the AND expression. This is a description in equation form of the oper-ation of a NAND gate with two inputs. Evaluating this expression for all possible values of the two input variables, you get the results shown in Table 2.10. The NAND expression can be extended to more than two input variables by including an additional letter to represent the other variables.

Helpful Hint: A bar over a variable or variables indicates an inversion.

TABLE 2.10

Boolean Expression in Truth Table for a 2-Input NAND Gate

A	B	Output ($X = \overline{AB}$)
0	0	$\overline{0.0} = \overline{0} = 1$
0	1	$\overline{0.1} = \overline{0} = 1$
1	0	$\overline{1.0} = \overline{0} = 1$
1	1	$\overline{1.1} = \overline{1} = 0$

Section 2.6 Review Quiz

The NAND gate is a _____ gate in the sense that any Boolean function can be implemented by NAND gates. (universal/computational/ basic/fundamental)

2.7 The NOR Gate

The NOR gate is the opposite of the digital OR gate and behaves in a manner that corresponds to the opposite of the OR gate. A logical 1 output results if both the inputs to the gate are 0. If one or more inputs are 1, the result is a 0 output. NOR is the result of the negation of the OR operator. The NOR gate, like the NAND gate, is a useful logic element because it can be used as a universal gate. NOR is a functionally complete operation such that combinations of NOR gates can be combined to generate any other logical function.

2.7.1 NOR Gate Symbol

The NOR gate is composed of two or more inputs and a single output, as indicated by the standard logic symbol shown in Figure 2.28. This figure shows the logic symbols for 2-, 3-, and 4- input gates. Inputs are on the left, and the output is on the right in each symbol.

2.7.2 Operation of a NOR Gate

A NOR gate produces a low output when any of its inputs is high. Only when all of its inputs are low is the output high. For the specific case of a 2-input NOR gate, as shown in Figure 2.28 with the inputs labeled A and B and the output labeled X, the operation can be stated as follows:

> *For a 2-input NOR gate, output X is LOW when either input A or input B is HIGH, or when both A and B are HIGH; X is HIGH only when both A and B are LOW.*

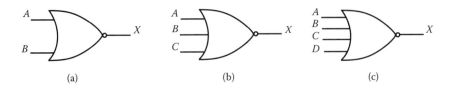

(a)   (b)   (c)

FIGURE 2.28
NOR gate symbols.

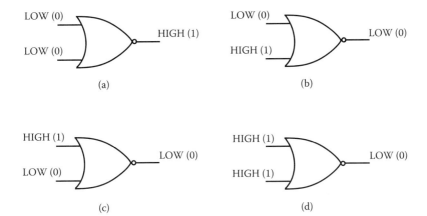

FIGURE 2.29
Operation of a 2-input NOR gate.

TABLE 2.11

Truth Table for a
2-Input NOR Gate

A	B	Output (X)
0	0	1
0	1	0
1	0	0
1	1	0

This operation results in an output level opposite that of the OR gate. In a NOR gate, the LOW output is the active or asserted output level as indicated by the bubble on the output. Figure 2.29 illustrates the operation of a 2-input NOR gate for all four possible input combinations, and Table 2.11 is the truth table for a 2-input NOR gate.

Helpful Hint: In a NOR gate, the only time a HIGH output occurs is when all the inputs are LOW.

2.7.3 Timing Diagram

The next two examples illustrate the operation of a NOR gate with pulse waveform inputs. Again, as with the other types of gates, we will simply follow the truth table operation to determine the output waveforms in the proper time relationship to the inputs.

Example 2.12

If the two waveforms shown in Figure 2.30 are applied to a NOR gate, what is the resulting output waveform?

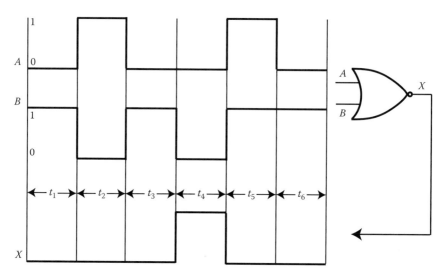

FIGURE 2.30
Timing diagram for a 2-input NOR gate.

SOLUTION

Whenever any input of the NOR gate is HIGH, the output is LOW as shown by the output waveform *X* in the timing diagram.

Example 2.13

What would be the effect on the output of the timing diagram in Example 2.12 if an input *C* was added that always remained at logical 0?

SOLUTION

Following the truth table of a 3-input NOR gate, it can be easily established that the output waveform *X* will not have any effect if an input *C* that always remains at 0 is added to the timing diagram of Example 2.12.

Helpful Hint: To solve a timing analysis problem, it is useful to look at the gate's truth table to see what the unique occurrence is for the gate. In the case of the NOT gate, the odd occurrence is when the output goes high due to all low inputs.

2.7.4 Negative-AND Equivalent Operation of the NOR Gate

NAND or NOR gates can be combined to create any type of gate. This enables a circuit to be built from just one type of gate, either NAND or NOR. For example, an AND gate is a NAND gate, followed by a NOT gate (to undo the inverting function). A NOT gate can be built from a NAND gate

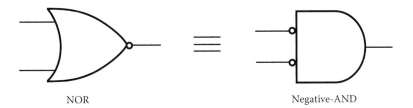

FIGURE 2.31
Standard symbols representing the two equivalent operations of a NOT gate.

by combining 2 inputs into one. Note that AND and OR gates cannot be used to create other gates because they lack the inverting (NOT) function. A NOR gate, like the NAND, has another aspect of its operation that is inherent in the way it logically functions. Table 2.11 shows that a 1 is produced on the gate output only when all of the inputs are 0. Therefore, a NOR gate can be used for an AND operation that requires all 0 inputs to produce a 1 output. This aspect of the NOR operation is called negative-AND. The term negative in this context means that the inputs are defined to be in the active or asserted state when 0. For a 2-input NOR gate performing a negative-AND operation, output X is 1 only when both inputs A and B are 0. When a NOR gate is used to detect all LOWs on its inputs rather than one or more HIGHs, it is performing the negative-AND operation and is represented by the standard symbol in Figure 2.31. Remember that the two symbols in Figure 2.31 represent the same physical gate and serve only to distinguish between the two modes of its operation.

Helpful Hint: Reducing a NAND or NOR gate to just one input creates a NOT gate.

Example 2.14

For a 2-input NOR gate operating as a negative-AND in Figure 2.32, determine the output relative to the inputs.

SOLUTION

Any time all the input waveforms are 0, the output is 1 as shown by output waveform X in the timing diagram.

2.7.5 Logic Expressions for a NOR Gate

The logic NOR gate is a combination of the digital logic OR gate with that of an inverter or NOT gate connected together in series. The NOR gate has an output that is normally at logical 1, and only goes to logical level 0 when any of its inputs are at logical 1. The NOR gate is the reverse or "complementary"

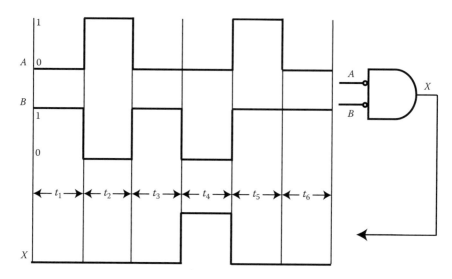

FIGURE 2.32
Timing diagram for Example 2.14.

TABLE 2.12

Boolean Expression in Truth
Table for a 2-Input NOR Gate

A	B	Output ($X = \overline{A + B}$)
0	0	$\overline{0 + 0} = \overline{0} = 1$
0	1	$\overline{0 + 1} = \overline{1} = 0$
1	0	$\overline{1 + 0} = \overline{1} = 0$
1	1	$\overline{1 + 1} = \overline{1} = 0$

form of the OR gate we have seen previously. The Boolean expression for
the output of a 2-input NOR gate can be written as $X = \overline{A + B}$. This equation
says that the two input variables are first ORed and then complemented, as
indicated by the bar over the OR expression. Evaluating this expression, you
get the results shown in Table 2.12. The NOR
expression can be extended to more than two
input variables by including additional letters
to represent the other variables.

Helpful Hint: Some students find it easier
to analyze a NOR gate by solving it as an
OR gate and then inverting the result.

Section 2.7 Review Quiz

The NOR gate is a _____ gate in the sense that any Boolean function
can be implemented by NOR gates. (universal/computational/basic/
fundamental)

2.8 The Exclusive-OR Gate

The logical operation Exclusive-OR (abbreviated as XOR, EOR, or EXOR), is a type of logical disjunction on two operands that results in a value of true if exactly one of the operands has a value of true. An OR gate recognizes words with one or more 1s. The Exclusive-OR gate is different; it recognizes only words that have an odd number of 1s. A simple way to state this is "one or the other but not both." The XOR gate performs modulo-2 addition.

2.8.1 XOR Gate Symbol

The XOR gate is composed of two or more inputs and a single output, as indicated by the standard logic symbol shown in Figure 2.33. This figure shows the logic symbols for 2-, 3-, and 4- input gates. Inputs are on the left, and the output is on the right in each symbol.

2.8.2 Operation of XOR Gate

The output of an XOR gate is high only when the two inputs are at opposite logic levels. This operation can be stated as follows with reference to input A and B, and output X:

For an XOR gate, output X is high when input A is low and input B is high, or when input A is high and input B is low; X is low when A and B are both high or both low.

The four possible input combinations and the resulting outputs for an XOR gate are illustrated in Figure 2.34. The high level is the active or asserted output level and occurs only when the inputs are at opposite levels. Figure 2.35 shows one way to build an XOR gate. The upper AND gate forms the product $\overline{A}B$, and the lower gate gives $A\overline{B}$. Therefore, the Boolean equation is

$$Y = \overline{A}B + A\overline{B}$$

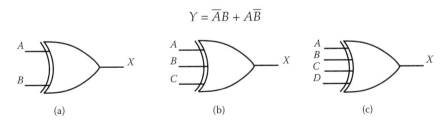

(a) (b) (c)

FIGURE 2.33
XOR gate symbols.

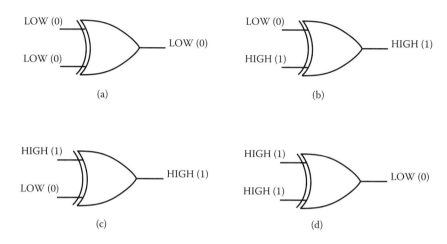

FIGURE 2.34
Operation of XOR gate.

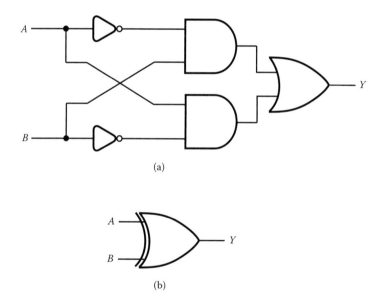

FIGURE 2.35
(a) XOR gate, (b) a 2-input XOR gate.

Let us discuss in detail the operation of this circuit. In Figure 2.35 (a), two low inputs mean both AND gates have low outputs. Thus, the final output is low. If *A* is set to low and *B* is set to high, the output of the top AND gate becomes high. Thus, the final output is high. Similarly, a high *A* and low *B* results in a final output that is high. If both inputs are high, both AND gates have low outputs, and the final out is low.

2.8.3 XOR Gate Truth Table

Truth table for a 2-input XOR gate is shown in Table 2.13. The output is high when A or B is high but not both. This is the reason why this circuit is called Exclusive-OR. In other words, the output is 1 only when the inputs are different, just like the previously stated phrase "one or the other but not both."

The logical XOR function of two variables is represented mathematically by a ⊕ between the two variables. The Boolean expression for the output of a 2-input XOR gate can be written as

TABLE 2.13

2-Input XOR Gate

A	B	Output $(\bar{A}B + A\bar{B})$
0	0	0
0	1	1
1	0	1
1	1	0

$$X = A \oplus B. \text{ The } \oplus \text{ sign is read as "XOR."}$$

2.8.4 Timing Diagram

The next two examples illustrate the operation of an XOR gate with pulse waveform inputs. Again, as with the other types of gates, we will simply follow the truth table operation to determine the output waveforms in the proper time relationship to the inputs.

Example 2.15

If the two waveforms shown in Figure 2.36 are applied to an XOR gate, what is the resulting output waveform?

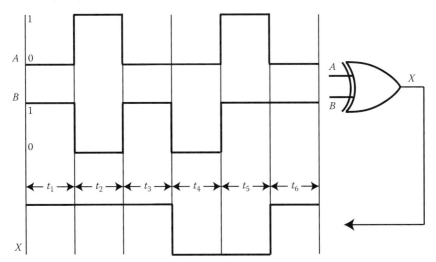

FIGURE 2.36
Timing diagram for XOR gate.

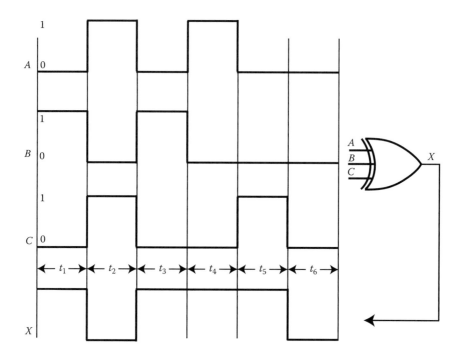

FIGURE 2.37
Timing diagram for XOR gate.

SOLUTION

The output waveform X always remains at 0 when an even number of inputs has 1s. The output becomes 1 when an odd number of inputs has 1s.

Example 2.16

If the two waveforms shown in Figure 2.37 are applied to an XOR gate, what is the resulting output waveform?

SOLUTION

The output waveform X always remains at 0 when an even number of inputs has 1s. The output becomes 1 when an odd number of inputs has 1s.

In Figure 2.38 (a), the upper gate produces $A\,B$, while the lower gate gives $C\,D$. The final gate XORs both of these sums to get

$$Y = (A \oplus B) \oplus (C \oplus D)$$

If we substitute input values into the question and solve for the output, we will get the result 1 for inputs of $A = 0$, $B = 0$, $C = 0$, and $D = 1$.

$$Y = (0 \oplus 0) \oplus (0 \oplus 1) = 1$$

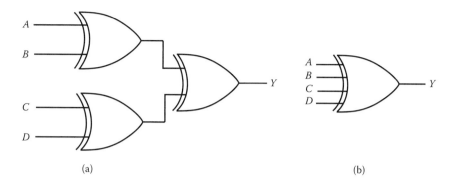

FIGURE 2.38
A 4-input XOR gate: (a) circuit with 2-input XOR gates, (b) equivalent logic symbol.

Another case is when an input of $A = 0$, $B = 1$, $C = 0$, and $D = 0$ produces an output of 1.

$$Y = (0 \oplus 1) \oplus (0 \oplus 0) = 1$$

We can plot the truth table by substituting the input values in this equation to get the output value. Another way is to analyze Figure 2.38 (a). If all inputs are 0s, the first two gates have 0 outputs. Thus, the final gate has a 0 output. If A to C are 0s and D is a 1, the upper gate has a 0 output, the lower gate has a 1 output, and the final gate has a 1 output. The circuit action can be analyzed in this way to come up with all the values in a truth table. We summarize the action in Table 2.14. The important point to note from this table is that each input word with an odd number of 1s produces a 1 output. For example, the first input word to produce a 1 output is 0001. This word has an odd number of 1's. The next word with a 1 output is 0010, which also has an odd number of 1's. A 1 output also occurs for the following words: 0100, 0111, 1000, 1011, 1101, and 1110, all of which have an odd number of 1's. The circuit basically recognizes words with an odd number of 1's and rejects words with an even number of 1's. This circuit is equivalent to a 4-input XOR gate.

Helpful Hint: XOR gate recognizes words with an odd number of 1's.

2.8.5 Parity

In telecommunications and computing, parity refers to the evenness or oddness of the number of bits with value 1 within a given set of bits. Thus parity is determined by the value of all the bits in a given set. Even parity means a word has an even number of 1's. For example, numbers 111001 and 110011 have even number of parity because they contain four 1's. Odd parity means a word has an odd number of 1's. For instance, 100011 and 110001 have odd parity because they contain three 1's. Here are two more examples:

TABLE 2.14

Truth Table for a 4-input XOR Gate

A	B	C	D	Output	Comments
0	0	0	0	0	Even
0	0	0	1	1	Odd
0	0	1	0	1	Odd
0	0	1	1	0	Even
0	1	0	0	1	Odd
0	1	0	1	0	Even
0	1	1	0	0	Even
0	1	1	1	1	Odd
1	0	0	0	1	Odd
1	0	0	1	0	Even
1	0	1	0	0	Even
1	0	1	1	1	Odd
1	1	0	0	0	Even
1	1	0	1	1	Odd
1	1	1	0	1	Odd
1	1	1	1	0	Even

1111 0000 0000 0011 = even parity

1111 0000 1111 0001 = odd parity

The first word has even parity because it contains six 1's, whereas the second word has odd parity because it contains nine 1's.

XOR gates are ideal for testing the parity of a word. XOR gates recognize words with an odd number of 1's. Therefore, even-parity words produce a low output, and odd-parity words produce a high output. This property of being dependent upon all the bits and changing value if any one bit changes allows for the XOR gate's use in error detection schemes.

2.8.5.1 Odd-Parity Generator

Figure 2.39 shows a 7-bit register that stores a letter in ASCII form. Let us take two examples, one with letter *A* and the other with letter *C*.

The ASCII code for letter *A* is 100 0001. Since this word has two 1s, an even parity, the XOR gate will have an output of 0. Additionally, there is a NOT gate in addition to the ASCII code *A* at the output. Therefore, the overall output of the circuit is the 8-bit word 1100 0001. The point to note here is that this word has an odd parity. Now let us take the example of the letter *C*. The ASCII code for letter *C* is 100 0011. Since this word has three 1's, an odd parity, the XOR gate will have an output of 1. Again, there is an inverter in addition to the ASCII code *C* at the output. Therefore, the overall output of

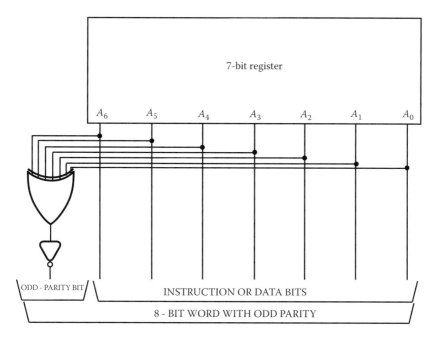

FIGURE 2.39
Odd-parity generator.

the circuit is the 8-bit word 0100 0011. Again, the point to note here is that this word also has an odd parity. The circuit is called an odd-parity generator because it produces an 8-bit output word with odd parity. If the register word has even parity, 0 is the output of the XOR gate, and the odd-parity bit is 1. In the other case, when the register word has odd parity, the XOR gate has a 1 output, and the odd-parity bit is set to 0. Irrespective of the register contents, the odd-parity bit and the register bits form a new 8-bit word that will always have odd parity.

One would be thinking as to what is the practical use for such a circuit. Data transmission is one of the areas that find such an application extremely useful. Because of transients and noise, a 1-bit error sometimes occurs in transmitted data. For example, the letter *A* may be transmitted over a phone line in ASCII form as 100 0001. If due to the transients the second bit from the right changes from 0 to 1, the received data will become 100 0011, which is actually *C* in ASCII. The solution is to transmit an odd-parity bit along with the data word and have an XOR gate test each received word for odd parity. If no error is detected, the XOR gate will recognize the word. If a 1-bit error has occurred, the XOR gate will disregard the received word, and the data can be rejected. In the specific case of *C* getting transmitted for *A*, the odd-parity bit after the inversion will be 1. This we have seen before. But the seven bits for *A* have now changed to 100 0011. The final word now becomes 1100 0011. This word has even parity and should be rejected.

Example 2.17

What is the output for each of these input words if they are the inputs to a 16-input XOR gate?

(a) 1111 1111 1111 1111
(b) 1010 1100 1001 1100
(c) 1010 1100 1011 1111
(d) 0000 0000 0000 0000

SOLUTION

(a) The word has sixteen 1's, an even number. Therefore, the output signal is

EVEN = 1

(b) The word has eight 1's, an even number. Therefore, the output signal is

EVEN = 1

(c) The word has eleven 1's, an odd number. Therefore, the output signal is

ODD = 0

(d) The word has zero 1's, an even number. Therefore, the output signal is

EVEN = 1

Section 2.8 Review Quiz

The XOR gate performs modulo-2 addition. (True/False)

2.9 The Exclusive-NOR Gate

The logical operation Exclusive-NOR (abbreviated as XNOR, ENOR, or EXNOR) is logically equivalent to an XOR gate followed by an inverter.

2.9.1 XNOR Gate Symbol

The XNOR gate is composed of two or more inputs and a single output, as indicated by the standard logic symbol shown in Figure 2.40. This figure shows the logic symbols for 2-, 3-, and 4- input gates. Inputs are on the left, and the output is on the right in each symbol. The bubble on the output of the XNOR symbol indicates that its output is opposite that of the XOR gate.

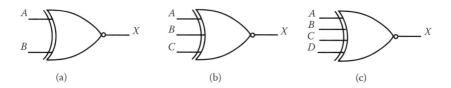

FIGURE 2.40
XNOR gate symbols.

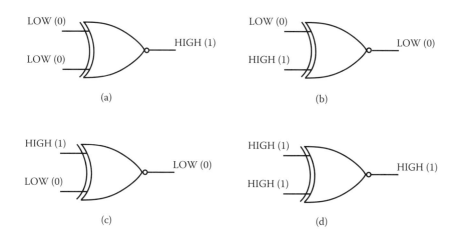

FIGURE 2.41
Operation of XNOR gate.

2.9.2 Operation of XNOR Gate

When the two input logic levels are opposite, the output of the exclusive-NOT gate is LOW. The operation can be stated as follows:

> *For a 2-input XNOR gate with inputs A and B, output X is low when input A is low and input B is high, or when A is high and B is low; X is high when X and B are both high or both low. The four possible input combinations and the resulting outputs for an XNOR gate are illustrated in Figure 2.41.*

2.9.3 XNOR Gate Truth Table

The truth table for a 2-input XNOR gate is shown in Table 2.15. Because of the inversion on the output side, the truth table of an XNOR gate is the complement of an XOR truth table. As shown in Table 2.15, the output is high when the inputs are the same. Therefore, the 2-input XNOR gate is ideally suited for bit comparison, recognizing when two input bits are identical.

Helpful Hint: Instead of recognizing odd-parity words, XNOR gates recognize even-parity words.

TABLE 2.15

2-Input XNOR Gate

A	B	Output
0	0	1
0	1	0
1	0	0
1	1	1

2.9.4 Timing Diagram

The next example illustrates the operation of an XNOR gate with pulse waveform inputs. Again, as with the other types of gates, we will simply follow the truth table operation to determine the output waveforms in the proper time relationship to the inputs.

Example 2.18

If the two waveforms shown in Figure 2.42 are applied to an XNOR gate, what is the resulting output waveform?

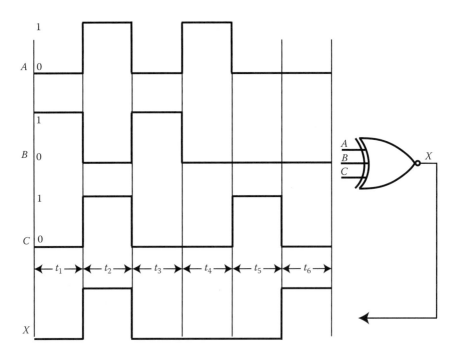

FIGURE 2.42
Timing diagram for XNOR gate.

SOLUTION

The output waveform X always remains at 1 when inputs are identical. The output becomes 0 when an odd number of inputs has 1's.

Section 2.9 Review Quiz

The logical operation XNOR is logically equivalent to an OR gate followed by an inverter. (True/False)

Common Practice: When interfacing memory to a microprocessor, logic gates are used to detect memory read-and-write cycles and to decode the address signals.

2.10 Summary

1. A signal that is normally low but pulses high when active is said to be active-high.
2. A signal that is normally high but pulses low when active is said to be active-low.
3. A logic circuit can be described by a truth table that lists all possible circuit inputs and the corresponding outputs.
4. Truth tables are important for testing and troubleshooting logic circuits.
5. The inverter output is the complement of the input.
6. The inverter is used to change an active-low signal into an active-high signal, and vice versa.
7. The AND gate output is high only when all the inputs are high.
8. The OR gate output is low when all the inputs are low.
9. Every AND gate has an OR gate equivalent, and every OR gate has an AND gate equivalent.
10. The NAND gate output is low only when all the inputs are high.
11. The NAND can be viewed as a negative-OR whose output is high when any input is low.
12. The NOR gate output is low when any of the inputs are high.
13. The NOR can be viewed as a negative-AND whose output is high only when all the inputs are low.
14. The exclusive-OR (XOR) gate output is high when the inputs are not the same.
15. Applications for the XOR gate include logic comparator and binary adder.
16. The exclusive-NOR gate output is low when the inputs are not the same.

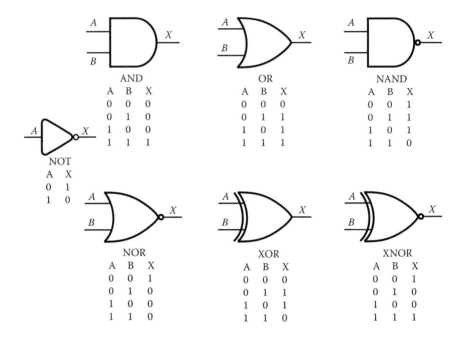

FIGURE 2.43
Summary of major gates.

17. Distinctive shape symbols and truth tables for various logic gates are shown in Figure 2.43.

18. An effective way to measure the precise timing relationships of digital waveforms is with an oscilloscope or a logic analyzer.

Glossary

AND gate: A logic gate that produces a high output only when all the inputs are high.

Bias: The voltage necessary to cause a semiconductor device to conduct or cut off current flow. A device can be forward biased or reverse biased, depending on what action is desired.

Boolean algebra: The mathematics of logic circuits.

Boolean equation: An algebraic expression that illustrates the functional operation of a logic gate or combination of logic gates.

CMOS: Complementary metal-oxide semiconductor; a class of integrated logic circuits that is implemented with a type of field-effect transistor.

Complement: The inverse or opposite of a number. Low is the complement of high, and 0 is the complement of 1.

Controller inverter: A logic circuit that produces the 1's complement of the input word.

Disabled: To disallow or deactivate a function or circuit.

Enabled: To allow or activate a function or circuit.

Even parity: An even number of 1s in a binary word.

Exclusive-NOR gate: A logic gate that produces a logical 0 when the two inputs are at opposite levels.

Exclusive-OR (XOR) gate: A logic gate that produces a logical 1 when the two inputs are at opposite levels.

Gate: A logic circuit with one or more input signals but only one output signal.

Inverter: Logic circuit that inverts or complement its inputs.

Logic: Study of arguments.

Logic circuit: A circuit whose input and output signals are 2-state, either low or high voltages. The basic logic circuits are OR, AND, and NOT gates.

NAND gate: A logic gate that produces a low output when all the inputs are high.

NOR gate: A logic gate that produces a low output when one or more of the inputs are high.

Odd parity: An odd number of 1's in a binary word.

OR gate: A logic gate that produces a high output when one or more of its inputs are high.

Parity generator: A circuit that produces either an odd- or even-parity bit to go along with the data.

Timing diagram: A graphical method of showing the exact output behavior of a logic circuit for every possible set of input conditions.

Truth table: A mean for describing how a logic circuit's output depends on the logic levels present at the circuit's input.

Word: A string of bits that represents a coded instruction or data.

Answers to Section Review Quiz

2.2 True

2.3 NOT

2.4 Logical

2.5 Addition

2.6 Universal

2.7 Universal

2.8 True

2.9 False

True/False Quiz

1. Inverters can be drawn using two different symbols but electrically the two are the same gate.

2. An exclusive-OR gate will produce a high output whenever either or both of its input are high.

3. An AND gate and an active-low input NOR gate each have the same truth table.

4. An inverter performs the NOR operation.

5. If any input of an OR gate is 1, the output is 1.

6. Gates like AND and OR can have only two inputs.

7. A NAND gate has an output that is opposite to the output of an AND gate.

8. If all inputs to an AND gate are 1, the output is always 0.

9. The NOR gate can be viewed as an OR gate followed by an inverter.

10. If the inputs to a 2-input XOR gate are opposite, the output is always 0.

11. CMOS stands for complementary metal-oxide semiconductor.

12. An odd-parity generator can never be used in data transmission.

13. The Boolean expression for the output of a 2-input XOR gate can be written as $X = A \oplus B$.

14. A timing diagram is a representation of a set of signals in the space domain.

15. Boolean addition is the same as the OR function.

Questions

QUESTION 2.1

What are some of the applications of truth tables in digital electronics?

QUESTION 2.2

What are the benefits of universal gates?

QUESTION 2.3

Describe Boolean addition and multiplication and their relationship with logic gates.

QUESTION 2.4

Is it possible to implement an XOR gate by only using a NAND gate? If so, how?

QUESTION 2.5

Describe how parity is used in communication links.

Problems

PROBLEM 2.1

Sketch the output voltage for the circuit of Figure 2.44. Use the ideal diode approximation.

PROBLEM 2.2

What is the output waveform of the circuit in Figure 2.45?

PROBLEM 2.3

Show how to complement each bit in Figure 2.46 such that the output from A to F is 100111.

PROBLEM 2.4

Describe the truth table of an 8-input AND gate.

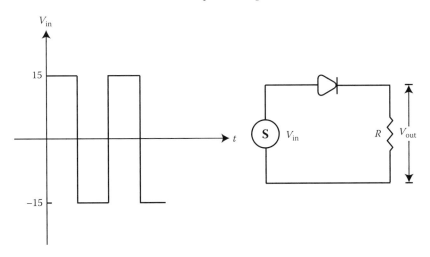

FIGURE 2.44
Circuit diagram for Problem 2.1.

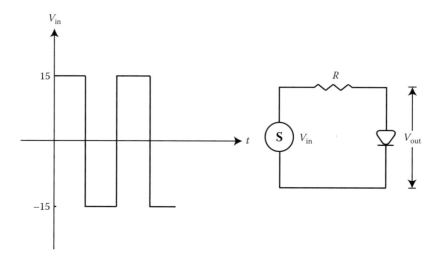

FIGURE 2.45
Circuit diagram for Problem 2.2.

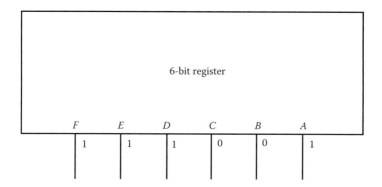

FIGURE 2.46
Basic setup for Problem 2.3.

PROBLEM 2.5

A NOR gate has 6 inputs.

(a) How many input words are in its truth table?
(b) What is the only input word that produces a 1 output?
(c) If each input is connected to an inverter (NOT gate), describe the change in the behavior of the circuit.

PROBLEM 2.6

Design a digital circuit to input two numbers S0 and S1 and produce the outputs SUM and CARRY representing the sum of the two bits and the carry (if any).

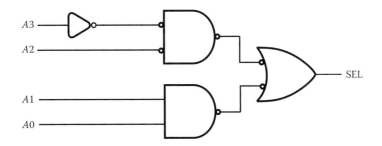

FIGURE 2.47
Logic circuit for Problem 2.7.

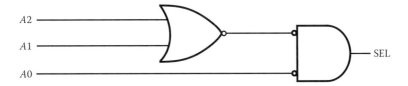

FIGURE 2.48
Logic circuit for Problem 2.8.

PROBLEM 2.7

Determine the word description and truth table for the logic circuit shown in Figure 2.47.

PROBLEM 2.8

Determine the word description and truth table for the logic circuit shown in Figure 2.48.

PROBLEM 2.9

A seven-segment display is a form of electronic display device for displaying decimal numerals. Seven-segment displays are widely used in digital clocks, electronic meters, and other electronic devices for displaying numerical information. Each digit of these displays is comprised of seven bars of light-emitting semiconductor (or light-absorbing liquid crystal) material arranged as shown in Figure 2.49. If the decimal number 2 is to be displayed, then segments a, b, g, and d are energized. You are required to create the truth table for logic that receives a *BCD* digit as input and provides seven outputs to drive a corresponding display digit.

PROBLEM 2.10

Show how a truth table may be recreated from Boolean equations by constructing the truth table that corresponds to the following equation:

$$X = A'B'C' + A'BC' + AB'C' + AB'C$$

$$Y = A'BC + ABC'$$

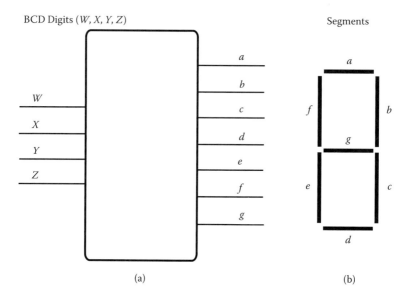

BCD Digits (W, X, Y, Z)   Segments

(a)   (b)

FIGURE 2.49
Basic setup for Problem 2.9.

PROBLEM 2.11

Determine the output waveform for the AND gate shown in Figure 2.50.

PROBLEM 2.12

What will happen to the X output waveform in Figure 2.50 if the B input is kept at the 0 level?

PROBLEM 2.13

How many input words are in the truth table of an 8-input OR gate? Which input words result in a high output?

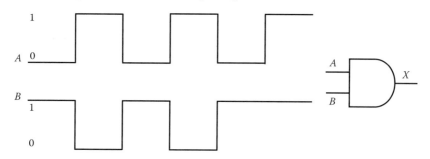

FIGURE 2.50
Input signals for Problem 2.11.

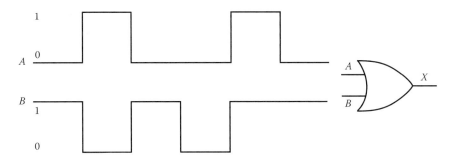

FIGURE 2.51
Input signals for Problem 2.14.

PROBLEM 2.14

Determine the OR gate output in Figure 2.51.

PROBLEM 2.15

For the situation depicted in Figure 2.52, determine the waveform at the OR gate output.

PROBLEM 2.16

What would happen to the glitch in the output in Figure 2.52 if input C sat in the 1 state during the first half of the timing diagram?

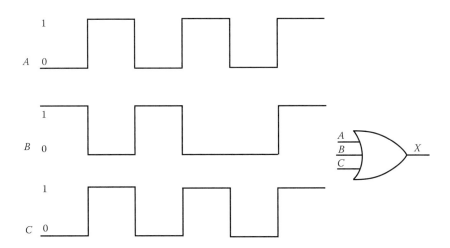

FIGURE 2.52
Input signals for Problem 2.15.

PROBLEM 2.17

Certain chemicals are stored in two tanks at a manufacturing plant equipped with wireless sensor network. Each tank has a wireless sensor mote that detects when the chemical level drops to 20% of full. The wireless sensor sends a signal with logical 1 when the tanks are more than 20% of full. When the volume of chemical in a tank drops to 20% of full, the sensor mote sends a signal with logical 0 to the base station. The base station has a light-emitting diode (LED) on an indicator panel to show when both tanks are more than 20% of full. Demonstrate with the help of a logic diagram how a NAND gate can be used to implement this function.

PROBLEM 2.18

Describe the behavior of a circuit where an inverter is connected to one of the inputs of an XOR gate.

PROBLEM 2.19

Figure 2.53 illustrates a controlled inverter, sometimes referred to as a programmed inverter. Analyze the circuit and show how it transmits a binary word or its 1's complement.

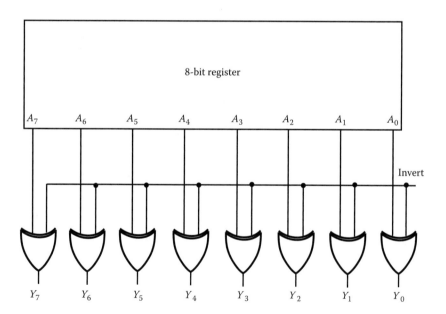

FIGURE 2.53
A controlled inverter.

PROBLEM 2.20

What is the output of a 16-input XNOR gate for each of these input words?

(a) 0000 0000 0000 0000
(b) 0000 0000 0000 1111
(c) 1111 0101 1111 1100
(d) 0101 1100 0111 0011
(e) 1111 0000 1010 0110
(f) 1111 1111 1111 1110.

3

Microcontroller Hardware

> Man is a wonderful creature; he sees through the layers of fat (eyes), hears through a bone (ears) and speaks through a lump of flesh (tongue).

> **—Ali ibn Abu Talib (7th Century Saint)**

OUTLINE

OBJECTIVES

Upon completion of this chapter, you should be able to

1. Explain the operation of a common-emitter transistor circuit used as a digital inverter switch.
2. Read and understand digital IC terminology as specified in the manufacturer's data sheets.
3. Make basic comparisons between the major IC technologies—TTL and CMOS.
4. List specific fixed-function integrated-circuit devices that contain the various logic gates.
5. Describe the operation of seven-segment displays.
6. Appreciate the basic operation of liquid-crystal displays and keypads.

7. Understand the operational role of the different types of buses and their signals in a microcontroller.
8. Understand the major functional units of a processor.
9. List the HC11 processor registers and processor modes.
10. Compare and contrast various memory types.
11. Understand how to relocate RAM and systems registers.
12. Understand how to enable and disable ROM and EEPROM through CONFIG.
13. List the HC11 I/O ports.
14. Use some basic BUFFALO commands to control the EVBU.

Key Terms: Accumulator, Address Bus, Arithmetic Logic Unit (ALU), Bipolar, BUFFALO, Bus, Central Processing Unit (CPU), Chip, CMOS, Control Bus, Control Unit, Cutoff, Data Bus, Dual In-Line Package (DIP), EEPROM, Embedded Systems, Enable, EPROM, EVBU, Hardware, Input/Output (IO), Integrated Circuit (IC), Keypad, Light-Emitting Diode (LED), Liquid-Crystal Display (LCD), Microcontroller, Monitor Program, MOSFET, Peripheral, Port Number, Processor Register (or General-Purpose Register), Program Counter, Programmable Logic Controller (PLC), Saturation, Seven-Segment Display, Stack Pointer, Surface-Mounted Device (SMD), TTL, VLSI

3.1 Introduction

A microcontroller is a small computer on a single integrated circuit (IC) containing a processor core, memory, and programmable input/output peripherals. Our world is full of microcontrollers. We find several of them in automatically controlled products and devices, such as automobile engine control systems, implantable medical devices, remote controls, home and office appliances, power tools, and toys. In the last two decades, the advent of technology has made them smaller in size and relatively less expensive compared to a design that uses a separate microprocessor, memory, and input/output devices. In general, ICs are cheaper to manufacture and, for that reason, have found a home in almost nearly all modern electrical products such as cars, television sets, CD players, cellular phones, etc. A monolithic IC is an electronic circuit that is constructed entirely on a single small chip of silicon. All the components that make up the circuit, that is, transistor, diodes, resistor, and capacitors, are an integral part of that single chip. In fixed-function logic, the logic functions are set by the manufacturer and cannot be altered. The IC package can be classified in many ways. One such classification is based on the way they are mounted on printed circuit boards as

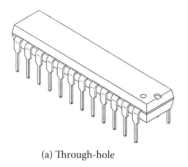

(a) Through-hole

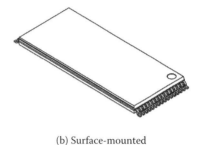

(b) Surface-mounted

FIGURE 3.1
(a) Through-hole and (b) surface-mounted device.

either through-hole mounted or surface mounted. The first type, that is, the through-hole type package, has pins (leads) that are inserted through holes in the printed circuit board, and these can be soldered to conductors on the opposite side. The most common type of through-hole package is the dual in-line package (DIP) shown in Figure 3.1 (a).

Figure 3.1 (b) illustrates the second type of IC package that uses surface-mounted technology (SMT). The surface-mount technology was developed in the 1960s and became widely used in the late 1980s. In the industry, it has largely replaced the through-hole technology construction method of fitting components with wire leads into holes in the circuit board. Surface mounting is a space-saving alternative to through-hole mounting. The holes through the PC board are unnecessary for SMT. The pins of surface-mounted packages are soldered directly to conductors on one side of the board, leaving the other side free for additional circuits. Also, for a circuit with the same number of pins, a surface-mounted package is much smaller than a dual in-line package because the pins are placed closer together.

3.2 A Transistor as a Switch

We discussed in Chapter 2 how an inverter is used to complement the input to the opposite state at its output. Figure 3.2 shows how a common-emitter-connected transistor switch can be used to perform the function of an inverter. When V_{in} equals 1 (or +5 V), the transistor goes into saturation state, and so the transistor is turned on. At this time, the V_{out} equals 0 (0 V). On the other hand, when V_{in} equals 0 (0 V), the transistor goes into the cutoff state and turns off. Now the V_{out} equals 1 (or approximately +5 V), assuming that R_L is much greater than R_C ($R_L > R_C$).

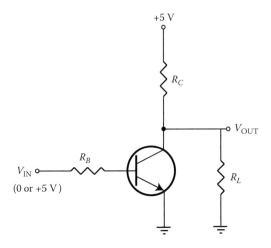

FIGURE 3.2
Common-emitter configuration.

Example 3.1

If R_C = 1 kΩ and R_L = 10 kΩ in Figure 3.1, determine the value of V_{out} when the input voltage is zero, that is, V_{in} = 0. What is the effect of adding more loads in parallel with R_L such that R_L decreases to 1 kΩ?

SOLUTION

With R_C = 1 kΩ and R_L = 10 kΩ

$$V_{out} = (5 \text{ V} \times 10 \text{ k}\Omega)/(1 \text{ k}\Omega + 10 \text{ k}\Omega) = 4.55 \text{ V}$$

When the load is increased such that R_L decreases to 1 kΩ, the V_{out} drops to

$$V_{out} = (5 \text{ V} \times 1 \text{ k}\Omega)/(1 \text{ k}\Omega + 1 \text{ k}\Omega) = 2.5 \text{ V}$$

Example 3.2

Let R_C = 1 kΩ and R_L = 10 kΩ in Figure 3.1. Determine the value of I_C when input voltage is high, that is, V_{in} = 1 (or + 5 V). What is the value of I_C if the R_C is set to a large value?

SOLUTION

With R_C = 1 kΩ and R_L = 10 kΩ,

$$I_C = 5 \text{ V}/R_C = 5 \text{ mA}$$

The I_C decreases with the increase in the value of R_C.

From Example 3.1, it can be concluded that when the transistor is cut off ($V_{out} = 1$), a small value of R_c is desirable to ensure that V_{out} is close to 5 V. But, as observed in Example 3.2, when the transistor is saturated, to avoid excessive collector current, a large value of R_c is desirable. This idea of needing a variable R_c resistance is accommodated by the transistor–transistor logic (TTL) integrated circuit.

The types of transistors with which all ICs are implemented are either MOSFETs (metal-oxide semiconductor field-effect transistors) or bipolar junction transistors. A circuit technology that uses MOSFETs is CMOS (complementary MOS). TTL uses a bipolar transistor in place of R_c to act like a varying resistance. BiCMOS uses a combination of both CMOS and bipolar.

Helpful Hint: If you appreciate the idea that V_{out} varies depending on the size of the connected load, it will help you realize the reason why in actual circuits the output voltage is not exactly 0 or 5 V.

Section 3.2 Review Quiz

Both the NPN and PNP type bipolar transistors can be made to operate as an "ON/OFF" type solid-state switches. (True/False)

3.3 The TTL Integrated Circuit

Integrated circuits are generally called ICs or chips. They are complex circuits that have been etched onto tiny chips of semiconductor. The chip is packaged in a plastic holder with pins spaced on a grid that will fit the holes on breadboards. Delicate wires inside the package provide the connections from the chip to the pins. The transistor-transistor logic (TTL) is a very popular family of ICs. It is much more widely used than RTL (resistor-transistor logic) or DTL (diode-transistor logic) circuits, which were the forerunners of TTL. It is called transistor-transistor logic because both the logic gating function (e.g., AND) and the amplifying function are performed by transistors. TTL is notable for being a widespread integrated-circuit family used in many applications such as computers, instrumentation and controls, consumer electronics, etc. One basic function of a TTL integrated circuit is as a complementing switch, or inverter.

The pin configuration of the 7404 is shown in Figure 3.3. The power supply connections to the IC are made to pin 14 (+5 V) and pin 7 (ground), which supply power to all six logic circuits. Although never shown in the pin configuration top view of the digital ICs, each gate is electrically tied internally to both V_{cc} and ground. In the case of the 7404, the logic circuits are called inverters. The symbol for each inverter is the same as the NOT gate, that is, a triangle with a circle at the output. The circle is used to indicate the inversion function.

Helpful Hint: In the later chapters, when we design some of the microcontroller-based projects, we will often use the inverters of 7404 IC.

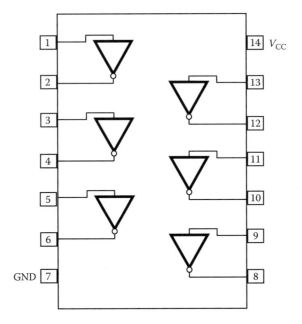

FIGURE 3.3
A 7404 hex inverter pin configuration.

Section 3.3 Review Quiz

Transistor-transistor logic (TTL) is a class of digital circuits built from bipolar junction transistors (BJT) and resistors. (True/False)

3.4 The CMOS Integrated Circuit

In the world of ICs, another common integrated-circuit technology used in digital logic is the CMOS (complementary metal-oxide semiconductor). CMOS uses a complementary pair of metal-oxide semiconductor field-effect transistors (MOSFETs) instead of the bipolar transistors used in TTL chips. CMOS is also sometimes referred to as complementary–symmetry metal-oxide-semiconductor (or COS-MOS). The words "complementary–symmetry" refer to the fact that the typical digital design style with CMOS uses complementary and symmetrical pairs of p-type and n-type metal-oxide semiconductor field-effect transistors (MOSFETs) for logic functions. CMOS allows a high density of logic functions on a chip. It was for primarily this reason that CMOS won the race in the 1980s and became the most used technology to be implemented in very large-scale integration (VLSI) of chips.

Two important advantages of CMOS devices are high noise immunity and low static power consumption. Significant power is only consumed while the transistors in the CMOS device are switching between ON and OFF states. Consequently, CMOS devices do not produce as much waste heat as other forms of logic devices, for example TTL. For that reason, CMOS technology is used in microprocessors, microcontrollers, static RAM, and other digital logic circuits. It is also used for several analog circuits such as image sensors, data converters, and highly integrated transceivers for many types of communication. The disadvantage of using CMOS is that, generally, its switching speed is slower than that of TTL, and it is susceptible to burnout due to electrostatic charges if not handled properly. Figure 3.4 shows the pin configuration for a 4049 CMOS hex inverter.

Common Practice: The channel in MOSFET can be of n-type or p-type, and is accordingly called an nMOSFET or a pMOSFET (also commonly nMOS, pMOS).

Helpful Hint: MOSFETs are by far the most common transistors in both digital and analog circuits, though the bipolar junction transistors (BJTs) were at one time much more common.

Section 3.4 Review Quiz

    (a) MOSFET stands for _____. (b) CMOS stands for _____.

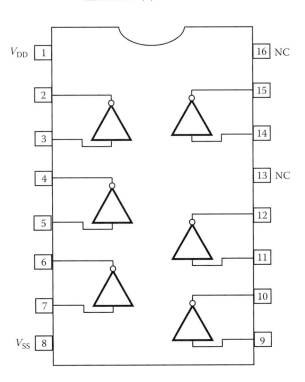

FIGURE 3.4
A 4049 CMOS hex inverter pin configuration.

3.5 Using Integrated-Circuit Logic Gates

We studied logic gates in great detail in Chapter 2. We know by now that logic gates are the basic components in digital electronics. They are used to create digital circuits and even complex ICs. All the logic gates are available as ICs. The IC pin layout, logic gate type, and technical specification are all contained in the logic data manual supplied by the manufacturer of the IC. Generally, the pins are numbered anticlockwise around the IC (chip) starting near the notch or dot. Referring to a TTL or a CMOS logic data manual, we can see that there are several logic gate ICs. We list just a few:

- The 7400 TTL and the 4011 CMOS are quad two-input NAND ICs.
- The 7404 TTL and the 4049 CMOS are hex inverter ICs.
- The 7402 TTL and the 4001 CMOS are quad two-input NOR ICs.
- The 7408 (74HC08) and 7432 (74HC32) are quad two-input AND gate.
- The 7411 (74HC11) is a triple three-input AND gate.
- The 7421 (74HC21) is a dual four-input AND gate.

HC stands for high-speed CMOS. For example, the 7408 is a TTL AND gate, and the 74HC08 is the equivalent CMOS AND gate. The terms quad (four), triple (three), and dual (two) refer to the number of separate gates on a single IC.

Figure 3.5 illustrates in detail the 7408 IC. The 7408 is a 14-pin dial-in-line package (DIP) IC. The power supply connections are made to pins 7 and 14. This supplies the operating voltage for all four AND gates on the IC. Pin 1 is identified by a small indented circle next to it or by a notch cut out between pin 1 and 14. The IC pin configuration can also be summarized as in Table 3.1.

Example 3.3

The top view of an IC package is shown in Figure 3.6. Label the pin numbers.

SOLUTION

The pins are numbered counterclockwise around the IC starting near the notch or dot. Figure 3.7 shows the pin numbers.

Common Misconception: Students often assume that gate output obtains its high and low voltage levels from its input pin, but the fact is that it receives its voltage from V_{CC} or ground.

Section 3.5 Review Quiz

The pins are numbered clockwise around the IC (chip) starting near the notch or dot. (True/False)

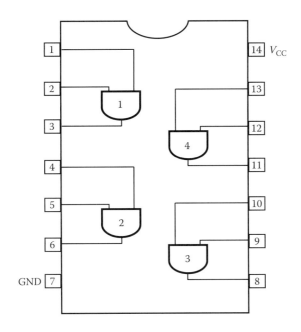

FIGURE 3.5
The 7408 IC.

TABLE 3.1

The 7408 IC Pin Configuration

Pin Number	Pin Description
1	Input gate 1
2	Input gate 1
3	Output gate 1
4	Input gate 2
5	Input gate 2
6	Output gate 2
7	Ground
8	Output gate 3
9	Input gate 3
10	Input gate 3
11	Output gate 4
12	Input gate 4
13	Input gate 4
14	Positive supply

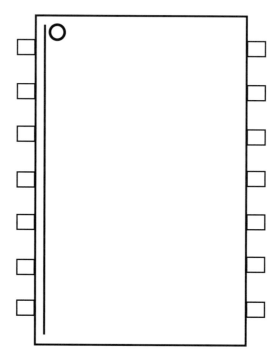

FIGURE 3.6
Top view of an IC.

3.6 Seven-Segment Displays

Often one common requirement for many digital devices is a visual numeric display. Individual light-emitting diodes (LEDs) are only capable of displaying the binary states of a set of latches or flip-flops. As an output device, the LED is cheap and easy to wire to a microcontroller. However, users are far more used to thinking and dealing with decimal numbers. To meet this requirement of users, the display must be capable of clearly representing decimal numbers without any requirement of translating binary to decimal or any other format. A seven-segment display is a form of electronic display device for displaying decimal numerals that is an alternative to the more complex dot-matrix displays. In addition to the numbers from 0 to 9, the seven-segment display can show certain letters. Seven-segment displays can be found in patents as early as 1908 but did not achieve extensive use until the advent of LEDs in the 1970s. They are widely used in digital clocks, electronic meters, and other electronic devices for displaying numerical information. Two types of seven-segment displays are the LED and the LCD. A

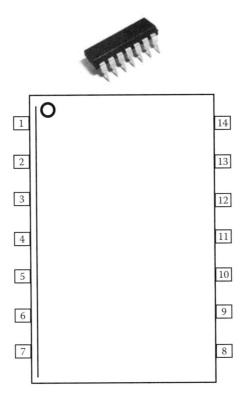

FIGURE 3.7
Pin numbers of an IC.

seven-segment display, as its name indicates, is composed of seven elements. Individually on or off, they can be combined to produce simplified representations of Arabic numerals. Often, the seven segments are arranged in an oblique or slanted arrangement, which aids readability. Each of the seven segments in an LED display uses a light-emitting diode to produce a colored light when there is current through it and can be seen in the dark. An LCD or liquid-crystal display operates by polarizing light so that when a segment is not activated by a voltage, it reflects incident light and appears invisible against its background; however, when a segment is activated, it does not reflect light and appears black. LCD displays cannot be seen in the dark.

Figure 3.8 illustrates the decimal digits 0 through 9 as displayed on a seven-segment display. To display a decimal digit using seven segments, input bits of the seven-segment display are set to cause it to light up the appropriate segments to show the proper digit. Consider, for example, Figure 3.9. In order to display digit '9', the bit at 'e' must be LOW (0) and remaining bits must be set to HIGH (1). Note that the optional DP decimal point (an "eighth segment") is used for the display of non-integer numbers. Some seven-segment

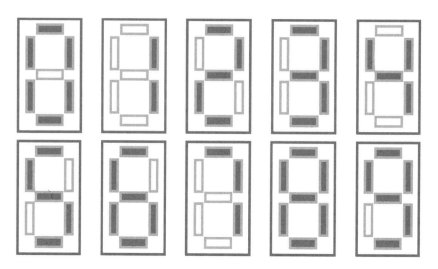

FIGURE 3.8
Decimal digits (0–9) as displayed on a seven-segment display.

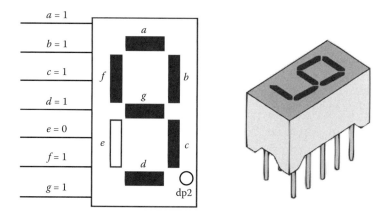

FIGURE 3.9
Logical state of a seven-segment display when displaying the digit '9.'

displays have two decimal points viz dp1 and dp2. The display of digit '9' is illustrated in Figure 3.9 and Table 3.2.

Following the methodology from Table 3.2 for all the digits from 0 to 9, we can obtain the logical states and seven-segment display status as shown in Table 3.3. This table provides the display pattern for digits from 0–9.

From Tables 3.3 and 1.4 we can easily create Table 3.4 to include the ASCII equivalent for each digit.

TABLE 3.2

Logical States of a Seven-Segment
Display When Displaying the Digit '9'

Input Pins	Logical State to Display "9"	Corresponding Segment Label
a	1	a
b	1	b
c	1	c
d	1	d
e	0	e
f	1	f
g	1	g
h	1	dp2

Input	h	g	f	e	d	c	b	a
State	0	1	1	0	1	1	1	1
HEX		6				F		

TABLE 3.3

Display Pattern for Digits 0 to 9

Character	DP2	g	f	e	d	c	b	a	Hexadecimal
0	0	0	1	1	1	1	1	1	3F
1	0	0	0	0	0	1	1	0	06
2	0	1	0	1	1	0	1	1	5B
3	0	1	0	0	1	1	1	1	4F
4	0	1	1	0	0	1	1	0	66
5	0	1	1	0	1	1	0	1	6D
6	0	1	1	1	1	1	0	1	7D
7	0	0	0	0	0	1	1	1	07
8	0	1	1	1	1	1	1	1	7F
9	0	1	1	0	1	1	1	1	6F

Example 3.4

Calculate the total power supplied to a three-and-a-half digit LED display when it indicates 1888. A 5 V supply is used, and each LED has a 15 mA current.

SOLUTION

Let N denote the total number of segments.

$$N = [3 \times (\text{segments for 8})] + [1 \times (\text{segments for 1})]$$

$$= (3 \times 7) + (1 + 2) = 23$$

TABLE 3.4

Mapping the LED Display, ASCII Equivalent, and Input Bits (hex)

LED Display	ASCII Equivalent	DP2 Binary	Input Bits (hex)	LED Display	ASCII Equivalent	DP2 Binary	Input Bits (hex)
0	30	0	3F	A	41	0	77
1	31	0	06	b	62	0	7C
2	32	0	5B	C	43	0	39
3	33	0	4F	d	64	0	5E
4	34	0	66	E	45	0	79
5	35	0	6D	F	46	0	71
6	36	0	7D	S	53	0	6D
7	37	0	07	g	67	0	6F
8	38	0	7F	H	48	0	76
9	39	0	6F	P	50	0	73

Let I_t denote the total current.

$$I_t = N \times \text{(current per segment)} = 23 \times 15 \text{ mA} = 345 \text{ mA}$$

Let P denote the total power supplied LED display.

$$P = I_t \times V_{cc} = 345 \text{ mA} \times 5 \text{ V} = 1.725 \text{ W}$$

Section 3.6 Review Quiz

In addition to the ten numerals, seven-segment displays can be used to show some letters including punctuation. (True/False)

Team Discussion: In addition to the ten numerals, seven-segment displays can be used to show letters of the Latin, Cyrillic, and Greek alphabets including punctuation, but only a few representations are unambiguous and intuitive at the same time. Discuss a few of these unambiguous and intuitive symbols.

3.7 Liquid-Crystal Displays

The earliest discovery leading to the development of the liquid-crystal display (LCD) technology, the discovery of liquid crystals, dates from 1888. An LCD is a thin, flat electronic visual display that uses the light-modulating properties of liquid crystals (LCs). Figure 3.10 shows a low-cost LCD that can display 2 lines by 16 characters for a hobby project. An LCD controls the reflection of available light, whereas an LED display generates or emits light energy as current is passed through the individual segments. The available light may simply be ambient light such as sunlight or normal room lighting; reflective LCDs use ambient light. Alternatively, the available light might be

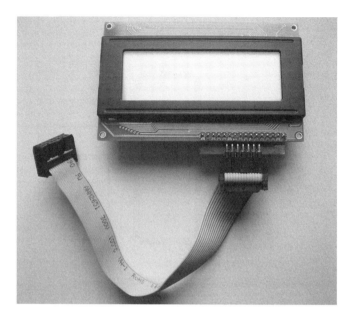

FIGURE 3.10
LCD for a hobby project.

provided by a small light source that is part of the display unit; backlit LCDs use this method. LCDs are used in a wide range of applications including computer monitors, televisions, instrument panels, aircraft cockpit displays, etc. They are common in battery-operated consumer devices such as video players, gaming devices, clocks, digital watches, calculators, and telephones due to very low power consumption compared to LEDs. On the other hand, LEDs have the advantage of a much brighter display, which, unlike reflective LCDs, is easily visible in dark or poorly lit areas.

LCDs are more energy efficient and offer safer disposal than CRTs (Cathode Ray Tubes). They operate from a low-voltage (typically 3 to 15 V rms), low-frequency (25 to 60 Hz) a.c. signal and draw very little current. They are frequently arranged as seven-segment displays for numerical read-outs as shown in Figure 3.11. The a.c. voltage required to turn on a segment is applied between the segment and the backplane, which is common to all segments. The segment and the backplane form a capacitor that draws very little current as long as the a.c. frequency is kept low. To avoid visible flicker, it is generally not lower than 25 Hz.

LCDs can add a lot to your application in terms of providing a useful inter-face for the user, debugging an application, or just giving it a professional look. LCs are available as multidigit seven-segment decimal numeric dis-plays. A wide range of LCDs come with many special characters such as colons (:) for clock displays, +/− indicators for digital voltmeters; decimal points for calculators, and battery-low indicators since many LCD devices

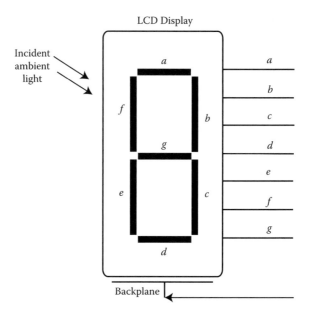

FIGURE 3.11
Basic arrangement of LCD.

are battery powered. A more complex but readily available LCD display is the alphanumeric LCD module. A connector can plug them directly into the decoder/driver chip or the starter kits. These are available from many companies in numerous formats such as 1 line by 16 characters up to 4 lines by 40 characters. The LCD in Figure 3.10 is a 2 lines by 16 characters type. For uniformity, the interface for these modules has been standardized so that an LCD module from any manufacturer will use the same signal and data format. Eight data lines are used to send the ASCII code for whatever the user wants to display. These data lines also carry special control codes to the LCD command register. Three other inputs (Register Select, Read/Write, and Enable) are used to control the location, direction, and timing of the data transfer. In the later chapters we will study how the instruction register (IR) and the data register (DR) of the LCD is controlled. We will create programs where, before starting the internal operation of the LCD, control information will be temporarily stored in these registers to allow interfacing with various peripheral control devices.

Section 3.7 Review Quiz

LCDs are less energy efficient and do not offer safer disposal than CRTs. (True/False)

FIGURE 3.12
A keypad.

3.8 Keypads

A keypad is a set of buttons arranged in a block or "pad" that usually bear digits and other symbols and usually a complete set of alphabetical letters. Switch matrix keyboards and keypads are really just an extension of the button concepts. Figure 3.12 shows a keypad with a connector that can plug the keypad directly into the decoder/driver chip or the starter kits. If a keypad only contains numbers, it is called a numeric keypad. Keypads are found on many alphanumeric keyboards and on other devices such as calculators, push-button telephones, combination locks, and digital door locks, which require mainly numeric input. They are part of mobile phones that are replaceable, and sit on a sensor board of the phone. Some multimedia mobile phones have a small joystick that has a cap to match the keypad. Keypads are also a feature of some combination locks. This type of lock is often used on doors such as that found at the main entrance to some offices.

A keypad is one device that is commonly used by a microcontroller. It is a matrix of switches that are organized as 4 input pins and 4 output pins. The microcontroller scans this matrix of switches using the keypad port,

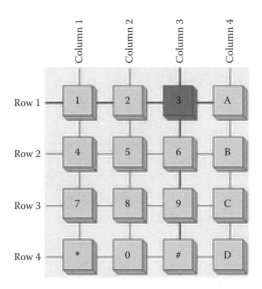

FIGURE 3.13
4×4 matrix in a keypad with 3 pressed.

driving a nibble (4 bits) and reading the results in a nibble (4 bits). Since it is a 4×4 matrix, the keypad port is only able to scan a maximum of 16 keys. A delay between writing a port and reading the result is provided through a scan program in order to delay operations in the microcontroller and the switch bounce in the keypad. In Figure 3.13, when the key "3" is pressed, the first row and third column are set to HIGH (1). The rest of the rows and columns remain at LOW (0). A unique combination of rows and columns can be mapped to 16 different symbols here.

Team Discussion: Discuss the reasons why the keypad of keyboards and calculators are different.

Section 3.8 Review Quiz

As a general rule, the keys on calculator-style keypads are arranged such that 123 is on the bottom row. (True/False)

3.9 The 68HC11/68HC12 Microcontroller

Figure 3.14 presents fundamental design philosophies for small-scale computers. Microcomputers are distinguished as bus-oriented computers, single-board computers, mixed designs, and single-chip microcomputers (or microcontrollers). Bus-oriented computers are distributed over a few

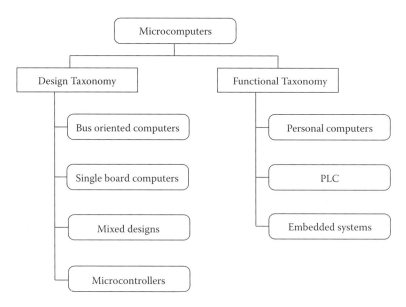

FIGURE 3.14
Fundamental design philosophies for small-scale computers.

separate boards and are best suited for adaptability. An example of adaptability is the addition of more memory by just plugging an additional board. A single-board computer is more attractive for the manufacturer from an economical point of view. Also, a single-board computer is usually more reliable. If we combine the bus-oriented and single-board computer designs, a mixed-design approach originates. Here, a motherboard contains the essential computer core and connectors for further expansion. Mixed designs have been widely used in the world of personal computers. A single-chip microcomputer (or microcontroller) came into existence only because IC manufacturing made possible the implementation of complex systems into a single chip. Microcontrollers are designed for embedded applications, in contrast to the microprocessors used in PCs or other general-purpose applications. Figure 3.14 also presents the functional taxonomy of microcomputers. Programmable Logic Controllers (PLC) are used for industrial control. Unlike PCs, the PLC is designed for multiple inputs and output arrangements, extended temperature ranges, immunity to electrical noise, and resistance to vibration and impact. Finally, a branch of the taxonomy tree called embedded systems presents computers that perform a specific function, usually as a part of a bigger system. A microcontroller can be considered a self-contained system with a processor, memory, and peripherals, and can be used as an embedded system. Embedded systems are, for example, the TV-set control system, the cruise control unit in a car, the laser-printer-embedded computer, the smart card, the sensor node in a wireless sensor network, and so on. Compared to PLCs, the embedded systems are more specialized both in hardware and software. The M68HC11E9 (we will call it HC11 in this book) is a computer system on a single chip. The version of the HC11 is used on the development board (EVBU) and in the examples in this book.

One of the common definitions of an embedded system includes reference to both hardware and software. An embedded system is defined as one that has computer hardware and software embedded in it as one of its most important components. It is a dedicated, computer-based system for an application or product. Embedded systems hardware consists of a processor, memory devices, I/O devices, and a basic hardware unit such as power supply, clock and reset circuit, I/O ports to access external devices, and other on-chip and off-chip units. Figure 3.15 shows the components of embedded system hardware, of which the processor is an important unit. A microcontroller is a very complex integrated chip that includes a great deal of logic circuitry, numerous registers with various functions, and several signal lines that connect it to the other elements in the microcontroller. For a beginner, it can be overwhelming and even intimidating to start out by considering a full-blown microcontroller. In order to reduce confusion, our approach here will be to introduce a simplified version of the HC11 by leaving out some of its more sophisticated elements and retaining only what is necessary to describe how the processor executes a stored program. The omitted portions will be added later as they are needed. A simplified arrangement of HC11 building blocks is shown in Figure 3.16. The address, data, and control buses are not shown in this figure.

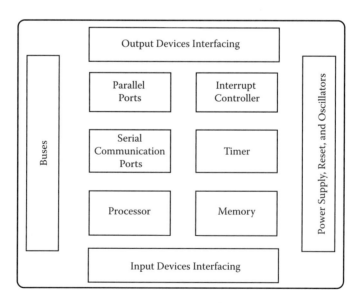

FIGURE 3.15
Hardware components of an embedded system.

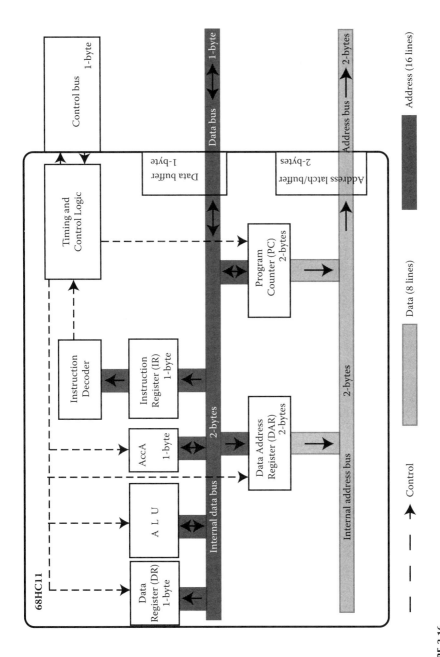

FIGURE 3.16
Simplified HC11 building blocks.

However, we will discuss buses later in this section. Readers are encouraged to refer to M68HC11E Series Programming Reference Guide for a complete block diagram of HC11 after reading this section.

In the next few subsections we will briefly get familiar with the terms that are used while dealing with these building blocks. In the later chapters we will go in depth if and when the need arises.

3.9.1 HC11 Processor

The processor is the device of a microcontroller that is responsible to execute instructions, manipulate data, and perform arithmetic functions. The term Central Processing Unit (CPU) is used to refer to the portion of a computing system that carries out the instructions of a microcontroller program, and is the primary element carrying out the microcontroller's functions. It is the unit that reads and executes program instructions. The HC11 supports four hardware modes, such as single-chip, expanded, special test, and bootstrap. The single-chip mode does not extend the address and data buses to external devices. Therefore, single-chip mode designs are confined to just the on-chip memory. On-chip memory refers to any memory that physically exists on the microcontroller itself. We will shortly discuss various on-chip memory types. The expanded mode, on the other hand, provides the flexibility to the HC11 to address memory devices that are not an integral part of the HC11 chip. In microcontrollers, the memory that is used externally when the microcontroller works in expanded mode for special cases is called expanded off-chip memory. The HC11 supports expanded off-chip memory. We will take a closer look at the HC11 memory in the next subsection. For now, let us discuss the processor building blocks. The following are the building blocks of the HC11 processor:

- Arithmetic Logic Unit (ALU): ALU is a digital circuit that performs arithmetic and logical operations such as addition and subtraction, logical AND, OR, and NOT operations, and data shifting.

- Processor registers: The processor contains several registers that are used to store various kinds of information needed by the processor as it performs its functions. These registers serve as dedicated memory locations inside the processor chip. A basic register block configuration is shown in Figure 3.17. Table 3.5 outlines briefly various processor registers, and Table 3.6 describes the HC11 processor registers. We will discuss in detail these registers in Chapter 4.

- Control unit: A control unit is primarily responsible for reading the instruction from memory and ensuring the execution of the instructions. The memory address register (MAR) holds the address of the

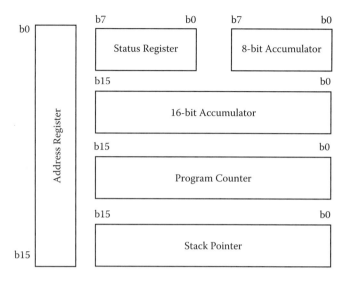

FIGURE 3.17
Processor register block configuration.

TABLE 3.5

General Processor Registers

Processor Register	Description
Accumulators	An accumulator is a special register linked to the ALU in which intermediate arithmetic and logic results are stored. The size of the accumulator is related to the size of the data bus.
Program counter	A program counter is a special register that keeps track of the address of the next location in memory that will be accessed. The program counter is the same size as the address bus. It is also called instruction pointer.
Stack pointer	A stack pointer is a special register that contains an address pointer that indicates the next available memory location on the stack. A stack is a temporary memory area that performs the job of temporary storage and retrieval of data from memory during the processing of instructions.
Address register	An address register is a special register that is used specifically by the instructions to address memory. It helps in simpler instructions, because it does not need to be concerned about the address.
Status register	A status register is a special register dedicated to reporting status through bits that indicate certain results of the last operation. The most common status bits are sign flags, carry/borrow flags, zero flags, and overflow flags.

TABLE 3.6

HC11 Processor Registers

Processor Register	Description
Accumulator A	Referred to as "A" register or as "AccA," it is an 8-bit register that is used as the primary data-processing register.
Accumulator B	Referred to as "B" register or as "AccB," it is an 8-bit register identical to accumulator A in function.
Accumulator D (A:B)	Referred to as "D" register or as "AccD," it is a 16-bit register such that two 8-bit accumulators are combined in it.
Index register X	Referred to as "IX," "[X]," or "X" register, it is a 16-bit address register that is used by the indexed addressing mode instructions. It can also be used as a general-purpose 16-bit register.
Index register Y	Referred to as "IY," "[Y]," or "Y" register, it is a 16-bit address register identical to Index Register X.
Stack pointer	Referred to as the "S" or the "SP" register, it is a 16-bit register that always contains the address of the next available stack memory location.
Program counter	Referred to as the "PC," it is a 16-bit address register that always contains the address of the next location in the memory that will be addressed.
Condition code register	Referred to as "P," "C," "CCR," or "Status" register, it is an 8-bit status and control register. Out of the eight bits, five bits are H, N, Z, V, and C. These are called status flags. The remaining three bits are S, X, and I. These are called control bits.

memory location of the next instruction to be executed. While the first instruction is being executed, the address of the next memory location is held by it. The memory data register (MDR) is the register of a control unit that contains the data to be stored in the memory. It acts like a buffer and holds anything that is copied from the memory ready for the processor to use it. The instruction register (IR) is a special register that always contains the opcode for the current instruction. The instruction decoder of a processor is a combinatorial circuit often in the form of a read-only memory, and sometimes in the form of an ordinary combinatorial circuit. Its purpose is to translate an instruction code into the address in the memory where the code for the instruction starts.

3.9.2 HC11 Memory

Previously, we defined the terms on-chip and off-chip memory. Recall that an on-chip memory refers to any memory that physically exists on the microcontroller itself. Also, the memory that is used externally when the microcontroller works in expanded mode for special cases is called expanded off-chip memory. But what is a memory? A memory refers to computer components

```
$ 0000
    512 B RAM
$ 01FF
                            $ 0200
              Expanded Memory
                            $ 0FFF
$ 1000
    64 B REGISTER BLOCK
$ 103F
                            $ 1040
              Expanded Memory
                            $ B5FF
$ B600
    512 B EEPROM
$ B7FF
                            $ B800
              Expanded Memory
                            $ CFFF
$ D000
    12 KB ROM
$ FFFF
```

SINGLE CHIP

FIGURE 3.18
Memory map on the HC11.

and recording media that retain digital data used for computing for some interval of time. As any storage has a location and an address to that location, a memory also has its location and address on the hardware. A memory address is an identifier for a memory location at which a microcontroller program or a hardware device can store data and later retrieve it. Generally, this is a binary number from a finite, monotonically ordered sequence that uniquely describes the memory itself. The HC11 uses a 16-bit address such that numbers of a unique address are equal to 2^n, where n is the number of address bits in the system. Thus, with $n = 16$, 64K unique memory locations are possible. Figure 3.18 illustrates the memory map on the HC11.

The memory unit stores groups of bits (words) that represent instructions that a microcontroller is to execute (that is, a program), and data that are to be operated on by the program. The memory also serves as temporary storage of the results of operations performed by the CPU. RAM, ROM, and EEPROM are the three types of memory that are included on-chip.

3.9.2.1 RAM

RAM stands for Random Access Memory and is often referred to as read/write memory. The RAM is a general-purpose volatile type of memory, where the information (data or programs) is lost after the power is switched off. By default, the HC11E9 RAM is located at $0000–$01FF of the memory map. To achieve the flexibility with on-chip and off-chip resources, the user is given the ability to relocate the RAM in the memory map by altering the RAM map position control bits in the INIT register. Figure 3.19 shows the structure of the 8-bit INIT register where the four bits, such as RAM3, RAM2, RAM1, and RAM0 specify the most significant hex digit of the 16-bit RAM address in the memory map. As shown in Table 3.7, the location of RAM can be moved to

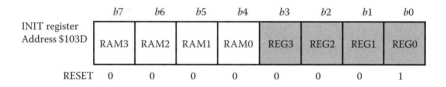

FIGURE 3.19
The INIT register (RAM map position).

TABLE 3.7

RAM Mapping Using INIT Control Bits

Bits in INIT Register				HEX Equivalent	Start of Address Range	End of Address Range
RAM3	RAM2	RAM1	RAM0			
0	0	0	0	$0	$0000	$01FF
0	0	0	1	$1	$1000	$11FF
0	0	1	0	$2	$2000	$21FF
0	0	1	1	$3	$3000	$31FF
0	1	0	0	$4	$4000	$41FF
0	1	0	1	$5	$5000	$51FF
0	1	1	0	$6	$6000	$61FF
0	1	1	1	$7	$7000	$71FF
1	0	0	0	$8	$8000	$81FF
1	0	0	1	$9	$9000	$91FF
1	0	1	0	$A	$A000	$A1FF
1	0	1	1	$B	$B000	$B1FF
1	1	0	0	$C	$C000	$C1FF
1	1	0	1	$D	$D000	$D1FF
1	1	1	0	$E	$E000	$E1FF
1	1	1	1	$F	$F000	$F1FF

the start of any 4K memory block. Recall that the HC11 uses a 16-bit address. Therefore, since 4 bits are already used, only 12 bits of the 16-bit address remain to be used. The maximum number of memory locations that can be addressed with the remaining 12 bit is $2^{12} = 4K$. Before we move on to the next type of memory in HC11, let us take an example to see how the RAM mapping can be changed using the INIT register.

Example 3.5

How can a user move the location of the RAM in the memory map from the address range $0000–$01FF to $F000–$F1FF?

SOLUTION

The control bits of the register INIT located at $103D specifies the most significant hex digit of the 16-bit RAM address in the memory map. Since we require $F to be that hex digit, we will have to set the most significant nibble (RAM3 to RAM0 bits) of INIT register to $F or 1111_2.

3.9.2.2 ROM

ROM stands for Read-Only Memory. It is nonvolatile memory and always retains its data. ROM is a type of memory that normally can only be read, as opposed to RAM that can be both read and written. There are two main reasons that read-only memory is used for certain functions within the microcontroller. The first is permanence of information. The information (data or programs) stored in ROM are always there, whether the power is on or not. For this reason, it is called nonvolatile storage. The second important reason to use ROM is information security. The fact that ROM cannot easily be modified provides a measure of security against accidental (or malicious) changes to its contents. Read-only memory is used to store the BUFFALO monitor program during manufacturing. Having this in a ROM means it is available when the power is turned on so that the processor can use it to boot up the system. The BUFFALO monitor program on the EVBU is located in the ROM from $E000 to $FFFF. The ROM address range $D000–$D000 is unused and unavailable. The ROM address range $FFD6–$FFFF holds with the Interrupt Vector Table. We will study more about interrupts later in Chapter 9.

The CONFIG register located at memory location $103F is used to disable the on-chip ROM. Why would one want to disable on-chip ROM? A user may want to disable on-chip ROM when the user programs are located in an external memory chip. The version M68HC11E1 has ROM permanently disabled. Let us look into how the on-chip ROM can be disabled using the CONFIG register. Figure 3.20 shows the internal structure of the CONFIG register. The ROMON bit is the control bit that can be set or cleared to enable

	b7	b6	b5	b4	b3	b2	b1	b0
CONFIG register Address $103F	-	-	-	-	NOSEC	NOCOP	ROMON	EEON
RESET	0	0	0	0	*u*	*u*	1	*u*

FIGURE 3.20
The CONFIG register.

and disable the on-chip ROM. When the HC11 is reset or powered on, the ROMON control bit is set to enable the on-chip ROM.

3.9.2.3 EEPROM

An EPROM, or erasable programmable read-only memory, is a type of nonvolatile memory chip that retains its data when its power supply is switched off. Once programmed, an EPROM can be erased by exposing it to strong ultraviolet light from a mercury-vapor light source. EPROMs are easily recognizable by the transparent fused quartz window in the top of the package, through which the silicon chip is visible, and which permits exposure to UV light during erasing. Microcontrollers with EPROM are rare and have been replaced by EEPROM, which is easier to use and cheaper to manufacture. EEPROM stands for electrically erasable programmable read-only memory and is a type of nonvolatile memory used to store small amounts of data that must be saved when power is removed, for example, calibration tables or device configuration. EEPROM can be erased with electrical signals and then reprogrammed.

The property of an EEPROM to erase data electronically has given microcontrollers enhanced versatility. HC11E9 contains 512 bytes of on-chip EEPROM mapped to $B600–$B7FF. The logical arrangement of EEPROM on HC11 is in 32 rows of 16 bytes each. The memory location $B600–$B60F is the first row. The next row is $B610–$B61F. In this way, 32 rows constitute the EEPROM in the HC11. Recall from our discussion of RAM that a user has the ability to remap the RAM from one page to another. In the next subsection we will study how system registers can be remapped also. The EEPROM, on the other hand, cannot be remapped. But, like the ROM, the EEPROM can be disabled, too.

Recall that the CONFIG register located at memory location $103F is used to disable the on-chip ROM. This register CONFIG has also the control bit that can disable the on-chip EEPROM. Let us look into how the on-chip EEPROM can be disabled using the CONFIG register. Figure 3.20 shows the internal structure of the CONFIG register. The EEON bit is the control bit that can be set or cleared to enable and disable the on-chip EEPROM. When the HC11 is reset or powered on, the EEON control bit is undefined. The Appendix contains more information about HC11 EEPROM that is useful for the design of EEPROM-based programs.

3.9.2.4 System Registers

The system registers in HC11 are held by default from $1000 to $103F in the memory map. This is a 64-byte block of memory space dedicated to these registers. System registers contain a wide range of control bits, status bits, hardware-related information, and information common to the user software and the on-chip hardware system. To achieve flexibility with on-chip and off-chip resources, the user is given the ability to relocate the system registers in the memory map by altering the REG map position control bits in the INIT register. Figure 3.21 shows the structure of the 8-bit INIT register where the four bits such as REG3, REG2, REG1, and REG0 specify the most significant hex digit of the 16-bit register block address in the memory map. As shown in the Table 3.8, the location of the register block can be moved to the start of any 4K memory block. Recall that the HC11 uses a 16-bit address.

	b7	b6	b5	b4	b3	b2	b1	b0
INIT register Address $103D	RAM3	RAM2	RAM1	RAM0	REG3	REG2	REG1	REG0
RESET	0	0	0	0	0	0	0	1

FIGURE 3.21
The INIT register (register map position).

TABLE 3.8

Register Mapping Using INIT Control Bits

Bits in INIT Register				HEX Equivalent	Start of Address Range	End of Address Range
REG3	REG2	REG1	REG0			
0	0	0	0	$0	$0000	$003FF
0	0	0	1	$1	$1000	$103F
0	0	1	0	$2	$2000	$203F
0	0	1	1	$3	$3000	$303F
0	1	0	0	$4	$4000	$403F
0	1	0	1	$5	$5000	$503F
0	1	1	0	$6	$6000	$603F
0	1	1	1	$7	$7000	$703F
1	0	0	0	$8	$8000	$803F
1	0	0	1	$9	$9000	$903F
1	0	1	0	$A	$A000	$A03F
1	0	1	1	$B	$B000	$B03F
1	1	0	0	$C	$C000	$C03FF
1	1	0	1	$D	$D000	$D03F
1	1	1	0	$E	$E000	$E03F
1	1	1	1	$F	$F000	$F03F

Therefore, since 4 bits are already used, only 12 bits of the 16-bit address remain to be used. The maximum number of memory locations that can be addressed with the remaining 12 bits is $2^{12} = 4K$. Before we move on to the next section, let us take an example to see how the register mapping can be changed using the INIT register.

Example 3.6

How can a user move the location of the system register block in the memory map from the address range $0000–$003F to $F000–$F03F?

SOLUTION

The control bits of the register INIT located at $103D specify the most significant hex digit of the 16-bit register block address in the memory map. Since we require $F to be that hex digit, we will have to set the least significant nibble (REG3 to REG0 bits) of INIT register to $F or 1111_2.

Helpful Hint: Students are encouraged to read Section 4 and 3.3.1 of the HC11 Reference Manual, and Section 4 of the Technical Data Manual, when designing for a technical solution of an HC11 application.

Helpful Hint: The HC11 is an 8-bit microcontroller family. The HC12 is an enhanced 16-bit version of the HC11.

3.9.3 HC11 Advanced On-Chip Input/Output (I/O) Capabilities

The input unit consists of devices that allow data and information from the outside world to be entered into the microcontroller's internal memory of the CPU. These devices are often referred to as peripherals because they are physically separated from the processor of the microcontroller. Typical input peripherals include keypads, analog-to-digital converters (ADCs), etc. On the other hand, the output unit consists of peripheral devices that transfer data and information from the internal memory or CPU to the outside world. Typical output peripherals include LED/LCD displays, video monitors, digital-to-analog converters (DACs), etc. Different versions of the HC11 have different numbers of external ports, labeled alphabetically. The most common version has five ports, A, B, C, D, and E. Each port is eight bits wide, except for D, which is usually six bits. With external memory, B and C are used as address and data bus, respectively. In this mode, port C is multiplexed to carry both the lower byte of the address and data. We will take a closer look on these I/O ports from Chapter 8 onward. The HC11 I/O ports are summarized in Table 3.9. The HC11 also includes an ADC and an advanced timing system to support various event-driven functions.

Before we end this section, it is worth mentioning that, in general, a bus consists of wires that are used to transfer data either in serial or parallel transmission. A unidirectional bus allows information flow only in one direction, whereas a bidirectional allows it in both directions. Figure 3.22 illustrates a possible bus arrangement in an HC11. There are three buses external to the processor and two buses within the processor. The external

TABLE 3.9

Summary of HC11 I/O Ports

Port	Input Pins	Output Pins	Bidirectional Pins	Total Number of Pins	Broad Area of Usage	Address	RESET
A	3	3	2	8	Timer event	PORTA: $1000 PACTL: $1026	00000000
B	0	8	0	8	Output only	PORTB: $1004	00000000
C	0	—	8	8	Programmable I/O	PORTC: $1003 PORTCL:$1005 DDRC: $1007	UUUUUUUU
D	0	—	6	6	Serial communication	PORTD: $1008 DDRD: $1009	00000000
E	8	—	0	8	Analog capture	PORTE: $100A	UUUUUUUU
Total	11	11	16	38			

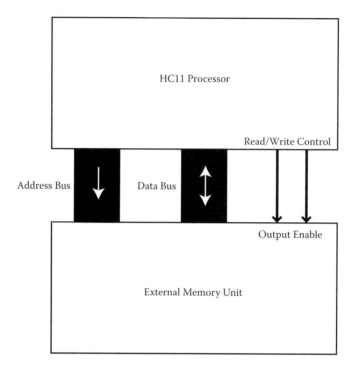

FIGURE 3.22
A simplified bus arrangement.

address and data buses are extensions of the same buses inside the processor, so there are really only three different buses:

- Address bus: An address bus is a unidirectional, 16-bit bus that carries the 16-bit address code from the processor to the memory unit to select the memory location that the processor is accessing for a READ or WRITE operation.
- Data bus: Though the name of this bus is data bus, in reality it often carries the information like instruction codes. A data bus is a bidirectional, 8-bit bus over which 8-bit words can be sent from the processor to memory (that is, a WRITE operation) or from the memory to the processor (a READ operation).

  Helpful Hint: The program counter always holds the instruction address that refers to the program portion of the memory, and the memory data register holds data (operands) addresses that refer to the data portion of the memory.

- Control bus: A control bus is a grouping of all the timing and control signals needed to synchronize the operation of the processor with the other units of the microcontroller.

  Helpful Hint: The default processing capability of HC11 is of one byte. Most 8-bit processors such as the HC11 will have an 8-bit data bus and a 16-bit address bus.

Section 3.9 Review Quiz

The 68HC11 has two ____-bit accumulators *A* and *B*, two ____-bit index registers *X* and *Y*, a condition code register, a ____-bit stack pointer, and a program counter. (8/16/32/64)

3.10 EVBU/BUFFALO

The EVBU is the Motorola M68HC11 Universal Evaluation Board, a development tool for HC11 microcontroller-based designs. Since Motorola 68HC11 is a popular microcontroller and several evaluation boards are available (including Motorola's original board, known as the EVBU) for exploring the capabilities of this microcontroller, we will create programs for execution on this EVBU. Figure 3.23 is a photograph of the EVBU built by Axiom Manufacturing, Inc.

The hardware components of EVBU are

- The HC11 chip
- M68HC68 real-time clock chip
- Standard serial communications port
- Breadboard area

FIGURE 3.23
An evaluation board.

A monitor program with a User Interface (UI) called BUFFALO (Bit User Fast Friendly Aid to Logical Operations) is software that provides a controlled environment for the HC11 chip to operate.

BUFFALO is the standard boot-loader for the HC11. Not all HC11 models come with the BUFFALO boot-loader. The 68HC11A0 and A1 do not, but the A8 does. The UI facilitates the EVBU to run programs, enter simple commands, and monitor the HC11. The BUFFALO monitor program on the EVBU uses a substantial piece of the RAM for temporary storage. This storage can be used for calculations, variables, data, the user program, the interrupt vector jump table, or work in progress. Often, programmers refer to this area of RAM as the scratchpad due to its usage similar to the way people take notes on a scrap of paper. Figure 3.24 shows the layout of the scratchpad memory. We will come back to the EVBU in later chapters.

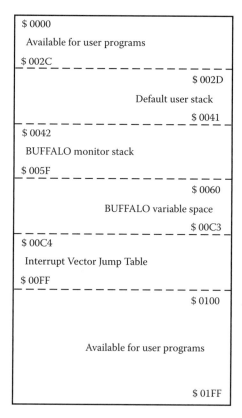

RAM MEMORY MAP

FIGURE 3.24
EVBU scratchpad memory.

A complete list of BUFFALO commands is provided in the EVBU User's Manual. Readers are encouraged to go through the manual in order to get a good grasp of those commands. In the next few examples, we explore a few of the frequently used commands.

Self-Learning: Connect the EVBU, attach the serial cable, and run the BUFFALO terminal program. Perform each BUFFALO command as you read along the EVBU User's Manual.

Example 3.7

Show how you will load "FF" in the memory range ($0030 to $003F) on the BUFFALO monitor. Note that the $ symbol is used to indicate that the data is hex.

SOLUTION

We will use the BF<start><end><data>, that is, Block Fill command. This command fills successive memory locations with the data starting with the start address and ending with the row that contains the ending address.

>BF 0030 003F FF <cr>

Helpful Hint: The $ symbol is used to indicate that the data is hex.

Example 3.8

After executing the command in Example 3.7, show what command you will use on the BUFFALO monitor to display them.

SOLUTION

We will use MD<start><end>, that is, Memory Display command.

>MD 0030 003F <cr>
>0030 FF FF FF FF FF FF FF FF FF FF FF FF FF FF FF FF
>

Example 3.9

Describe in words what the user is trying to achieve as follows:

>MD 100 16 <cr>
0100 FF FF FF FF FF FF FF FF FF FF FF FF FF FF FF FF

>MM 100 <cr>
0100 FF AA <cr>

>MD 100 16 <cr>

0100 AA FF FF FF FF FF FF FF FF FF FF FF FF FF FF FF

>MM 100 <cr>

0100 AA FF <cr>

>MD 100 16 <cr>

0100 FF FF FF FF FF FF FF FF FF FF FF FF FF FF FF FF

SOLUTION

Here the user first looks at the contents of the memory range $0100–$010F. This is accomplished by the first MD command. The MD command provides a dump of a range of memory locations.

Next, the contents of the memory location $0100 are changed to $AA. This is done by the first MM command. The MM command first lists the memory contents of the given address and then gives the user an opportunity to change that value. If the user enters a new value and presses the enter key, the new value is inserted. The contents remain unaffected if no new value is entered.

By the second MD command, the user looks again at the contents of the same memory range $0100–$010F. Next, the user changes the contents of the memory location $010F back to $FF. This is accomplished by the second MM command. Finally, to verify if the contents have changed back, the user views the contents of the same memory range again. The last MD command is used to look at the contents.

Helpful Hint: Once you have entered the MM command, enter it again but do not press the enter key to accept the value. Rather, press the equal sign key "=", so the same address is repeated. Pressing the "+" key (or shift and "+") advances to the next address, and pressing the "-" changes to the preceding address.

Section 3.10 Review Quiz

BUFFALO stands for _____.

3.11 Summary

1. The transistor is the basic building block of the modern digital integrated circuit. It can be switched on or off, applying the appropriate voltage at its base connection.
2. CMOS is made with MOS field-effect transistors.
3. Bipolar TTL is made with bipolar junction transistors.
4. As a rule, CMOS has lower power consumption than bipolar.
5. TTL and CMOS integrated circuits are formed by integrating thousands of transistors in a single package.

6. TTL and CMOS are the most popular ICs used in the digital world today.

7. SMD-style ICs are gaining popularity over the through-hole style DIP ICs because of their smaller size and reduced manufacturing costs.

8. A seven-segment display is a popular form of electronic display device for displaying decimal numerals. Individually on or off, they can be combined to produce simplified representations of Arabic numerals.

9. The interface for alphanumeric LCD modules is standardized with eight data lines for sending the ASCII code. These data lines also carry special control codes to the LCD command register. Three other inputs (Register Select, Read/Write, and Enable) are used to control the location, direction, and timing of the data transfer.

10. A keypad is a matrix of switches that are organized as 4 input pins and 4 output pins. The microcontroller scans this matrix of switches using the keypad port, driving a nibble (4 bits) and reading the results in a nibble (4 bits).

11. Microcomputers are distinguished as bus-oriented computers, single-board computers, mixed-designs computers, and single-chip microcomputers (or microcontrollers).

12. The M68HC11E9 is a computer system on a single chip.

13. The processor is the device of a microcontroller that is responsible for executing instructions, manipulating data, and performing arithmetic functions.

14. The HC11 supports four hardware modes such as single chip, expanded, special test, and bootstrap.

15. The building blocks of the HC11 processor are ALU, processor registers, and control unit.

16. The definition of various HC11 processor registers can be found in Table 3.6.

17. Typically, processor registers are not considered part of the normal memory range for a system.

18. RAM, ROM, and EEPROM are the three types of memory that are included on-chip.

19. The RAM is a general-purpose volatile memory located by default at $000–$01FF of the memory map.

20. The BUFFALO monitor program on the EVBU uses a substantial portion of the RAM for temporary storage.

21. The BUFFALO monitor program on the EVBU is located in the ROM from $E000 to $FFFF.

22. ROM, burnt during manufacturing process, is located at $B600–$B7FF in the memory map.

23. The entire EEPROM is available to the user for information manipulation.
24. The HC11 has a 64-byte block of memory space for system registers located by default at $1000–$103F in the memory map.
25. Unlike RAM and system registers, the ROM and EEPROM cannot be relocated within the memory map, but they can be disabled with control bits in the CONFIG register.
26. The HC11 uses a 16-bit address and, consequently, 64K unique memory locations are possible.
27. The HC11 is an 8-bit microcontroller family. The 68HC12 is an enhanced 16-bit version of the 68HC11.
28. Different versions of the HC11 have different numbers of external ports, labeled alphabetically. The most common version has five ports, A, B, C, D, and E.
29. The three types of buses are address bus, data bus, and control bus.
30. The hardware components of EVBU are the HC11 chip, M68HC68 real-time clock chip, standard serial communications port, and breadboard area.
31. BUFFALO is a monitor program that provides a controlled environment for the HC11 chip.

Glossary

Accumulator: A parallel register in a microcontroller that is the focal point of all arithmetic and logic operations.

Address Bus: An address bus is a unidirectional, 16-bit bus that carries the 16-bit address code from the processor to the memory unit to select the memory location that the processor is accessing for a READ or WRITE operation.

Arithmetic Logic Unit (ALU): An ALU is a digital circuit that performs arithmetic and logical operations like addition and subtraction; logical AND, OR, and NOT operations; and data shifting.

Bipolar: A class of integrated logic circuits implemented with bipolar transistors; also known as TTL.

BUFFALO: Stands for Bit User Fast Friendly Aid to Logical Operations. It is a monitor program that provides a controlled environment for the HC11 chip.

Bus: A bus consists of wires that are used to transfer data either in serial or parallel transmission.

Chip: The term given to an IC.

CMOS: Complementary metal-oxide semiconductor; a class of integrated logic circuits that is implemented with a type of field-effect transistor.

Central Processing Unit (CPU): The term is used to refer to the portion of a computing system that carries out the instructions of a microcontroller program. It is the primary element carrying out the microcontroller's functions.

Control Bus: This is a grouping of all the timing and control signals needed to synchronize the operation of the processor with the other units of the microcontroller.

Control Unit: Primarily responsible for reading the instruction from memory and ensuring the execution of the instructions.

Cutoff: A term used in transistor switching that signifies that the collector-to-emitter junction is turned off or is not allowing current flow.

Data Bus: A data bus is a bidirectional, 8-bit bus over which 8-bit words can be sent from the processor to memory (that is, a WRITE operation), or from the memory to the processor (a READ operation).

Dual-In-Line Packages (DIP): The most common pin layout for integrated circuits. The pins are aligned in two straight lines, one on each side of the IC.

EEPROM: A type of nonvolatile programmable link based on electrically erasable programmable read-only memory cells that can be turned on or off repeatedly by programming.

Embedded Systems: A computer system designed to perform one or a few dedicated functions often with real-time computing constraints.

Enable: To activate or put into an operational mode; an input on a logic circuit that enables its operation.

EPROM: A type of nonvolatile programmable link based on electrically programmable read-only memory cells that can be turned either on or off once with programming.

EVBU: This is the Motorola M68HC11 Universal Evaluation Board, a development tool for HC11 microcontroller-based designs.

Hardware: The integrated circuits and electronic devices that make up a computer system.

Input/Output (I/O): The communication between a processor and the outside world.

Integrated Circuit (IC): The fabrication of several semiconductor and electronic devices such as transistors, diodes, and resistors onto a single piece of silicon crystal.

Keypad: A set of buttons arranged in a block or "pad" that usually bear digits and other symbols and is usually a complete set of alphabetical letters.

Light-Emitting Diode (LED): A semiconductor light source used as indicator lamp in many devices.

Liquid-Crystal Display (LCD): An LCD is a thin, flat electronic visual display that uses the light-modulating properties of liquid crystals (LCs).

Microcontroller: A small computer on a single IC containing a processor, memory, and I/O peripherals.

Monitor Program: The computer software program initiated at power-up that supervises system-operating tasks, such as reading the keyboard and driving the computer monitor.

MOSFET: Metal-oxide semiconductor field-effect transistor.

Peripheral: A device such as a printer or model that provides communication with a computer.

Port Number: An 8-bit number used to select a particular I/O port.

Processor Register (or general-purpose register): A processor register is a small amount of storage available on the CPU whose contents can be accessed more quickly than storage available elsewhere.

Program Counter: A 16-bit internal register that contains the address of the next program instruction to be executed.

Programmable Logic Controller (PLC): A programmable logic controller is a digital computer used for automation of electromechanical processes, such as control of machinery on factory assembly lines, amusement rides, or lighting fixtures.

Saturation: A term used in transistor switching that signifies that the collector-to-emitter junction is turned on or conducting current heavily.

Seven-Segment Display: This is a form of electronic display device for displaying decimal numerals that is an alternative to the more complex dot-matrix displays.

Stack Pointer: A 16-bit internal register that contains the address of the last entry on the RAM stack.

Surface-Mounted Device (SMD): The newest style of integrated circuits, soldered directly to the surface of a printed circuit board. They are much smaller and lighter than the equivalent logic constructed in the DIP through-hole style logic.

Transistor–Transistor Logic (TTL): The most common integrated circuit used in the digital world today.

Very Large-Scale Integration (VLSI): The process of creating IC by combining thousands of transistors into a single chip.

Answers to Section Review Quiz

3.2 True

3.3 True

3.4 (a) Metal-oxide semiconductor field-effect transistor, (b) Complementary metal-oxide semiconductor

3.5 False

3.6 True

3.7 False

3.8 True

3.9 8, 16, 16

3.10 Bit User Fast Friendly Aid to Logical Operation

True/False Quiz

1. In a common-emitter transistor circuit, when V_{out} is 0, R_c should be large.

2. In a common-emitter transistor circuit, when V_{out} is 1, R_c should be large.

3. The HC11 supports 16-bit-by-16-bit divide instructions.

4. An inverter performs the NOR operation.

5. The sign flag indicates the sign of the last data processed.

6. DIP are the most common pin layout for ICs with pins aligned in a single straight line only on one side of the IC.

7. HC11 uses a 16-bit address bus.

8. Seven-segment displays receive all the inputs only from the V_{cc}.

9. The BUFFALO monitor program controls the HC11 within the environment created by the EVBU hardware.

10. Accumulators A and B are identical in every sense other than the names.

11. The data bus in HC11 is an 8-bit bidirectional bus over which 8-bit words can be transferred.

12. Analog-to-digital converters (ADC) and digital-to-analog converters (DAC) can never be integrated on a chip.

13. The embedded control applications are characterized by high production volumes.

14. Unlike EEPROM chips, EPROMs do not need to be removed from the computer to be modified.

15. Programmable Logic Controller (PLC) is used for industrial control.

Questions

QUESTION 3.1

Explain how a Bipolar Junction Transistor (BJT) can act like a switch.

QUESTION 3.2

What are the advantages of TTL over CMOS?

QUESTION 3.3

What are the most common methods for constructing a seven-segment display? Which one consumes the maximum power?

QUESTION 3.4

In seven-segment displays, does the common-anode configuration have any advantage over the common-cathode configuration? List the factors that decide the selection of a configuration in a circuit.

QUESTION 3.5

How many accumulators does HC11 have? Describe them with their sizes in bits.

QUESTION 3.6

What is the function of the BUFFALO command "asm <address>"?

QUESTION 3.7

Your colleague assumed that "BUFFALO is case sensitive and only understands hexadecimal numbers." Discuss if his assumption is correct or not.

QUESTION 3.8

The following is a snapshot of work performed by your colleague.

```
>T 1
DEX                    P-010C Y-FF00 X-09C2 A-99 B-F8 C-90 S-0041
>
BNE    $010A           P-010A Y-FF00 X-09C2 A-99 B-F8 C-90 S-0041
>
NOP                    P-010B Y-FF00 X-09C2 A-99 B-F8 C-90 S-0041
>
DEX                    P-010C Y-FF00 X-09C1 A-99 B-F8 C-90 S-0041
```

The first command was a trace command. How was your colleague able to proceed without typing any command after that?

QUESTION 3.9

Buffalo often initializes the stack pointer register to $41, but you can manually set the stack pointer register. How?

QUESTION 3.10

Most manufacturers release multiple versions of the microcontroller chip. For example, we have 68HC11A1 as well as 68HC811E2 in the 68HC11 family. Why is it important for the programmer to know the version of the chip?

Problems

PROBLEM 3.1

Determine V_{out} for the common-emitter transistor inverter circuit of Figure 3.1 using $V_{in} = 0$ V, $R_B = 1$ MΩ, $R_C = 330$ Ω, and $R_{load} = 1$ MΩ.

PROBLEM 3.2

The design engineer in the lab suggested modifying the configuration of Problem 3.1. If the load resistor (R_{load}) used in Problem 3.1 is changed from 1 MΩ to 470 Ω, describe what happens to V_{out}.

PROBLEM 3.3

In the circuit of Figure 3.1 with $V_{in} = 0$ V, V_{out} will be almost 5 V as long as R_{load} is much greater than R_C. An intern newly appointed at the design lab suggested that R_C be made very small to ensure that the circuit will work for all value of R_{load}. How will you respond to his suggestion?

PROBLEM 3.4

In Figure 3.1, if $R_C = 100$ Ω, determine the collector current when $V_{in} = +5$ V.

PROBLEM 3.5

Use the 7408 TTL IC to build a clock-enabled circuit. Use a clock oscillator to pass the signal on to the receiving device when the switch is in the enable (1) position, and block the signal when in the disable (0) position.

PROBLEM 3.6

We saw how a single byte can encode the full state of a seven-segment display. The most popular bit encodings are gfedcba and abcdefg—both usually assume 0 is off and 1 is on. We discussed one of the encodings in this chapter. Create a table that will provide the hexadecimal encodings of both types for displaying the digits 0 to 9.

PROBLEM 3.7

Which rows and columns will get enabled when 123 are pressed in a keypad?

PROBLEM 3.8

Show how you will load '33' in the memory range ($0010 to $001F) on the BUFFALO monitor.

PROBLEM 3.9

If '33' is loaded in the memory range ($0010 to $001F), show what command you will use on the BUFFALO monitor to display them.

PROBLEM 3.10

Using the BUFFALO monitor commands, perform the following:

(a) Modify the data stored in the memory location $0000 for a new value of $FF.
(b) Display this memory location for verification.
(c) Again modify the data stored in the memory location $0000 to its original value.
(d) Finally, display this memory location for verification.

PROBLEM 3.11

How can a user move the location of the RAM in the memory map from the address range starting at $0000 to the address range ending at $A1FF?

PROBLEM 3.12

How can a user move the location of the system register block in the memory map from the address range ending at $003F to the address range starting at $A000?

4

Microcontroller Software

Divide each of the difficulties under examination into as many parts as possible, and as might be necessary for its adequate solution.

—René Descartes (father of modern philosophy)

OUTLINE

4.1 Introduction

4.2 Programming Concepts

4.3 System Software

4.4 Developing a Program

4.5 Flow and State Diagrams

4.6 HC11 Programming Model

4.7 HC11 Memory-Addressing Modes

4.8 Summary

OBJECTIVES

Upon completion of this chapter you should be able to

1. Make comparisons between assembly language, machine language, and higher-level languages.
2. Describe a simple assembly language program.
3. Define opcode, operand, and address of an operand.
4. Explain the function of software program instructions in a microcontroller-based system.
5. Appreciate the role of editors, assemblers, compilers, interpreters, debuggers, and operating systems.
6. Cite and use the various steps in the software development process.
7. Understand the meaning of flowchart.
8. Draw the basic parts of the flowchart such as flowchart symbols.
9. Appreciate the advantages and limitations of the flowchart.
10. Fully understand all of the flags within the condition code register of the HC11.
11. Explain the purpose of memory-addressing modes.
12. Demonstrate the working of memory-addressing modes on the HC11.

Key Terms: Address, Assembler, Assembly Language, Carry Flag (C), Comment, Compiler, Condition Code, Register (CCR), Control Bits, Debugger, Direct Addressing (DIR), Extended Addressing (Ext), Flowchart, Half-Carry Flag (H), Hand Assembly, Higher-Level Language, Immediate Addressing (Imm), Implementation, Indexed Addressing (Indx, Indy), Inherent Addressing (Inh), Integrated Development Environment (Ide), Interrupt Mask Flag (I), Machine Language, Machine Code, mnemonics, Monitor Program, Negative Flag (N), Nonmaskable Interrupt Flag (X), Offset, Opcode, Operand, Operand Address, Operating System, Overflow Flag (V), Prebyte, Program, Programmer, Relative Addressing (Rel), Requirements Analysis, Single-Byte Instruction, Software, Software Design, Software Testing, Source Program, State Diagram, Statement Label, Status Flags, Stop Disable Flag (S), Three-byte instruction, Two-byte instruction, User Interface (Ui), Validation, Verification, Zero Flag (Z)

4.1 Introduction

René Descartes (1596–1650) was a French philosopher, mathematician, physicist, and writer, one of the key figures in the Scientific Revolution. One of Descartes' influences in mathematics was the Cartesian coordinate system, which is named after him. Descartes published a philosophical and mathematical treatise titled *Discourse on the Method* in 1637. (Its full name is *Discourse on the Method of Rightly Conducting One's Reason and of Seeking Truth in the Sciences.*) Descartes notes down in the *Discourse on the Method* the quote that begins this chapter. In today's software engineering world, the best practice of decomposing a problem is what Descartes was talking about. Best practices in software engineering employ the divide-and-conquer strategy in order to decompose the problem or the customer requirements into smaller and easy-to-handle problems or product requirements. At another place in the same thesis, Descartes points to the importance of enumeration as "In every case to make enumerations so complete, and reviews so general that I might be assured that nothing was omitted." In software engineering's state of the art practices, enumeration holds the key to problem-solving methodologies. It is essential since it not only helps in project planning but also aids in keeping track of the completeness of the proposed solution.

Descartes suggested drawing boxes on a paper, and connecting them. This idea has led to a multitude of graphic thinking aids that we use today. One such well-known approach is flowcharting. A flowchart is a collection of characteristically shaped boxes that are connected by line segments. Each box represents a type of activity. Specific details required to perform the various activities are entered within the boxes according to agreed-upon conventions. The line segments connecting the boxes are annotated with arrowheads, implying a sense of direction. The implication is that when one activity has been performed, the next step in the procedure is found by moving along a directed line segment to the activity described by the next box. The goal of this chapter is to achieve proficiency in problem-solving and design development approaches with respect to embedded software development, especially the HC11 programming. We will now start learning what operations HC11 can do on its own, and gain knowledge of how some of these can be put together to solve a problem. It can be expected that after reading the next few chapters, readers should be able to pick up a manufacturer's documentation and understand enough of it to harvest the details they need to design and implement a specific program.

Helpful Hint: The divide-and-conquer strategy requires that means for communicating data and controlling program flow be shared between elements.

4.2 Programming Concepts

An algorithm may be loosely defined as a set of instructions for solving a problem. In other words, it is an effective method for solving a problem expressed as a finite sequence of steps. Proficiency with algorithms is of strategic value in using the computer as a problem-solving tool, since a computer can solve a problem only after it has been told how to solve it. This means that human effort is required to develop detailed solution procedures that can subsequently be communicated to the computer for implementation. A solution is said to be computer implemented when instructions have been prepared that enable the computer to carry out the procedure. These instructions must be communicated to the computer in a language that it can "understand." Such a language is often referred to as a *programming language*. A hierarchy diagram of computer programming languages relative to the computer hardware is shown in Figure 4.1. At the lowest level is the computer hardware (CPU, memory, disk drive, input/output). Next is the *machine language* that the hardware understands because it is written with 1s and 0s. Recall from Chapter 2 that high and low levels were denoted by 1s and 0s. Machine code is commonly referred to as *object code*, and a machine code program is referred to as an *object code program*. The machine code consists of the following:

- *Operational code (opcodes)*: An opcode is a unique multibit code given to identify each instruction to the microcontroller.
- *Operands (data)*: The parameters that follow the assembly language mnemonic to complete the specification of the instruction. An operand is data that is operated upon by the instruction. For example, an

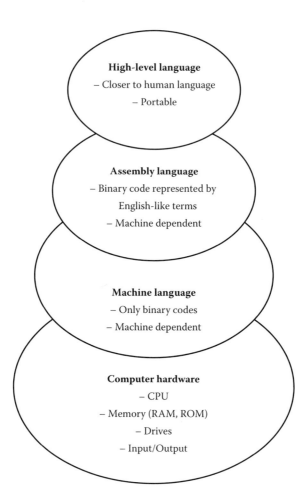

FIGURE 4.1
Hierarchy of programming languages relative to computer hardware.

addition instruction adds operands, and a load instruction reads an operand from an address in memory and loads it into a processor register. The operand field is defined as a group of up to three bytes following the opcode.

- *Addresses of the operands*: An address is a pointer into memory. Each memory location has a unique address by which it is identified. An address is an identifier for a memory location, at which a computer program or a hardware device can store data and later retrieve it. An absolute address is a complete address, whereas an offset is a value that is added to a base address to reference another memory location.

At the level above machine language is *assembly language* where the 1s and 0s are represented by English-like words. Assembly languages are considered low level because they are closely related to machine language and are machine dependent, which means a given assembly language can only be used on a specific microcontroller or microprocessor. At the level above assembly language is *higher-level language,* which is closer to human language and further from machine language. An advantage of higher-level language over assembly language is that it is portable, which means that a program can run on a variety of computers. Also, higher-level language is easier to read, write, and maintain than assembly language. Most system software like Windows, and applications software like word processors and spreadsheets, are written with higher-level languages. As mentioned earlier, higher-level languages contain English-like commands that are readily understandable by the programmer. They normally combine a number of assembly-level statements into a single high-level statement. A compiler is used to translate the higher-level languages such as C, C++, or JAVA into machine language.

Example 4.1

Identify the following from the load instruction given in the following text:

```
86    01    LDAA    #$01
```

(a) Machine code
(b) Source code
(c) Complete instruction
(d) Operand field
(e) Opcode
(f) Mnemonic
(g) Immediate data or actual operand

SOLUTION

(a) Machine code: 86 01
(b) Source code: LDAA #$01
(c) Complete instruction: "86 01" and "LDAA #$01"
(d) Operand field: 86 01
(e) Opcode: 86
(f) Mnemonic: LDAA
(g) Immediate data or actual operand: 01

Example 4.2

Find all of the fields asked in Example 4.1 for the following:

```
B7    01    00    STAA    $0100
```

SOLUTION

(a) Machine code: B7 01 00
(b) Source code: STAA $0100
(c) Complete instruction: "B7 01 00" and "STAA $0100"
(d) Operand field: 01 00
(e) Opcode: B7
(f) Mnemonic: STAA
(g) Extended mode effective address: 0100

Example 4.3

Find all of the fields asked in Example 4.1 for: 1B   ABA

SOLUTION

(a) Machine code: 1B
(b) Source code: ABA
(c) Complete instruction: "1B" and "ABA"
(d) Operand field: empty
(e) Opcode: 1B
(f) Mnemonic: ABA

Helpful Hint: The advantages of higher-level languages over assembly language are ease of readability and maintainability.

An 8-bit processor like HC11 cannot provide the opcode and operand address in a single 8-bit word. With a one-byte word size, there are three basic instruction formats: single-byte, two-byte, and three-byte. These are illustrated in Figure 4.2. The single-byte instruction contains only an 8-bit opcode. There is no operand address specified. The instruction ABA in Example 4.3 is a single-byte instruction. The first byte of the two-byte instruction shown in Figure 4.2 is an opcode, and the second byte is an 8-bit address code that specifies the operand address. These two bytes are always stored in memory in this order. The instruction LDAA $10 is a two-byte instruction. The meaning of this instruction is to take the data currently stored in memory address 10 (or $0010) and load it into accumulator A (AccA). The contents of address 10 (or $0010) remain unchanged. Here, 96 is the opcode for LDAA and 10 is the address containing data to be loaded in AccA.

Finally, the three-byte instruction is shown in Figure 4.2. In a three-byte instruction, the first byte is the opcode, and the second and third bytes for a 16-bit operand address. An example of a three-byte instruction is ADDA $00FE. This instruction indicates that the data stored in memory address 00FF should be taken, added to the current contents of AccA, and the result placed into AccA. Here, BB is the opcode for ADDA, 00 is the high-order byte of operand address, and FF is the low-order byte of operand address.

The single-byte, two-byte, and three-byte instruction formats that we have just discussed always begin with an opcode byte. But this rule has a few exceptions. For example, all of the instructions that use the index register Y will have a specific prebyte that precedes the opcode byte. This prebyte indicates

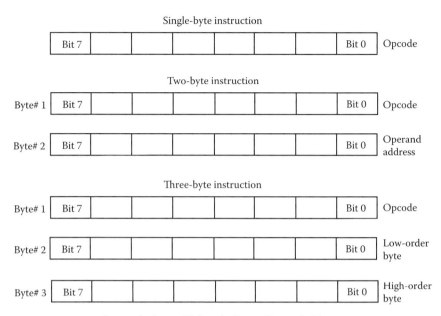

FIGURE 4.2
Instruction formats used in HC11.

that the instruction uses the internal Y register of the processor. Therefore, some instructions will have a two-byte opcode such that the first opcode is the prebyte and the second is the actual opcode for the instruction to be executed.

Group Discussion: Discuss where higher-level languages could be unsuitable as compared to assembly languages.

Section 4.2 Review Quiz

JAVA and C# are two examples of higher-level languages. (True/False)

4.3 System Software

Typical microcontroller system software provided in microcontroller development systems such as the HC11 includes editors, assemblers, compilers, interpreters, debuggers, and an operating system.

The editor is used to create or change source programs. Many editors for software developers include source program syntax highlighting and automatic completion to make programs easier to read and write. Source programs can be written in assembly language or a higher-level language such as C or C++. The editor has commands to change, delete, or insert lines or characters. Text editors are often provided with operating systems or software development packages, and can be used to change configuration files and programming language source programs. Programming editors often permit one to select the name of a subprogram or variable, and then jump to its definition and back.

The microcontroller is run by software instructions to perform particular tasks. Usually, the instructions are first written in assembly language using mnemonics abbreviations and then converted to machine language so that they can be interpreted by the microcontroller. The conversion from assembly language to machine language involves translating each mnemonic into the appropriate hexadecimal machine code and storing the codes in specific memory addresses. This can be done by a software package called an *assembler*, provided by the microcontroller manufacturer as shown in Figure 4.3. Another way to do this is by the programmer such that he or she looks up the codes and memory addresses. This process is called *hand assembly*.

A compiler is a program that compiles or translates a program written in a higher-level language and converts it into machine code that can be executed later, as shown in Figure 4.4. The compiler examines the entire source program and collects and reorganizes the instructions. Every higher-level language comes with a specific compiler for a specific computer, making the higher-level language independent of the computer on which it is used.

Group Discussion: What are some differences between a complier and an assembler?

Like a compiler, an interpreter usually processes a higher-level language program. Unlike a complier, an interpreter actually executes the higher-level

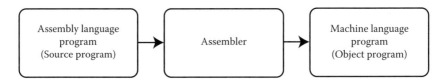

FIGURE 4.3
Assembly-to-machine conversion using an assembler.

FIGURE 4.4
High-level-to-machine conversion using a compiler.

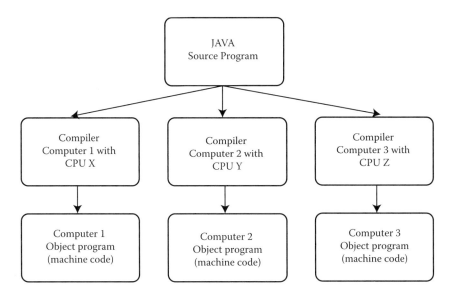

FIGURE 4.5
Machine independence of a program written in a higher-level language.

language program one statement at a time, rather than translating the whole program into a sequence of machine instructions to be run later.

All programs written in higher-level languages, such as JAVA and C++, will run on any computer. A given higher-level language is valid for any computer, but the compiler that goes with it is specific to a particular type of CPU. This is illustrated in Figure 4.5, where the same higher-level language program written in JAVA is converted by different machine-specific compilers.

A debugger provides an iterative method of executing and debugging the user's software, one or a few instructions at a time, allowing the user to see the effects of small pieces of the program and thereby isolate programming errors. Typically, debuggers offer sophisticated functions such as running a program step by step (single-stepping), stopping (breaking or pausing the program to examine the current state) at some event or specified instruction by means of a breakpoint, and tracking the values of some variables. Some debuggers have the ability to alter the state of the program while it is running, rather than merely observing it.

An *operating system* performs resource management and human-to-machine translation functions. A *resource* may be the processor, memory, or an I/O device. An operating system is another program that tells the machine what to do under a variety of conditions. Major operating system functions

Best Practice: Experienced software developers always utilize an integrated development environment (IDE) for software development. An IDE is a software application that provides comprehensive facilities to computer programmers for software development.

include efficient sharing of memory, I/O peripherals, and the microcontroller among several users. An operating system is the interface between the hardware and users. It manages the system resources in accordance with system policy to achieve system objectives. Operating systems for microcontrollers became available when microcontrollers moved from process control applications to the general-purpose computer applications. It was appropriate to write a process control program in assembly language because the microcontroller was required to perform dedicated real-time control functions. But when the microcontrollers evolved to the point of controlling several I/O devices, a need for organizing the operating system was felt. As mentioned in Chapter 3, the EVBU is the Motorola M68HC11 Universal Evaluation Board, a development tool for HC11 microcontroller-based designs. The hardware components of EVBU are the HC11 chip, M68HC68 real-time clock chip, standard serial communications port, and breadboard area. A monitor program with a user interface (UI) called BUFFALO (Bit User Fast Friendly Aid to Logical Operations) is software that provides a controlled environment for the HC11 chip to operate. BUFFALO is the standard boot-loader for the HC11. A user interface (UI) facilitates the EVBU to run programs, enter simple commands, and monitor the HC11. We saw a few BUFFALO commands in Chapter 3 through worked-out examples.

Team Discussion: An IDE normally consists of a source code editor, a compiler and/or an interpreter, build automation tools, and a debugger. Discuss your experience in using an IDE for programming in higher languages such as JAVA or C#.

Helpful Hint: BUFFALO is the standard boot-loader for the HC11.

Section 4.3 Review Quiz:

A program written in assembly language consists of a series of mnemonic statements and meta-statements, comments, and data. (True/False)

4.4 Developing a Program

The process of program development is divided into four phases:

- Problem analysis (requirements analysis)
- Design development (technical solution)
- Coding (implementation involving product integration)
- Testing (verification)

Since in this book we will be dealing with programs that are relatively smaller in size, one person will be performing all the just listed tasks.

However, in a project of relatively larger size involving numerous personnel, this process is often referred to as a software development process. Perhaps the most crucial task in creating a software product is extracting the requirements. This is called *requirements analysis*. Customers (end users of embedded system applications) typically have an abstract idea of what they want as an end result, but not what software should do. Incomplete, ambiguous, or even contradictory requirements are recognized by skilled and experienced software engineers at this point. Customer and product requirements are enumerated in a bidirectional traceability matrix. Whereas *customer requirements* define the problem to be solved, *product requirements* define the sensible features of the solution. A bidirectional traceability matrix maps requirement, design elements, code snippets, and test steps. Good traceability practices allow for *bidirectional traceability*, meaning that the traceability chains can be traced in both the forward and backwards direction as illustrated in Figure 4.6. *Forward traceability* looks at tracing the requirements sources to their resulting product requirements to ensure the completeness of the product requirement specification. Forward traceability is also a mechanism for tracing each unique product requirement forward into the design that implements that requirement, the code that implements that design and the tests that verify that requirement and so on. The goal is to ensure that each requirement is implemented in the product and that each requirement is thoroughly tested. The *backwards traceability* looks at tracing each unique work product (e.g., design element, code segment or unit, test procedures, etc.) back to its corresponding requirement. Backward traceability can verify that the requirements have been kept current with the design, code, and test. It is a mechanism that traces each requirement back to its source.

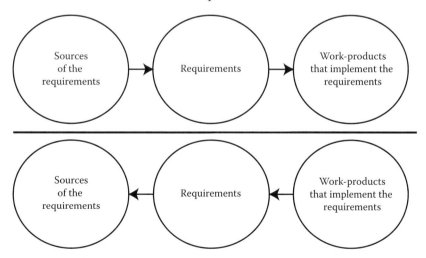

FIGURE 4.6
Bidirectional (forward and backward) traceability.

Before designing, coding, testing, and fielding the product, it might be prudent to determine whether satisfying the requirements also satisfies the original needs. Requirements are validated to ensure the resulting product will perform as intended in the user's environment. This may be done through prototype development, demonstrating presentation slides with dummy screenshots in meetings with customers, etc.

Software design is a process of problem-solving and planning for a software solution. After the purpose and specifications of software are determined, software developers have the daunting task to develop a plan for a solution. This may include low-level component and algorithm implementation issues as well as the architectural view. After the design development comes the implementation. One important point to note here is that the sequence of the phases like design development, coding, verification through test procedures, etc. is taken from a typical waterfall lifecycle of software development. For a graphical representation of the stages in waterfall lifecycle model, refer to the Appendix.

Implementation is the part of the process where computer programmers actually program the code for the project. Implementation using the detailed design implementation is carried out by the programmers according to the software requirements and specifications agreed upon between the development team and customer. Work products in the form of program code, databases, etc., are created to accomplish information or events processing as per the software design. Each module of every component is first implemented independently with unit tests performed by the developer and code walkthrough performed by peer reviewers. Later, as a product integration phase, these components are coupled and interfaced.

Software verification is an integral and important part of the software development process. This part of the process ensures that defects are recognized as early as possible. Peer review of work products like software design and source code, etc., constitutes a vital part of verification. The objective of testing is to find errors and mistakes in order to verify that the implemented software is as per the requirements specification. Test plan, design, and procedures are created as a part of software testing.

There are many other phases after the verification phase like the deployment phase, maintenance phase, etc., but due to the relevance to HC11 solution development, we will limit our discussion to just the four phases recognized above.

4.4.1 Problem Analysis

Before an efficient program can be developed to solve a problem, it is essential to:

- Define the problem.
- Analyze all the facts to state the parts of the problem that can be solved separately.

- Enumerate all the parts of the problem.
- Reduce the problem to a mathematical expression.

The mathematical expression should state the problem in its simplest form. Complex statements can be reduced through the process of numerical analysis. Computers are frequently used to reduce the labor of extensive numerical analysis.

Best Practice: The Z notation is a formal specification language used for describing and modeling computing systems. Experience software developers and requirements analysts use Z notation to make requirements specification unambiguous.

Helpful Hint: Eliciting needs is the first opportunity for failure.

4.4.2 Design Development

Often referred to as a *technical solution*, the purpose of design development is to design, develop, and implement solutions to requirements. Design development with respect to HC11 is usually the process of sequencing the operations into an order that will simplify coding, minimize execution time, and conserve storage space. It is not always possible to attain all three objectives. It is frequently necessary to compromise between minimum time and maximum conservation of storage space. However, the most efficient program is the one that solves the problem with the minimum number of instructions and in a minimum amount of time. This requires the designer to come up with design alternatives and choose the best-suited solution for a specific problem. Probably one of the hardest and most discipline-challenging aspects of writing software is to develop alternative solutions. A program flowchart is a useful tool in program organization. It helps to keep the entire program in view and aids in developing the proper sequence of operations.

4.4.3 Coding

Working from the flowchart, it is relatively straightforward to code the program if the person performing the coding is well versed in programming language.

Some of the fundamental properties that are most sought from a piece of code are

1. *Efficiency and performance*: It is highly desirable from a program to minimize the amount of system resources it consumes. These system resources can be processor time, memory space, slow devices such as disks, network bandwidth, and to some extent, even user interaction.

2. *Reliability*: Presence of logic errors such as division by zero or off-by-one errors can be a nightmare for debugging the program. The ultimate requirement from a program is for its results to come out correctly on every run. We have seen before that the conceptual correctness of algorithms is vital for any program to be successful. A systematic approach by a programmer is also important in order to

minimize many of the common programming mistakes such as mistakes in resource management, etc.

3. *Robustness*: Not taking into account the problems due to programmer error, the program should be written such that it accounts for various other problems such as incorrect, inappropriate or corrupt data, unavailability of needed resources such as memory, operating system services and network connections, and user error. Data validation is one of the important requirements management activities that programmers often utilize to accomplish robustness.

4. *Usability*: The more intuitive the user interface (UI) of a program, the better are scores on its usability. A wide range of textual, graphical, and sometimes hardware elements that improve the clarity, intuitiveness, cohesiveness, and completeness of a program's UI play a vital role in its success.

5. *Maintainability*: Maintainability has a relationship to the longevity of software. It is highly desirable that a program should be easily modifiable by its present or future developers in order to make improvements or customizations, fix bugs and security holes, or adapt it to new environments. Good practices during initial development make the difference in this regard. One such practice to improve maintainability is adding comments to the source code. Comments are usually added with the purpose of making the source code easier to understand. Comments have a wide range of potential uses: from augmenting program code with basic descriptions, to generating external documentation. Comments are also used for integration with source code management systems and other kinds of external programming tools. In the assembly language for HC11, a *comment* is any text after all operands for a given mnemonic have been processed. It is a line beginning with * up to the end of line. An empty line may also be considered as a comment.

Those of you who have heard of Visual Studio, NetBeans, and Eclipse know that these are IDEs. IDEs are generally used for debugging computer programs. As mentioned earlier in this chapter, the debugger provides an iterative method of executing and debugging the user's software, one or a few instructions at a time. This facilitates the programmer's seeing the effects of small pieces of the program and by this means isolating programming errors. Debugging is one of the key tasks in the software development process, because an incorrect program can have significant consequences for its users. Some programming languages are more prone to some kinds of faults because their specifications do not require compilers or assemblers to perform as much checking as other languages. Use of a static analysis tool can help detect some possible problems. Debuggers can vary from sophisticated IDEs to quite simpler ones. An example of a standalone debugger that is less of a visual environment and uses a command line is the gdb.

The types of instructions and programming techniques that are used with the 8-bit HC11 are similar to those used by all 8-bit microcontrollers. Thus, once you become proficient at programming, the fairly sophisticated HC11, it should be relatively easy to learn how to program other 8/16/32-bit microcontrollers. The result of the coding phase should be a properly sequenced, symbolically coded program that is ready for testing.

Example 4.4

Rewrite the following code snippet with appropriate comments.

```
Loop    LDAA    #$FF
        DECA
        BNE     Loop
```

Internet Search: Do some Internet research to write one page report on the Product Integration (PI) Phase in Software Development.

SOLUTION

```
Loop    LDAA #$FF        ; Load AccA with 25510
        DECA             ; Decrement AccA register
        BNE      Loop    ; Branch to Loop if not zero
```

4.4.4 Program Testing

A program with any degree of complexity seldom, if ever, operates satisfactorily on the first trial run. The more complex the program, the greater is the possibility of errors. Errors occur in coding, in interpretation of machine functions, in input data, and in machine operation. These errors are expensive because the programmer's time and effort are wasted until the errors are located and corrected. Errors must be kept to a minimum by checking and rechecking every step in the programming process. Peer review of selected work products like design specification, test procedures, etc., is vital in verification during the software development. Usually, all the product requirements are tested before releasing a product using test procedures. Before we end this discussion on verification, it is important to note that verification and validation have different meanings. Validation demonstrates that the product, as provided, will fulfill its intended use; whereas, verification addresses whether the work product properly reflects the specified requirements. In other words, verification ensures that "you built it right," whereas validation ensures that "you built the right thing." The depth of software testing depends on the size of the program. Test planning, test design, and test procedures with test steps are created by the test engineer in programs with relatively large size. We will omit these for now and visit them later if and when needed in later chapters.

Section 4.4 Review Quiz

Inspection and peer review are generally considered a part of software verification. (True/False)

4.5 Flow and State Diagrams

Human language provides a less acceptable means for communicating easily grasped descriptions of algorithms than does a programming language. This is the case because, unless the person who undertakes the explanation of a complicated procedure is extremely careful with his choice of words, ambiguities and misinterpretations can easily arise. These complications can be avoided by adopting certain conventions enabling algorithms to be described in a graphic form known as flow diagram or flowchart. The principal value of a flowchart is due to the fact that it shows a lot at a glance. It graphically represents organized procedures and data flow. The broad essentials and many details are readily apparent.

4.5.1 Flowchart Symbols

In flowcharting, symbols and words support one another for maximum clarity. Standard symbols enhance the graphical clarity of flowchart functions. These symbols are shown in Figure 4.7. Table 4.1 presents the details of these symbols. In a program flowchart, the emphasis is on computer decisions and processing. The programmer uses the flowchart to develop each step of his program, and starts with symbols representing major functions. As a program develops, the programmer extracts large segments and describes them in detail on subsidiary flowcharts. The finished flowchart is a programmer's guide to coding the program.

4.5.2 Flowcharting Techniques

We will use a simple problem to illustrate the use of flowcharts.

Example 4.5

Select the largest of three numbers.

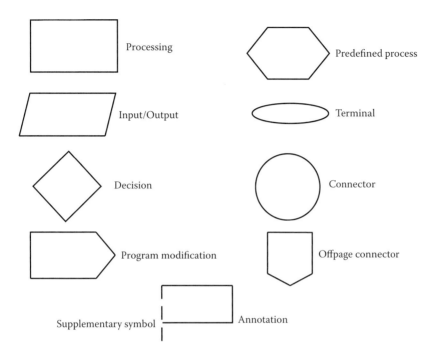

FIGURE 4.7
Program flowchart symbols.

TABLE 4.1

Program Flowchart Symbols

Symbol	Description
Annotation	The addition of descriptive comments or explanatory notes as clarification. The broken line may be drawn on either the left or right, and connected to a flow-line where applicable
Connector	An entry from, or an exit to, another part of the program flowchart
Decision	Point in the program where a branch to alternate paths is possible, is based upon variable conditions
Input/Output	Any function of an I/O device (making information available for processing, recording processing information, etc.)
Offpage connector	Used instead of the connector symbol to designate entry to or exit from a page
Predefined process	A group of operations not detailed in the particular set of flowcharts
Processing	A group of program instructions that performs a processing function of the program
Program modification	An instruction or group of instructions that changes the program
Terminal	The beginning, end, or a point of interruption in a program

SOLUTION

With this basic information, we can construct the chart in Figure 4.8. A decision block follows each comparison because the largest number is not known. The programmer must follow the decision blocks to the ultimate conclusion because he must provide for all selections and still select the largest number.

Team Discussion: One of the elements in the Technical Solution or Design Development phase is to come up with alternative designs. Discuss another method to find the largest of three numbers.

The chart seems to indicate that this process will take three random numbers and select the largest of the three. Now it is a simple matter to code the program from our finalized flowchart.

Flowcharts are a better way of communicating the logic of a system to all concerned. With the help of a flowchart, a problem can be analyzed in a more effective way. Additionally, program flowcharts serve as good program documentation, which is needed for various purposes. The flowcharts act as a guide or blueprint during the systems analysis and program development phase. Often, the flowchart helps in the debugging process. Furthermore, the maintenance of an operating program becomes easy with the help of a flowchart. It helps programmers put their efforts more efficiently on what needs to be done. But flowcharts have their own limitations. Often, in the real world the program logic to solve a difficult problem is quite complicated. In that case, a flowchart becomes complex and clumsy. Another limitation of a flowchart is that if alterations are required, the flowchart may require a complete redrawing. A graphical flowcharting utility can be used to overcome this limitation. The most important factor of all to keep in mind is that the essentials of what is done can easily be lost in the technical details of how it is done.

4.5.3 State Diagrams

The flowchart views a system as a series of events that are passed through, one after another. Many systems, however, act in a different sort of way. They tend to move from one state to another, maybe spending a considerable period of time in that state and departing from it only when a time period is completed or a specific event or action occurs. Such systems are best represented by a state diagram. The distinction between state diagrams and flowcharts is important because these two concepts represent two completely opposed programming paradigms. State diagrams are used for event-driven programming, whereas flowcharts are for structured programming. A designer of state diagrams has to constantly think about the available events in order to translate an accurate system behavior in state diagrams. In contrast, events are only a secondary concern for flowcharts (Figure 4.9).

Helpful Hint: State diagrams are used to give an abstract description of the behavior of a system.

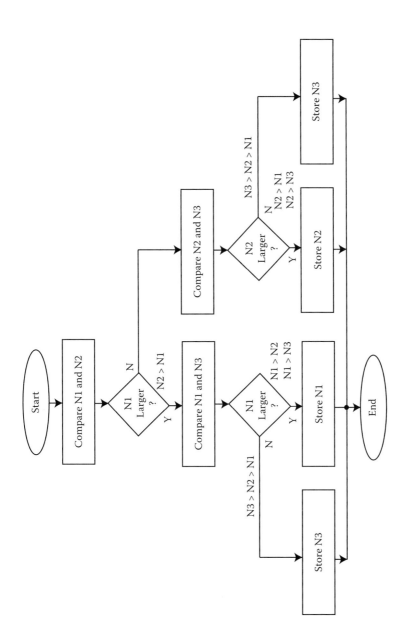

FIGURE 4.8
Flowchart to determine largest of three numbers.

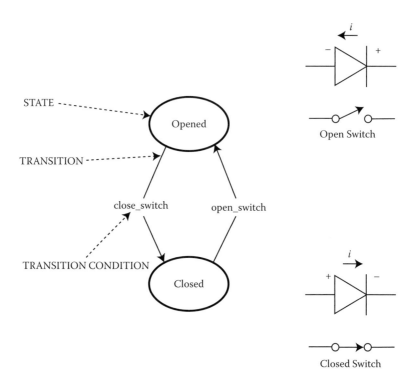

FIGURE 4.9
State diagram.

Section 4.5 Review Quiz

A state diagram is a type of diagram used in computer science to describe the behavior of systems. (True/False)

4.6 HC11 Programming Model

A microcontroller is often represented by its programming model. The programming model has significance due to its role in the representation of registers that the programmer can directly control through the microcontroller instruction set. The HC11 contains several registers that are used to store various kinds of information needed by the processor as it performs its functions. These registers serve as dedicated memory locations inside the

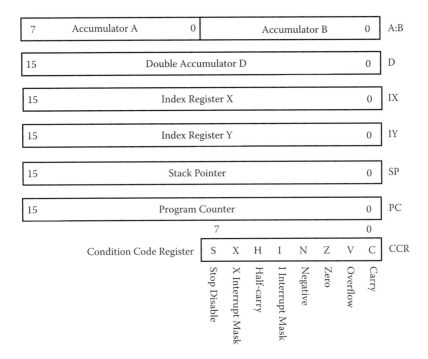

FIGURE 4.10
HC11 programmer's model.

processor chip. Recall that a basic register block configuration was shown in Figure 3.17. The important fact is that these are the internal registers that the programmer has to deal with throughout the coding phase, or even during the design development. Figure 4.10 illustrates the programming model for HC11. It contains two 8-bit accumulators (AccA and AccB), a 16-bit double accumulator D (AccD) that is a combination of AccA and AccB, 16-bit X and Y index registers, a 16-bit stack pointer, a 16-bit program counter, and an 8-bit condition code register.

4.6.1 The Condition Code Register (Flags)

CCR stands for condition code register. Recall from Table 3.6, where we briefly described HC11 processor registers, that this register contains 8 bits. These 8 bits are called *flags*, and they are used by the processor to monitor different conditions that exist during the execution of a program. The programmer can use conditional branch instructions to check the logic state of these flags to decide whether to branch or to continue in sequence. Let us examine the function of each flag.

4.6.1.1 Zero Flag (Z)

Whenever the processor executes an instruction that transfers a byte of data or performs an arithmetic, logic, or shift operation, it will automatically set or clear the Z flag according to whether or not the result of the operation is zero (00_{16} or 0000_{16}). The action can be summarized as "if the result is equal to zero, the processor set Z = 1, whereas if the result is not equal to zero, the processor clears Z = 0." The Z flag is affected whenever any instruction processes data or transfers data between registers or between a register and memory. The processor sets or clears the Z flag so that it can keep track of the "value being zero or not" of the last operation it performed.

Example 4.6

What is the effect on Z when the following operations occur?

(a) Subtraction of an operand from AccA, producing a result of $0000\ 0000_2$.
(b) Loading AccB with a data byte from memory that is equal to $0000\ 0000_2$.
(c) Decrementing the contents of the register X such that $[X] = 0000_{16}$.
(d) Incrementing the contents of the register Y such that $[Y] = 0000_{16}$.

SOLUTION

All of these operations will cause the processor to set Z = 1.

4.6.1.2 Negative Flag (N)

The negative flag (N) is set or cleared by the processor to indicate the sign of the result of any operation that transfers or processes data. Therefore, N is always equal to the sign bit (MSB) of the result. The processor will make N the same as the MSB of the result even when the program is not using signed numbers.

Common Misconception: Often, students think that the processor will keep track of whether the data represent unsigned numbers, signed numbers, or non-numerica information. This is something that the programmer has to keep track of.

Example 4.7

What is the value of N if the result of an addition instruction that is placed in:

(a) AccA is 01100111_2.
(b) X, Y or AccD is 7979_{16}.

SOLUTION

The processor will make:

(a) N = 0 since bit 7 (the MSB) in 01100111_2 is 0.
(b) $7979_{16} = 0111100101111001_2$. Therefore, N = 0 since bit 15 (the MSB) is 0.

Example 4.8

What will happen to N if the above result in:

(a) Part (a) is changed to 11000110_2.
(b) Part (b) is changed to $F3F3_{16}$.

SOLUTION

The processor will make

(a) N = 1 since bit 7 in 11000110_2 is 1.
(b) $F3F3_{16} = 1111001111110011_2$. Therefore, N = 1 since bit 15 (the MSB) is 1.

4.6.1.3 Carry Flag (C)

This flag is used to indicate a carry out of the MSB position when two 8-bit numbers or two 16-bit numbers are added. There are, however, many uses of this flag. Let us explore some of them through a few examples.

Example 4.9

What will be the value of the Carry Flag (C) when AccA which is 10101000_2 and the operand 01010110_2 from memory are added.

SOLUTION

$$
\begin{array}{r}
10101000_2 \\
+01010110_2 \\
\hline
11111110_2 \text{ (there is no carry at the MSB position)}
\end{array}
$$

Since there is no carry produced at the MSB position, the processes will clear the C flag to 0.

Example 4.10

What will be the value of the Carry Flag (C) when AccA which is 10101000_2 and the operand 01100110_2 from memory are added?

SOLUTION

$$
\begin{array}{r}
10101000_2 \\
+01100110_2 \\
\hline
1\ 00001110_2 \text{ (there is a carry of 1 at the MSB position)}
\end{array}
$$

Since there is a carry of 1 produced at the MSB position, the processes will set the C flag to 1.

When a binary number is subtracted from a smaller binary number, the MSB will have to "borrow" a 1 to complete the subtraction. When this happens, the processor will set $C = 1$ to indicate that a borrow has occurred. If a subtraction operation does not require a borrow, the MPU will clear $C = 0$. Thus, the C flag is also used to keep track of "borrows" when a subtract operation is performed on unsigned binary numbers.

Other than the previously mentioned uses, the C flag also participates in the various shift and rotate instructions that will be described in later chapters. The HC11 has specific single-byte instructions that can be used by the programmer to set or clear the C flag at any time. The flag C is set to 1 when the SEC instruction (opcode OD) is executed. The CLC instruction (opcode OC), when executed, will clear C to 0. There are no corresponding instructions for the Z or N flags. The HC11 also uses the C flag during its multiply instruction and its divide instructions.

Example 4.11

Branching was successful with the following instructions. What were the critical values (1 or 0) of Z or C flags for this to happen?

  (a) BEQ: branch if equal to zero
  (b) BNE: branch if not equal to zero:
  (c) BCC: branch if carry is clear
  (d) BCS: branch if carry is set

SOLUTION

  (a) Branch is made if Z flag is 1 (indicating a zero result).
  (b) Branch is made if Z flag is 0 (indicating a non-zero result).
  (c) Branch is made if C flag is 0, indicating that a carry did not result from the last operation.
  (d) Branch is made if C flag is 1, indicating carry occurred.

4.6.1.4 Overflow Flag (V)

By now we know that for signed numbers, the MSB is the sign bit with 0 = positive and 1 = negative. This leaves only 7 bits for the magnitude, with 2's complement used for negative values. With a sign bit and 7 magnitude bits, the numbers that can be represented will range from -128_{10} to $+127_{10}$. With this in hand, we define the V flag. The V flag is used to indicate overflow when signed numbers are added or subtracted by the processor. The processor will set $V = 1$ when an addition or subtraction operation results in an answer that falls outside the range. The HC11 instruction SEV can be used to set the V flag. Similarly, the instruction CLV can be used by the programmer to clear the V flag.

Example 4.12

What is the value of overflow flag (V) when $+96_{10}$ and $+64_{10}$ are added?

SOLUTION

$+96_{10} = 0\ 1100000_2$ (signed bit is zero)
$+64_{10} = 0\ 1000000_2$ (signed bit is zero)

$\overline{\phantom{+64_{10} = }}\ 1\ 0100000_2$ (signed bit is 1)

The signed bit is 1 which means that the result looks like a negative number. This does not look correct since the addition of 96 and 64 is 160, which is a positive number. The signed bit came into picture due to overflow from the addition of the magnitude bits into the sign bit position. When this overflow happens, the processor sets V = 1 to reflect the correct result.

Example 4.13

What is the value of overflow flag (V) when $+96_{10}$ and $+4_{10}$ are added?

SOLUTION

$+96_{10} = 0\ 1100000_2$ (signed bit is zero)
$+4_{10}\ = 0\ 0000100_2$ (signed bit is zero)

$\overline{\phantom{+4_{10}\ = }}\ 0\ 1100100_2$ (signed bit is zero)

The signed bit is 0, and no overflow from the addition of the magnitude bits into the sign bit position has happened. With no overflow, the processor clears V = 0 to reflect the correct result. Note that the result 100_{10} is not out of the range (-128_{10} to $+127_{10}$).

Common Misconception: Students often get confused between the V and C flags. V is used for signed arithmetic operations and indicates an overflow of the 7-bit magnitude capacity whereas C is used for unsigned operations and indicates an overflow of the 8-bit magnitude capacity.

4.6.1.5 Interrupt Mask Flag (I)

This flag can be set or cleared by using the SEI or CLI instructions, respectively. In Chapter 9, we will study the interrupts in detail where we will discuss more about the I flag.

4.6.1.6 Nonmaskable Interrupt Flag (X)

The X flag can only be set by two hardware conditions, system reset and the detection of an $\overline{\text{XIRQ}}$ signal. This flag can only be cleared by two software instructions, TAP (transfer from AccA to CCR) and RTI (return from interrupt). In Chapter 9, we will study the interrupts in detail where we will discuss more about the X flag.

TABLE 4.2

Summary of HC11 CCR (flags)

Flag	Name	Description
Z	Zero	Indicates that the result of a mathematical or logical operation was zero
C	Carry	Indicates that the result of an operation produced an answer greater than the number of available bits
X	XIRQ Interrupt Mask	Masks the XIRQ request when set. It is set by the hardware and cleared by the software, as well being set by unmaskable XIRQ
N	Negative	Indicates that the result of a mathematical operation is negative
V	Overflow	Indicates that the result of an operation has overflowed, according to the processor's word representation, similar to the carry flag but for signed operations
I	IRQ Interrupt Mask	Interrupts can be enabled or disabled by either setting or clearing this flag
H	Half-Carry	The H flag indicates whether or not a carry is produced from bit position 3 to bit position 4 during the addition of two 8-bit binary numbers
S	Stop Disable	The Stop Disable flag when set to 1 will prevent the STOP instruction from being executed

4.6.1.7 Half-Carry Flag (H)

The processor uses this flag whenever the 8-bit binary numbers represent BCD digits. The H flag indicates whether or not a carry is produced from bit position 3 to bit position 4 during the addition of two 8-bit binary numbers.

Helpful Hint: There are no specific instructions for clearing or setting the H flag.

4.6.1.8 Stop Disable Flag (S)

When set to 1, the Stop Disable flag (S) prevents the STOP instruction from being executed. The STOP instruction causes the oscillator and all of the HC11 clocks to stop. Table 4.2 summarizes all the CCR (flags).

Section 4.6 Review Quiz

List all the HC11 Flags.

4.7 HC11 Memory-Addressing Modes

One way to classify instructions is the fashion in which the operand address portion of the instruction is specified. In the instruction LDAA $2E00, the opcode byte is followed by a two-byte operand address $2E00. In Example 4.3, we saw an instruction with only opcode and no operand address. Thus, one type of instructions has an opcode byte followed by a two-byte operand address. The other type consists only of an opcode, since no operand address is required.

The HC11 has seven different address modes. This means that there are seven ways in which the operand address portion of an instruction can be specified. In this section we will discuss each of these address modes. Apart from the inherent mode, all of these modes result in the processor generating a 16-bit effective address, which it places on the address bus during the execution portion of an instruction cycle.

4.7.1 Extended Addressing

This address mode uses a two-byte operand address following the opcode. The format for extended addressing is shown in Figure 4.11.

Example 4.14

Describe the operation performed by the following instruction

```
LDAA      $2E00
```

SOLUTION

LDAA means "load accumulator A." $2E00 specifies the hex address of the memory location or the input device from which data are to be taken and loaded into accumulator A. The instruction "LDAA $2E00" will cause the processor to take the data from the address location 2E00 and load them into accumulator A. When implemented, "LDAA $2E00" instruction will produce the opcode and operand address as: B6 2E00. Here, B6 is the opcode that tells the processor that it has to fetch the next two bytes to determine the operand address (2E00 in this example).

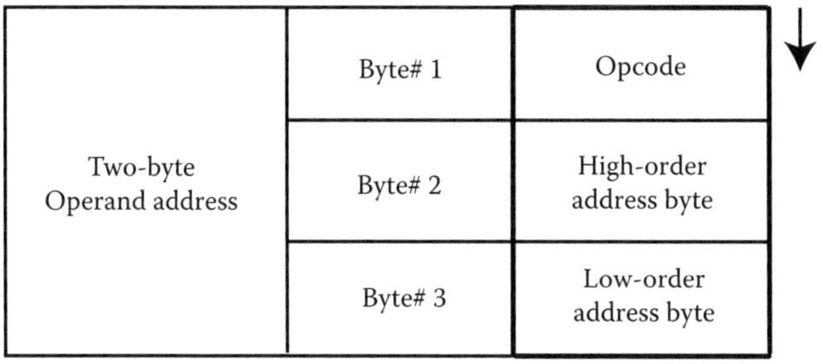

	Byte# 1	Opcode
Two-byte Operand address	Byte# 2	High-order address byte
	Byte# 3	Low-order address byte

FIGURE 4.11
Format for extended addressing.

Example 4.15

Describe the operation performed by the following instruction

```
ADDB   $00FF.
```

SOLUTION

ADDB means "Add the contents of memory to the contents of AccB and place the result in AccB." $00FF specifies the hex address of the memory location or the input device from which data are to be taken and added to the contents of accumulator B. The instruction "ADDB $00FF" will cause the processor to take the data from the address location 00FF and add them to the contents of accumulator B. When implemented, "ADDB $00FF" instruction will produce the opcode and operand address as: The opcode FB tells the processor what operation to perform (ADDB) and what address mode to use.

Example 4.16

Describe the operation performed by the following instruction

```
STY   $00FE.
```

SOLUTION

Recall our discussion earlier in this chapter about prebytes. Some HC11 instructions require a two-byte opcode; the prebyte followed by the opcode. Since this instruction involves the Y index register, the prebyte comes into play. For this particular instruction (STY) the prebyte is 18 and the opcode is FF. The instruction "STY $00FE" will cause the processor to store the most significant byte of the Y index register in memory location 00FE and the least significant byte of the Y index register in memory location 00FF, respectively. Remember, both index registers X and Y are 16-bit registers, and each memory location holds one byte, requiring two consecutive memory locations to store the entire contents of register Y.

4.7.2 Direct Addressing

Direct addressing requires only two bytes and typically takes only three clock cycles to execute. The format for direct addressing is shown in Figure 4.12.

The first and second bytes are the opcode and the lower-order address byte, respectively. There is no high-order byte because, in direct addressing, the high-order address byte is assumed to be $00000000_2 = 00_{16}$. Let us explore direct addressing through some examples.

Example 4.17

Describe the operation performed by the following instruction

```
LDAA   $FF.
```

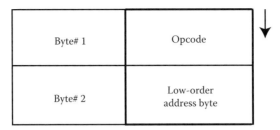

High-order address byte is always 00_{16}

FIGURE 4.12
Format for direct addressing.

SOLUTION

LDAA means "load accumulator A." $FF specifies the hex low-order address byte of the memory location or the input device from which data are to be taken and loaded into accumulator A. The high-order address byte is 00. Therefore, the combined address location is 00FF. The instruction "LDAA $FF" will cause the processor to take the data from the address location 00FF and load them into accumulator A. When implemented, "LDAA $FF" instruction will produce the opcode and operand address as: 96 00FF. Here, 96 is the opcode that tells the processor that it is direct addressing, and that it has to fetch the next two bytes to determine the operand address (00FF in this example).

Common Practice: Since it requires fewer instruction bytes and has a shorter execution time than extended addressing, programmers prefer to use direct addressing whenever possible.

4.7.3 Immediate Addressing

Consider the case when a programmer wants to load accumulator A with the value 10_{16}. Does the programmer need to specify an operand address? Absolutely not, because in this situation the data are known that are to be loaded into a processor register or to be operated on. The type of address mode where the operand itself is specified immediately following the opcode is called *immediate addressing*. Figure 4.13 presents the format of immediate addressing.

Example 4.18

Describe the operation performed by the following instruction

```
LDAA   #$FF
```

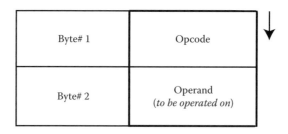

FIGURE 4.13
Format for immediate addressing.

SOLUTION

First observation here is that the # symbol appears in the #$FF. This symbol is used to indicate immediate addressing. The value that follows the # symbol is always data. Thus, FF is not an operand address, but is the operand itself. LDAA means "load accumulator A." #$FF specifies the data itself that is to be taken and loaded into accumulator A. The instruction "LDAA #$FF" will cause the processor to take the data FF and load it into accumulator A. When implemented, LDAA $FF will produce the opcode and operand address as: 86 FF. Here, 86 is the opcode that tells the processor that it is immediate addressing, and that it has to fetch, the operand (FF in this example).

Helpful Hint: The value that follows the # symbol is always data. The $ symbol is used to indicate that the data is hex.

4.7.4 Inherent Addressing

Inherent addressing occurs in the instructions that do not require any kind of memory access or I/O addresses. In HC11, many instructions will have one or more inherent or implicit addresses. What does inherent or implicit address mean? These are addresses that are implied by the instruction rather than explicitly stated by the programmer. Consider the example INX. Here INX means "increment the X register." The mnemonic is INX. The opcode, 08 here, notifies the processor that the operand is the contents of the X register. The processor then increments X by adding 1 to it.

Helpful Hint: All the instructions that set or clear flags in the CCR use inherent addressing.

4.7.5 Indexed Addressing

Indexed addressing is perhaps one of the most powerful address modes. It facilitates the handling of tables or blocks of data in memory. The format for indexed addressing is presented in Figure 4.14.

Usually, the indexed addressing is a two-byte instruction. The first byte is the opcode and the second byte is called *offset*. Offset is an unsigned

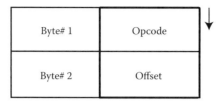

The offset is added to [X] or [Y] in order to obtain operand address

Operand address = [X] + offset
Operand address = [Y] + offset

FIGURE 4.14
Format for indexed addressing.

8-bit number. The effective operand address is obtained by adding the current contents of the index register X or register Y to this offset. Therefore, in indexed addressing mode, the process has to compute the operand address.

Helpful Hint: In indexed addressing mode, the operand address is not explicitly provided to the processor.

Example 4.19

Describe the operation performed by the instruction LDAA $FF, X.

SOLUTION

LDAA here means "load accumulator A with the data from address location given by adding FF to the X register." When implemented, "LDAA $FF,X" will produce the opcode and offset as: A6 FF. Here, A6 is the opcode that tells the processor that it is indexed addressing, and that it has to load the accumulator A with the data from the address location given by adding the offset (FF in this example) to the X register.

4.7.6 Relative Addressing

Branch instructions use the relative addressing mode. The significant point of interest is that relative addressing mode instructions do not process data. Instead, this mode controls the flow of the program. Table 4.3 summarizes all the HC11 addressing modes.

Section 4.7 Review Quiz

List all the HC11 addressing modes.

TABLE 4.3

Summary of HC11 Addressing Modes

Mode Name	Acronym	Operant Field	Operand	Remarks
Inherent	INH	Empty	No operand filed is empty. Operand is implied by the instruction	Memory access is not required
Immediate	IMM	ii	Operand found in memory locations following opcode (ii)	
Extended	EXT	hhll	Operand found in memory location pointed by the $hhll such that hh is the high-order and ll is the low-order byte of the 16-bit absolute address	Both hh and ll are found in the operand field
Direct	DIR	dd	Operand found in memory location pointed by $00dd such that dd is the direct mode address located in the operand field	
Indexed	INDX INDY	ff	Operand found in memory location addressed by the contents of the index register +ff such that ff is the 8-bit unsigned index mode offset located in the operand field	
Relative	REL	rr	In case the branch test is true, program execution will proceed at PC+$rr such that $rr is the signed 8-bit relative mode displacement located in the operand field. Else if the branch test is false, it will execute the next instruction	

4.8 Summary

1. The basic "language" of a computer is called *machine code* in which instructions are given as a series of binary codes.

2. Assembly language instructions are written using mnemonic abbreviations and then converted into machine language so that they can be interpreted by the microcontroller.

3. Higher-level languages like C or C++ are easier to write than assembly language, but they are not as memory efficient or as fast. All languages must be converted into a machine language matching that of the microcontroller before they can be executed.

4. Assembler is a software package that is used to convert assembly language into machine language.

5. HC11 machine code is made up of opcodes, operands, and addresses. The HC11 uses an 8-bit opcode and supports various ways of accessing operands in memory.

6. Compiler is a software package that converts a higher-level language program into machine language code.

7. A debugger provides an iterative method of executing and debugging the user's software, one or a few instructions at a time, allowing the user to see the effects of small pieces of the program and thereby isolate programming errors.

8. A software development process is a structure imposed on the development of a software product.

9. Requirements analysis encompasses those tasks that go into determining the needs or conditions to meet for a new or altered product.

10. Software design is a process of problem-solving and planning for a software solution.

11. Implementation is the realization of an application, or execution of a plan, idea, model, design, specification, standard, algorithm, or policy.

12. Software testing is an investigation conducted to provide stakeholders with information about the quality of the product or service under test.

13. A flowchart is a type of diagram that represents an algorithm or process showing the steps as boxes of various kinds, and their order by connecting these with arrows.

14. Flowcharts are used in analyzing, designing, documenting, or managing a process or program in various fields.

15. HC11 instructions can be of single, two, or three bytes type.

16. The condition code register (CCR) in the HC11 contains eight flags that indicate specific conditions that occur as the processor executes a program.

17. Each of the CCR flags conveys information to the processor or to the programmer.

18. By using a series of conditional branch instructions that are part of the HC11 instruction set, the programmer can test the values of some of these flags to determine what sequence of instructions to follow next.

19. CCR are the basis for all of the "decision-making" capabilities of any computer.

20. Each processor supports several methods it can use to access the memory called *memory-addressing modes*.

21. Immediate mode, direct mode, indexed mode, and relative mode are some of the examples of addressing modes.

22. Some modes allow memory to be accessed with an absolute address, while others use an address that is relative in nature.

23. The inherent mode is used for all instructions for which the operand is implied.

Glossary

Address: An address is an identifier for a memory location, at which a computer program or a hardware device can store data and later retrieve it.

Assembler: A software package that is used to convert assembly language into machine language.

Assembly Language: A low-level programming language unique to each microcontroller. It is converted, or assembled, into machine code before it can be executed.

Carry Flag (C): A bit in the CCR that indicates that the result of an operation produced an answer greater than the number of available bits.

Comment: A comment is a programming language construct used to embed programmer-readable annotations in the source code of a computer program.

Compiler: A software package that converts a higher-level language program into machine language code.

Condition Code Register (CCR): Referred to as "C," "CCR," or "Status" register, it is an 8-bit status and control register. Out of the 8 bits in CCR, three bits are S, X, and I. These are called *control bits*.

Control Bits: Out of the 8-bit bits in CCR, five bits are H, N, Z, V and C. These are called *status flags*. The remaining three bits are S, X, and I. These are called *control bits*.

Debugger: A debugger provides an iterative method of executing and debugging the user's software, one or a few instructions at a time, allowing the user to see the effects of small pieces of the program and thereby isolate programming errors.

Direct Addressing (DIR): Two bytes and typically takes only three clock cycles to execute.

Extended Addressing (EXT): Two-byte operand address following the opcode.

Flowchart: A diagram used by the programmer to map out the looping and conditional branching that a program must make. It becomes the blueprint for the program.

Half-carry Flag (H): A bit in the CCR that indicates whether or not a carry is produced from bit position 3 to bit position 4 during the addition of two 8-bit binary numbers.

Hand Assembly: The act of converting assembly language instructions into machine language codes by hand, using a reference chart.

Higher-level language: A type of computer language closest to human language that is a level above assembly language.

Immediate Addressing (IMM): The byte following the opcode is the operand.

Implementation: Implementation is the realization of an application, or execution of a plan, idea, model, design, specification, standard, algorithm, or policy.

Indexed Addressing (INDX, INDY): The operand address is obtained by adding the offset byte that follows the opcode to the contents of an index register.

Inherent Addressing (INH): Single-byte instruction that does not require an operand address.

Integrated Development Environment (IDE): An IDE also known as *integrated design environment* or *integrated debugging environment* is a software application that provides comprehensive facilities to computer programmers for software development.

Interrupt Mask Flag (I): A bit in the CCR that interrupts can be enabled or disabled by respectively setting or clearing this flag.

Lower-level Language: A lower-level programming language provides little or no abstraction from a computer's instruction set architecture. The word *lower* refers to the small or nonexistent amount of abstraction between the language and machine language.

Machine Language: Computer instruction written in binary code that is understood by a computer; the lowest level of programming language.

Machine Code: The binary codes that make up a microcontroller's program instructions.

Mnemonics: The abbreviated spellings of instructions used in assembly language.

Monitor Program: The computer software program initiated at power-up that supervises system operating tasks, such as reading the keyboard and driving the computer monitor.

Negative Flag (N): A bit in the CCR that indicates that the result of a mathematical operation is negative.

Nonmaskable Interrup Flag (X): A bit in the CCR that masks the XIRQ request when set. It is set by the hardware and cleared by the software as well is set by unmaskable XIRQ.

Offset: A byte that follows the opcode for a conditional branch instruction such as BEQ. The offset is added to the PC to determine the address to which the processor will branch.

Opcode: Operation code. It is the unique multibit code given to identify each instruction to the microcontroller.

Operand: The parameters that follow the assembly language mnemonic to complete the specification of the instruction.

Operand Address: Address in memory where an operand is currently stored or is to be stored.

Operating system: See Monitor Program.

Overflow Flag (V): A bit in the CCR that indicates that the result of an operation has overflowed according to the processor's word representation, similar to the carry flag but for signed operations.

Prebyte: First byte of a two-byte opcode.

Program: Complete sequence of instructions that directs a computer to perform a specific task or solve a problem.

Programmer: One who designs or writes a program.

Relative Addressing (REL): Used in conditional branch instructions to determine the branching address by adding the offset to the PC.

Requirements Analysis: Requirements Analysis encompasses those tasks that go into determining the needs or conditions to meet for a new or altered product.

Single-byte instruction: The single-byte instruction contains only an 8-bit opcode. There is no operand address specified.

Software: Computer program statements that give step-by-step instructions to a computer to solve a problem.

Software Design: Software design is a process of problem-solving and planning for a software solution.

Software Testing: Software testing is an investigation conducted to provide stakeholders with information about the quality of the product or service under test.

Source Program: In context to HC11, a program written in assembly language.

State Diagram: A state diagram is a type of diagram used to describe the behavior of systems.

Statement Label: A meaningful name given to certain assembly language program lines so that they can be referred to from different parts of the program.

Status Flags: Out of the 8-bit in CCR, five bits are H, N, Z, V, and C. These are called *status flags*.

Stop Disable Flag (S): A bit in the CCR that when set to 1 will prevent the STOP instruction from being executed.

Three-byte instruction: In a three-byte instruction, the first byte is the opcode, second and third bytes for a 16-bit operand address.

Two-Byte instruction: In a two-byte instruction, the first byte of the two-byte instruction is an opcode, and the second byte is an 8-bit address code that specifies the operand address. These two bytes are always stored in memory in this order.

User Interface (UI): An user interface is the space where interaction between humans and machines occurs.

Validation: Validation refers to an activity to ensure that the software that is being created is as per the requirements agreed upon at the analysis phase and to ensure product's quality.

Verification: Verification refers to an activity to ensure that specific functions are correctly implemented.

Zero Flag (Z): A bit in the CCR that indicates that the result of a mathematical or logical operation was zero.

Answers to Section Review Quiz

4.2 True

4.3 True

4.4 True

4.5 True

4.6 Refer to Table 4.2

4.7 Refer to Table 4.3

True/False Quiz

1. An assembly language is a high-level programming language for microcontrollers.

2. Typically a modern assembler creates object code by translating assembly instruction mnemonics into opcodes, and by resolving symbolic names for memory locations and other entities.

3. A program written in assembly language consists of a series of instructions–mnemonics that correspond to a stream of executable instructions that, when translated by an assembler, can be loaded into memory and executed.

4. Compilers enable the development of programs that are machine dependent.

5. Compiled languages are transformed into an executable form before running.

6. Interpreted languages are read and then executed directly, with no compilation stage. A program called an *interpreter* reads the program line by line and executes the lines as they are read.

7. A program flowchart is generally read from bottom to top.

8. Flowcharting symbols are connected together by means of bubbles.

9. An arrow coming from one symbol and ending at another symbol represents that this control passes to the symbol the arrow points to.

10. According to the best practices in software engineering, a comment in the source code of a computer program must always be avoided as much as possible because it wastes valuable space in a program.

11. Overflow can also occur when subtraction is performed between two numbers with opposite signs. The processor's internal logic can sense the overflow condition and will set or clear V accordingly.

12. Direct addressing typically takes only three clock cycles to execute.

13. In HC11 immediate addressing, the value that follows the # symbol is always data.

14. In HC11 immediate addressing, the $ symbol is used to indicate that the data is hex.

15. Z flag of CCR indicates that the result of a mathematical or logical operation was zero.

Questions

QUESTION 4.1

What is the basic difference between opcode and operand?

QUESTION 4.2

What are some of the limitations of using a flowchart in modern higher-level languages?

QUESTION 4.3

Describe the significance of the programming model for a microcontroller like HC11.

QUESTION 4.4

How many cycles do they need for execution?

QUESTION 4.5

A programmer wants to set the C flag. What kind of addressing mode will be used here? Explain the instruction in detail.

QUESTION 4.6

What is the difference between validation and verification in software development processes?

QUESTION 4.7

Describe some of the stages or phases in waterfall model of software development lifecycle?

QUESTION 4.8

A programmer defines a string as a local constant in a subroutine, and specifies the base address of the string as an address in the instruction. During iterations in a program, where should the programmer store the offset to access a specific character of the string?

QUESTION 4.9

Describe the operation performed by the following instruction.

```
LDAB #$A0
```

QUESTION 4.10

The following three instructions belong to one type of addressing mode. Name that addressing mode.

```
STX     $0011
ADDA    $0021
LDAB    $0034
```

Problems

PROBLEM 4.1

Draw a flowchart to find the sum of first 50 natural numbers.

PROBLEM 4.2

Draw a flowchart for computing factorial N (i.e., N!).

PROBLEM 4.3

Draw a flowchart to find the smallest of 10 numbers that are stored in sequential memory locations.

PROBLEM 4.4

Draw a flowchart to find the largest of 10 numbers that are stored in sequential memory locations.

PROBLEM 4.5

Write a program in any programming language to implement the algorithm presented in the flowchart to find the greatest of 3 numbers.

PROBLEM 4.6

What are the various addressing modes used in the assembly program code snippet given below?

```
Loop   LDAA   #$FF
       DECA
       BNE    Loop
```

PROBLEM 4.7

Identify the following from the load instruction 86 24   LDAA #$24

   (a) Machine code
   (b) Source code
   (c) Complete instruction
   (d) Operand field
   (e) Opcode
   (f) Mnemonic
   (g) Immediate data or actual operand
   (h) What is the meaning of # and $ in #$24?

PROBLEM 4.8

Correct the following pseudocode with respect to the Z flag of CCR. Explain your changes.

```
if (result = zero){
processor sets Z =0
}
elseif (result ≠ zero){
processor clears Z =0
}
```

PROBLEM 4.9

What is the effect on Z when the following operations occur?

(a) Clearing the contents of AccA, AccB, or any memory location such that the final result in any of these registers becomes $0000\ 0000_2$.

(b) Decrementing or incrementing the contents of the register X and Y such that [X] and [Y] are equal to 0000_{16}, respectively.

PROBLEM 4.10

Describe the operation with respect to addressing modes as performed by the HC11 instruction given below.

```
Instruction code   Assembly language
DB                     ADDB   $9C
9C
```

PROBLEM 4.11

What operation in terms of addressing modes will the HC11 perform in the following instruction?

```
Instruction code   Assembly language
      CE             LDX   #$0011
      00
      11
```

PROBLEM 4.12

Describe the sequence operation performed by the instruction sequence below. Keep your focus on the addressing modes rather than the data manipulation done by the instructions.

```
Instruction code   Assembly language
D6                     LDAB $55
55
CB                     ADDB #$01
02
F7                     STAB $CC00
CC
00
```

PROBLEM 4.13

Determine the contents of the HC11 internal Y register and the state of the Z flag at the completion of the instruction sequence below. Why does 18 appear in the instruction code?

```
Address   Instruction code    Assembly language
0000            18              LDY #$FFFE
0001            CE
0002            FF
0003            FE
0004            18              INY
0005            08
0006            18              INY
0007            08
```

PROBLEM 4.14

What are the addressing modes used in the following assembly language code?

```
LOOP:   LDAB   $00,X   ; Get a number from source list
        STAB   $00,Y   ; Put the number in destination list
        INX            ; Increment pointer of source list
        INY            ; Increment pointer of destination list
```

PROBLEM 4.15

Although the following three load accumulator A instructions look similar, there is a difference in the addressing mode that they employ. What are the addressing modes for each load accumulator A instruction?

```
LDAA   #$00
LDAA   $00
LDAA   $0000
```

5

Instructions

> When all else fails, read the instructions.
>
> **—Anonymous**

OBJECTIVES

Upon completion of this chapter you should be able to

1. Describe a few of the HC11 instructions.
2. Use and understand various aspects of HC11 instructions.
3. Understand the different instruction classifications of HC11 instruction set.
4. Determine addressing mode, the operand, and its effective addresses for HC11 instructions.
5. Write short programs that move data via load, store, and transfer instructions.
6. Perform simple arithmetic and logic operations.
7. Understand the concept of data masking.
8. Demonstrate the serial shifting of data.
9. Understand HC11 instructions to perform bit-level manipulation and comparisons.

Key Terms: AND Masking, Arithmetic, Arithmetic Shift, Carry Flag (C), Clear, Comment, Condition Code Register (CCR), Control Bits, Data Movement, Direct Addressing (DIR), Extended Addressing (EXT), Half-Carry Flag (H), Immediate Addressing (IMM), Indexed Addressing (INDX, INDY), Inherent Addressing (INH), Instruction, Interrupt Mask Flag (I), Logic, Logical Shifting, Negative Flag (N), Nonmaskable Interrupt Flag (X), Offset, Opcode, Operand, Operand Address, OR Masking, Overflow Flag (V), Rotate Operation, Status Flags, Set, Stop Disable Flag (S), Zero Flag (Z)

5.1 Introduction

Commands that control the actions of the processor are called *instructions*. In HC11, there are 145 mnemonic key words that give rise to 308 unique opcodes. The instruction LDAA performs the task of loading a byte of data from memory to Accumulator A (AccA). We know from Chapter 4 that there are many ways in which this loading of data can be performed. For example, the instruction LDAA can be implemented using the IMM, DIR, EXT, INDX, and INDY addressing modes. Therefore, the instruction LDAA gives rise to five opcodes, one for each addressing modes. The lower-case "x" is used in conjunction with the instruction mnemonics to indicate a wildcard. For example, LDAx is a set of all the mnemonics with the first three letters equal to LDA. Thus, LDAA and LDAB belong to the set of LDAx. Similarly, mnemonics STD, STS, STX, and STY belong to the set of mnemonics written as STx. In this chapter we will explore such HC11 instructions by working through several examples. The goal of this chapter is to develop a general understanding of the basic concepts of instructions, rather than fully define the HC11 instruction set. Fully defining the instruction set at this stage may result in your getting lost in details. A complete list of instructions is given in Appendix E. However, you are encouraged to refer to the Instruction Set Table in the M68HC11E Series Programming Reference Guide as and when needed. Furthermore, in this chapter, we will only describe a few of the hundreds of variations of instructions available to programmers. To simplify learning the instruction set, instructions can be divided into categories. One such categorization can be the following:

- Data movement
- Arithmetic
- Logic or bit manipulation
- Loops and jumps
- Subroutine and interrupts
- Processor control

TABLE 5.1

HC11 Processor Registers

Processor Register	Description
Accumulator A	Referred to as "A" register or as "AccA," it is an 8-bit register that is used as the primary data-processing register.
Accumulator B	Referred to as "B" register or as "AccB," it is an 8-bit register identical to accumulator A in function.
Accumulator D (A:B)	Referred to as "D" register or as "AccD," it is a 16-bit register such that two 8-bit accumulators are combined in it.
Index Register X	Referred to as "IX," "[X]," or "X" register, it is a 16-bit address register that is used by the indexed addressing mode instructions. It can also be used as a general-purpose 16-bit register.
Index Register Y	Referred to as "IY," "[Y]," or "Y" register, it is a 16-bit address register identical to Index Register X.
Stack pointer	Referred to as the "S" or the "SP" register, it is a 16-bit register that always contains the address of the next available stack memory location.
Program counter	Referred to as the "PC," it is a 16-bit address register that always contains the address of the next location in the memory that will be addressed.
Condition code register	Referred to as "P," "C," "CCR," or "Status" register, it is an 8-bit status and control register. Out of the eight bits, five bits are H, N, Z, V, and C. These are called *status flags*. The remaining three bits are S, X, and I. These are called *control bits*.

We will not adhere to the just-mentioned classification. Rather, we will split the elements of the above classification and put them in a natural sequence in order to achieve a rewarding learning experience. As an example, we will learn about multiplication and division instructions after we have understood data movement, addition and subtraction, logic, and shifting and rotating instructions. Learning in this order builds a strong foundation toward problem solving and assembly programming. Recall from Chapter 3 that HC11 has some registers called *processor registers*. These were summarized in Table 3.6. Since these processor registers are frequently used with almost all the instructions, we repeat the same table here as Table 5.1. Also, the condition code registers (CCRs) are frequently used throughout our journey of learning HC11 programming. CCR is shown in Figure 5.1. HC11 processor registers are listed in Table 5.1.

5.2 Data Movement

The simplest and most typical operation you can do within a computer system is to transfer data from one location to another and make a copy.

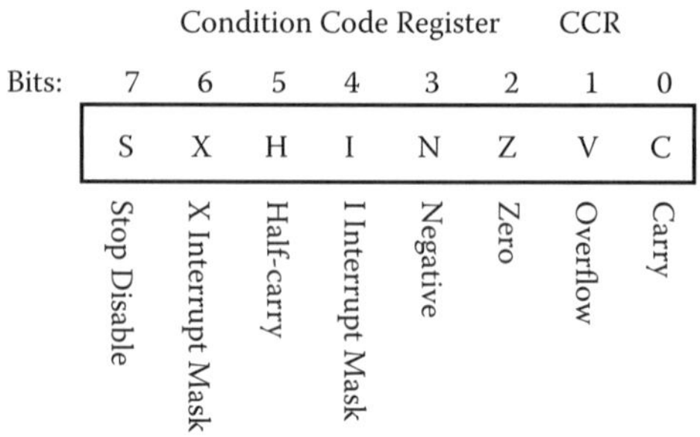

Condition Code Register   CCR

Bits: 7   6   5   4   3   2   1   0

| S | X | H | I | N | Z | V | C |

Stop Disable — X Interrupt Mask — Half-carry — I Interrupt Mask — Negative — Zero — Overflow — Carry

FIGURE 5.1
HC11 processor registers.

Inside the processor, there is a local data bus that is connected to all the registers in the processor. Recall from Chapter 3 that a bus is a subsystem that transfers data between computer components inside a computer or between computers. Each register is a latch whose outputs and inputs are attached to this bus. The data on the bus can be clocked into the inputs or enabled onto the bus from the outputs. Suppose that a programmer wants to transfer a number from Accumulator A (AccA) to Accumulator B (AccB). The output of AccA is enabled to the bus and AccB is clocked. This makes a clone of the contents of AccAs in AccB. Nothing happens to the contents of AccA, that is, the original value stored in AccA remains as it is. The same is true for all transfers from any place in a computer to any place else. When we say "any place," we mean either the processor register or memory. Thus, the movement of data can occur from the processor registers to memory, from memory to the processor register, or from one processor register to another processor register. Note that the HC11 does not support instructions that allow direct memory-to-memory transfers using a single instruction. The data movement in HC11 can be accomplished by load, store, clear, transfer, or exchange instructions. Let us look at these instructions in detail to get a better understanding of data movement in HC11.

5.2.1 Load Instructions

Figure 5.2 illustrates some of the load instructions. These instructions basically move data from memory to the processor registers. The set of load instructions is presented in Table 5.2.

Let us now look deeper into the interpretation of Table 5.2. Correct interpretation of such tables is critical for a better understanding of an instruction set. In Table 5.2, each row provides information about an instruction or mnemonic. The instruction or mnemonic is given in the first column. The next six columns are the mode of addressing. An "X" in a cell means that the mnemonic belongs to that addressing mode, while a "-" symbol means that it does not. Therefore, we can observe that LDAA belongs to IMM, DIR, INDX,

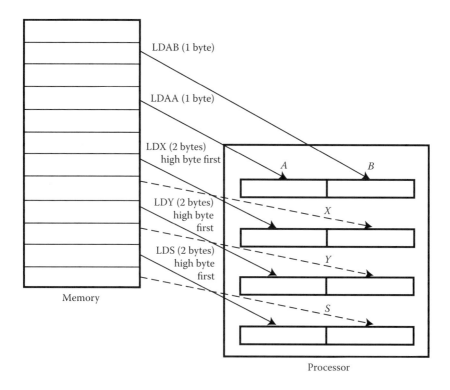

FIGURE 5.2
Visualizing HC11 load instructionsm,

TABLE 5.2

Set of Load Instructions

Instruction	IMM	DIR	INDX	INDY	EXT	INH	Function	H	I	N	Z	V	C
LDAA	X	X	X	X	X	—	(M) → A	—	—	X	X	0	—
LDAB	X	X	X	X	X	—	(M) → B	—	—	X	X	0	—
LDD	X	X	X	X	X	—	(M) → A						
							(M) → B	—	—	X	X	0	—
LDS	X	X	X	X	X	—	(M):(M+1) → s	—	—	X	X	0	—
LDX	X	X	X	X	X	—	(M):(M+1) → X	—	—	X	X	0	—
LDY	X	X	X	X	X	—	(M):(M+1) → Y	—	—	X	X	0	—

INDY, and EXT. It does not belong to the INH addressing mode. The function column contains the function performed by the mnemonic. Here, (M) → A means "load AccA with contents of memory." In the case of LDX, the function (M):(M+1) → X means "load 16-bit register with contents of memory." The knowledge of register size is critical here. Since index register X is a 16-bit register, LDX instruction loads two bytes into the index register X.

Each of the load instructions that copies two bytes of data follows the "high byte first" convention. In the final six columns, those flags of CCR that are affected are filled with the "X" symbol. A "-" symbol means that it does not get affected. A bit 0 or bit 1 would mean clear and set respectively. We can observe that all load instructions affect only three status flags in the CCR (N, Z, and V). The N and Z reflect the actual condition of the data that are loaded. The V bit is cleared, indicating that the sign (N flag) is correct. You are encouraged to refer to the Instruction Set Table in the M68HC11E Series Programming Reference Guide as and when needed.

Example 5.1

With the stored data and CCR as given here, answer the following for each of the instruction (I to III) listed.

(a) What is the addressing mode? How did you figure out?
(b) What are the effective addresses of the operand?
(c) What is the operand?
(d) Describe the instruction in plain English.
(e) What is the resulting status of each of the flags in the CCR? How did you figure them out?

Data:

0020	88	FF	00	00	FF	FF	00	00
0130	11	00	EE	00	00	00	00	00
CCR:	$D0							

	Memory Location	Machine Code		Source Code	
(I)	0000	86	10	LDAA	#10
(II)	0000	D6	20	LDAB	$20
(III)	0002	FE	01	33 LDX	$0133

SOLUTION

(I) (a) Looking at the # sign, we can easily figure out that we are dealing in IMM addressing mode.

(b) The effective address is the address immediately following the opcode in memory. To get that we add 1 to 0000. Therefore, the effective address is $0001.

(c) Whatever is next to # sign is the data. Therefore, 10 is the data (operand).

(d) The load instruction here makes the AccA value equal to 10.

(e) Original value of CCR is $D0 = 1101\ 0000_2$. Therefore, the sequence of SXHINZVC is 11010000. Keeping in mind that the operand is $10 (or 0001 0000_2$), N = 0 (MSB of operand is zero), Z = 0 (operand is not zero), and V = 0 (this is always clear in load instructions, see Table 5.2). Therefore, the value of CCR remains unchanged at $D0.

(II) (a) Since there is no # sign, we conclude that 20 is one byte of the address. Therefore, the addressing mode is DIR.

(b) The effective address can be obtained by putting 00 in front of 20. Thus, we get $0020.

(c) The operand is the data stored at the effective address. We can look at the data provided in the problem statement and see that $88 is located at the memory location $0020. Therefore, operand is $88.

(d) The load instruction here makes AccB value equal to contents of the memory location $0020.

(e) The original value of CCR is $D0 (or 1101 0000$_2$). Therefore, the sequence of SXHINZVC is 11010000. The operand $88 (or 1000 1000$_2$) makes N = 1 (MSB of operand is one), Z = 0 (operand is not zero), and V = 0 (this is always clear in load instructions; see Table 5.2). Therefore, the value of CCR changes from 1101 0000$_2$ to 1101 1000$_2$. Note that only N bit changed from 0 to 1. Thus, the new value of CCR is $D8.

(III) (a) The first observation in this instruction is that there is no # sign. The second observation is that there are two bytes of the address. Therefore, we conclude that it is an EXT mode of addressing.

(b) The effective addresses are two in number. The first one is the memory location $0133. The second one is obtained by adding 1 to the first one. Therefore, the second effective address is $0134.

(c) The operand (i.e., data stored at the effective address) is $0000 since $00 is located at each of the memory locations $0133 and $0134.

(d) The load instruction here makes the value of register X equal to the contents of the memory location $0133 and $0134.

(e) The original CCR value is $D0 (or 1101 0000$_2$). Therefore, the sequence of SXHINZVC is 11010000. We know from part (d) that the operand is $00 (0000 1000$_2$). The operand $00 makes N = 0 (since MSB of operand is zero), Z = 1 (operand is zero), and V = 0 (this is always clear in load instructions, see Table 5.2). Therefore, the value of CCR changes from 1101 0000$_2$ to 1101 0100$_2$. Note that only bit Z changed from 0 to 1. The new value of CCR is $D4.

5.2.2 Store Instructions

Figure 5.3 illustrates some of the store instructions. These instructions basically move data from the processor registers to memory. The set of store instructions is presented in Table 5.3.

The interpretation of Table 5.3 (store instructions) can be done in the same way we did for Table 5.2 (load instructions). Load and store instructions are complementary to each other.

Example 5.2

With the saved data and various processor registers set to the values given in the following text, answer the following for each of the instructions (I to III) listed.

(a) What is the addressing mode? How did you figure it out?
(b) What is the effective address of the operand?
(c) What is the operand?
(d) Describe the instruction in plain English.
(e) What is the resulting status of each of the flags in the CCR? How did you figure this out?

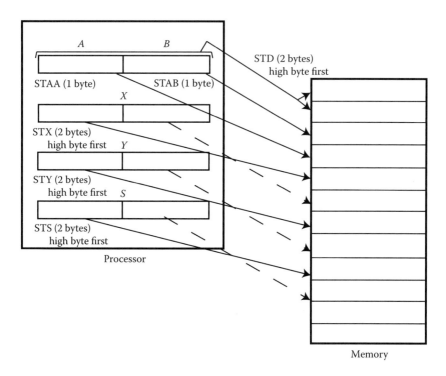

FIGURE 5.3
Visualizing HC11 store instructions.

TABLE 5.3

Set of Store Instructions

Instruction	IMM	DIR	INDX	INDY	EXT	INH	Function	H	I	N	Z	V	C
STAA	—	X	X	X	X	—	(A) → M	—	—	X	X	0	—
STAB	—	X	X	X	X	—	(B) → M	—	—	X	X	0	—
STD	—	X	X	X	X	—	(A) → M						
							(B) → M+1	—	—	X	X	0	—
STS	—	X	X	X	X	—	(S) → M:M+1	—	—	X	X	0	—
STX	—	X	X	X	X	—	(X) → M:M+1	—	—	X	X	0	—
STY	—	X	X	X	X	—	(Y) → M:M+1	—	—	X	X	0	—

Data:

0020	88	FF	00	00	FF	FF	00	00
0130	11	00	EE	00	00	00	8E	8F
AccA	$8E							
AccB	$F0							

X	$FF00
Y	$FF00
S	$00EE
CCR:	$D1

	Memory Location	Machine Code			Source Code	
(I)	0000	B7	01	36	STAA	$0136
(II)	0000	D7	A0		STAB	$A0
(III)	0002	DD	0A		STS	$0A

SOLUTION

(I) (a) Looking at no # sign and with two bytes of the address, we conclude that it is an EXT addressing mode.

(b) Since in EXT addressing mode, the effective address is $hhll (readers are encouraged to refer to the Instruction Set Table in the M68HC11E Series Programming Reference Guide), the effective address is $0136.

(c) The contents of AccA is the operand in this STAA instruction. AccA is given in the problem statement. Therefore, operand is $8E.

(d) The store instruction here makes the contents of the memory location $0136 equal to the contents of the AccA value.

(e) The original value of CCR value is $D1 = 1101 0001$_2$. Hence, the sequence of SXHINZVC is 11010001. From part (c), we know that the operand is $8E = 1000 1110$_2$. This operand makes N = 1 (MSB of operand is one), Z = 0 (operand is not zero), and V = 0 (this is always clear in store instructions; see Table 5.3). Therefore, the value of CCR changes from 1101 0001$_2$ to 1101 1001$_2$. Note that the N bit has changed from 0 to 1. Therefore, the new value of CCR is 1101 1001$_2$ = $D9.

(II) (a) Since the operand field of is a single byte ($A0), and there is no # sign, this store instruction is in DIR addressing mode.

(b) In DIR addressing mode, the effective address is obtained by putting 00 in front of the operand field of A0. Therefore, the effective address is $00A0.

(c) The contents of AccB is the operand in this store instruction. The contents of AccB is given in the program statement. Therefore, operand is $F0.

(d) The store instruction here makes the contents of the memory location $00A0 equal to the contents of AccB.

(e) The original value of CCR value is $D1 = 1101 0001$_2$. Hence, the sequence of SXHINZVC is 11010001. From part (c), the operand is $F0 = 1111 0000$_2$. This operand makes N = 1 (MSB of operand is one), Z = 0 (operand is not zero), and V = 0 (this is always clear in store instructions; see Table 5.3). Therefore, the value of CCR changes from 1101 0001$_2$ to 1101 1001$_2$. Note that only the N bit has changed from 0 to 1. Therefore, the new value of CCR is 1101 1001$_2$ = $D9.

(III) (a) Again, the operand field is a single byte (0A), and there is no # sign. Therefore, this store instruction is in DIR mode.
   (b) The instruction is STS. This is related to the stack pointer register of size 2 bytes. Since we are in DIR mode, the effective address is equal to 000A (putting 00 in front of 0A), and the following memory location 000B (adding 1 to 000A). Therefore, the effective addresses are $000A and $000B.
   (c) The data from stack pointer (S) is the operand in this STS instruction. S is given in the program statement. Therefore, the operand is $00EE.
   (d) The store instruction here makes the contents of the memory location $000A and $000B equal to the stack pointer value.
   (e) The original value of CCR value is $D1 = 1101 0001$_2$. Hence, the sequence of SXHINZVC is 11010001. From part (c), the operand is $00EE = 0000 1110$_2$. This operand makes N = 0 (MSB of operand is zero), Z = 0 (operand is not zero), and V = 0 (this is always clear in store instructions; see Table 5.3). Therefore, the value of CCR does not change from $D1.

5.2.3 Clear Instructions

A subset of the data transfer operations, clear operation, simply clears each bit of the operand to 0. The three types of clear instructions are shown in Table 5.4. The CLR means clear memory byte, whereas CLRA and CLRB mean clear AccA and AccB, respectively. The clear instruction is very similar to the store instruction such that $00 is stored in the memory location. The CLR instruction can use the extended and indexed address modes, while the CLRA and CLRB instructions use the inherent addressing mode. These instructions are often used to clear a memory location or an accumulator to zero before using it as a counter.

Example 5.3

What operations are performed in the following code snippet? Rewrite with appropriate comments.

```
CLRB
LDX     #DATA
LDAA    $00,X
```

TABLE 5.4

Set of Clear Instructions

Instruction	IMM	DIR	INDX	INDY	EXT	INH	Function	H	I	N	Z	V	C
CLR	—	—	X	X	X		$00 → M	—	—	0	1	0	0
CLRA	—	—	—	—	—	X	$00 → A	—	—	0	1	0	0
CLRB	—	—	—	—	—	X	$00 → B	—	—	0	1	0	0

SOLUTION

The first line has the instruction CLRB to clear the contents of AccB. Next, the address $0000 is loaded to the index register X. The third line has an offset of $00, which means that contents of the memory location pointed by the index register X has to be loaded to AccA. The current contents of X is $0000; therefore, the contents of memory location $0000 will be loaded to AccA. The code is rewritten with appropriate comments following:

Helpful Hint: In programming, a counter is used to keep track of the number of times some operation or event occurs, such as the number of times a program loop has executed.

```
CLRB              ; Clear AccB
LDX    #$0000     ; Initialize base address in X
LDAA   $00,X      ; Load contents at address in X to AccA
```

5.2.4 Transfer Instructions

Instructions that involve movement of data from one processor register to another fall into this category. All types of transfer instructions are shown in Table 5.5. The instruction TAB performs the function of transferring the contents of AccA to AccB. Thus, the TAB instruction copies data from AccA to AccB. The transfer instruction TBA is the reverse of TAB. TBA copies data from AccB to AccA. The CCR status flags, and the addressing modes can be interpreted from the table in a similar way we did previously for load instructions.

Common Practice: Often programmers new to HC11 keep the M68HC11E Series Programming Reference Guide handy for quick reference to the instruction set.

Example 5.4

Use the correct transfer instruction to write the following operation: Load AccA with the data 1111 1111$_2$ and then copy it to AccB.

TABLE 5.5

Set of Transfer Instructions

Instruction	IMM	DIR	INDX	INDY	EXT	INH	Function	H	I	N	Z	V	C
TAB	—	—	—	—	—	X	(A) → B	—	—	X	X	0	—
TAP	—	—	—	—	—	X	(A) → CCR	X	X	X	X	X	X
TBA	—	—	—	—	—	X	(B) → A	—	—	X	X	0	—
TPA	—	—	—	—	—	X	(CCR) → A	—	—	—	—	—	—
TSX	—	—	—	—	—	X	(S) + 1 → X	—	—	—	—	—	—
TSY	—	—	—	—	—	X	(S) + 1 → Y	—	—	—	—	—	—
TXS	—	—	—	—	—	X	(X) − 1 → S	—	—	—	—	—	—
TYS	—	—	—	—	—	X	(y) + 1 → s	—	—	—	—	—	—

SOLUTION

We will use LDAA for loading AccA with the data $1111\ 1111_2 = \$FF$. We will use
the # sign for IMM addressing since we are dealing with data. A TAB instruction
will be used to transfer data from AccA to AccB.

```
Address     Machine  Code    Source  Code   Comments
0000        86 FF            LDAA   #$FF    ; Load AccA with data
0002        16               TAB            ; Transfer data to AccB
```

5.2.5 Exchange Instructions

The instructions that involve swapping the contents of the double accumulator with the contents of one of the index register fall into this category. The
two exchange instructions are shown in Table 5.6. XGDX and XGDY swap the
contents of the D and X registers, and D and Y registers, respectively. From
the Table 5.5, we can observe that these instructions operate in the inherent
addressing mode and have no effect on the CCR.

Example 5.5

Use HC11 exchange instructions to transfer the contents of the index register X to
the index register Y. Illustrate with an example.

SOLUTION

If we use XGDX followed by XGDY, the contents of the index register X will be
first transferred to the register D with the first XGDX instruction. The subsequent
instruction XGDY will transfer the contents of register D to the index register Y. In
this way contents of the index register X will be transferred to index register Y. As an
example let us assume that the original contents of registers D, X, and Y are $0000, $FFFF, and $AAAA, respectively. With the first instruction XGDX, the contents of D, X, and Y will become $FFFF, $0000, and $AAAA, respectively. The next instruction XGDY will update the contents of D, X, and Y to $AAAA, $0000, and $FFFF, respectively. Therefore, we can see finally the data $FFFF in X has been transferred to Y.

Common Misconception: Students often get confused between transfer and exchange instructions. In transfer instructions, the source register does not get affected whereas exchange instructions act like a double transfer instruction because data is moved to and from each register. Thus, the source also gets affected in exchange instructions.

TABLE 5.6

Set of Exchange Instructions

Instruction	IMM	DIR	INDX	INDY	EXT	INH	Function	H	I	N	Z	V	C
XGDX	—	—	—	—	—	X	SWAP(D,X)	—	—	—	—	—	—
XGDY	—	—	—	—	—	X	SWAP(D,Y)	—	—	—	—	—	—

Section 5.2 Review Quiz

Load, store, clear, transfer, and exchange instructions are considered as data movement instructions. (True/False)

5.3 Arithmetic

The category of arithmetic instructions includes addition, subtraction, multiplication, and division instructions. However, we will delay in learning multiplication and division until we understand some logic, shifting, and rotating concepts. The HC11 has ten different types of addition (six) and subtraction (four) instructions. Each of these performs addition or subtraction operations on two operands. For four of these instruction types (ADDA, ADDB, SUBA, and SUBB), one of the operands comes from memory and the other from one of the 8-bit accumulators. In ADDD and SUBD, one of the operands comes from the memory and the other from AccD. For two of these instruction types (ABX and ABY), one of the operands comes from AccB and the other from either one of the index registers, X or Y. Finally, for two of these instruction types (ABA and SBA), one of the operands comes from AccA and the other from AccB.

5.3.1 Addition

From the Table 5.7, the instruction ABA adds the contents of AccA to AccB and places the result in AccA. The result will affect the H, N, Z, V, and C flags. For example, let $AccA = 68_{10}$ and $AccB = 50_{10}$. The ALU will add these two numbers as shown here:

$$AccA = 68_{10} = 0\ 100\ 0100_2$$
$$AccB = 50_{10} = 0\ 011\ 0010_2$$

Place result in $AccA = 118_{10} = 0\ 111\ 0110_2$ (note that no carry has occurred)

TABLE 5.7

Set of Addition Instructions

Instruction	IMM	DIR	INDX	INDY	EXT	INH	Function	H	I	N	Z	V	C
ADDA	X	X	X	X	X	—	(A) + (M) → A	X	—	X	X	X	X
ADDB	X	X	X	X	X	—	(B) + (M) → B	X	—	X	X	X	X
ADDD	X	X	X	X	X	—	(D) + (M):(M+1) → D	—	—	X	X	X	X
ABA	—	—	—	—	—	X	(A) + (B) → A	X	—	X	X	X	X
ABX	—	—	—	—	—	X	(X) + $00:(B) → X	—	—	—	—	—	—
ABY	—	—	—	—	—	X	(Y) + $00:(B) → Y	—	—	—	—	—	—

The addition does not produce a carry from bit 7 (MSP) position. Thus, the processor will make C = 0. The H flag is affected only by arithmetic addition of 8-bit operands. Whenever there is a carry-out from bit position 3 into bit position 4, the processor sets H = 1. In other words, if there is a carry from the lower nibble to higher nibble, the flag H is set. Therefore, since in this example there is no carry from lower nibble to higher nibble, H = 0. Also, the result is not exactly zero, so the processor clears Z. The processor also makes N = 0 since the N flag takes on the value of the sign bit of the result. The overflow flag, V, is cleared to 0 because the sign bit of the result is the same as the sign bit of the two operands. Recall that the V flag is used to indicate an overflow into the sign bit position when signed numbers are added or subtracted.

The ABX and ABY instructions are used to add the unsigned 8-bit number in AccB to the contents of index register X or Y, with the result placed in the same index register. The contents of AccB after the ABX and ABY execution remain unchanged. These instructions use the inherent address modes only, and neither one of them affects any of the status flags. The example here explains the operation of ADDA, ADDB, ADDD, and ADX.

Example 5.6

With the saved data and various processor registers set to the values given as follows, answer the following for each of the instruction (I to IV) listed.

(a) What is the addressing mode? How did you figure out?
(b) What is the effective address(es) of operand?
(c) What is the result?
(d) Describe the instruction in plain English.
(e) What is the resulting status of each of the flags in the CCR? How did you figure out?

Data:

0040	11	22	22	33	44	44	D6	21
0120	CC	FF	C5	00	FF	00	00	00

AccA	$90
AccB	$3F
X	$C600
Y	$FF00
S	$00EE
CCR:	$D3

	Address	Machine Code		Source Code	
(I)	0000	8B	55	ADDA	#$55
(II)	0005	FB	01 22	ADDB	$0122
(III)	0015	D3	44	ADDD	$44
(IV)	002A	3A		ABX	

SOLUTION

(I) (a) The # sign indicates that we are dealing in IMM addressing mode.

   (b) The effective address is the address immediately following the opcode in memory. The machine code 8B for ADDA resides at 0000. We can obtain the effective address by adding 1 to 0000. Therefore, the effective address is $0001.

   (c) $90 (AccA) + $55 (operand) = $E5. This is stored in AccA. The addition is shown in detail here:

$$AccA = 90_{HEX} = 1\ 001\ 0000_2$$
$$Operand = 55_{HEX} = 0\ 101\ 0101_2$$

Place result in $AccA = E5_{HEX} = 1\ 110\ 0101_2$ (note that no carry has occurred)

   (d) Add the contents of memory ($55) to the contents of AccA ($90). The result is stored in AccA.

   (e) Original value of CCR is $D3 = 1101\ 0011_2. Therefore, the sequence of SXHINZVC is 11010011. The result $E5 ($1110\ 0101_2$) makes H = 0 (no carry from the lower nibble of the result), N = 1 (MSB of the result is one), Z = 0 (result is not zero), V = 0 (sign of the result is correct), and C = 0 (there is no full carry in the result). Therefore, the value of CCR changes from 1101 0011_2 ($D3) to 1101 1000_2 ($D8).

(II) (a) In the absence of # sign and in the presence of two bytes of the address, we conclude that it is an EXT addressing mode.

   (b) Since in an EXT addressing mode, the effective address is $hhll (readers may refer to the Instruction Set Table in the M68HC11E Series Programming Reference Guide), the effective address is $0122.

   (c) $3F (AccB) + $C5 (from address $0122) = $04.

$$AccB = 3F_{HEX} = 0\ 011\ 1111_2$$
$$Operand = C5_{HEX} = 1\ 100\ 0101_2$$

Place result in $AccB = 04_{HEX} = 0\ 000\ 0100_2$

   Note that this addition has a carry from the lower nibble as well as a full carry.

   (d) Add contents of memory location $0122 to the contents of AccB ($3F). The result is stored in AccB. The contents of the memory location remain unchanged.

   (e) Original value of CCR is $D3 = 1101\ 0011_2. Therefore, the sequence of SXHINZVC is 11010011. The result $04 ($0000\ 0100_2$) makes H = 1 (there is a carry from the lower nibble of the result), N = 0 (MSB of the result is zero), Z = 0 (result is not zero), V = 0 (sign of the result is correct), and C = 1 (there is full carry in the result). Therefore, the value of CCR changes from 1101 0011_2 ($D3) to 1111 0001_2 ($F1).

(III) (a) Again, the operand field is a single byte ($44), and there is no # sign. Therefore, it is in DIR addressing mode.

   (b) The effective addresses are equal to $0044 (putting 00 in front of 44), and the following memory location 0045 (adding 1 to 0044).

(c) Register D is a combination of AccA and AccB. Therefore, the contents of D are \$90 and \$3F, i.e., \$903F. Also, from the data provided in the problem statement, \$44 is the content for each of the memory location \$0044 and \$0045.

The result = (D) + (\$0044):(\$0045)
= \$903F + \$4444
= \$D483.

(d) Add the contents of memory location \$0044 and \$0045 to the contents of the 16-bit register D and store the result in register D.
(e) Original value of CCR is \$D3 = 1101 0011_2. Therefore, the sequence of SXHINZVC is 11010011. The result \$D483 (1101010010000011_2) makes N = 1 (MSB of the result is one), Z = 0 (result is not zero), V = 0 (sign of the result is correct), and C = 0 (there is no full carry in the result). Therefore, the value of CCR changes from 1101 0011_2 (\$D3) to 1101 1011_2 (\$DB).

(IV) (a) The operand field is empty. Therefore, it is in INH addressing mode.
(b) There is no effective address in INH mode.
(c) Result = \$00:(B) + (X) = \$C500 + \$003F = \$C53F. This is stored in the index register X.
(d) Add contents of AccB to the index register X and store the result in the index register X.
(e) From Table 5.7 we can see that the status flags are not affected in this instruction. Therefore, CCR = \$D3.

5.3.2 Increment Instructions

An increment is an increase of some amount, either fixed or variable. Increment instructions (Table 5.8) are considered as arithmetic instructions due to their operation involving the addition of one to the operand. They overwrite the previous operand with the incremented result. No support is provided for a DIR addressing mode. Increment supports both 8-bit and 16-bit values. For example, INCA is increment A, which is an 8-bit value, whereas INX is increment index register X, which is a 16-bit value.

Helpful Hint: In loops, the idea of increment is to add 1 to a variable that is usually acting as a counter. The existing value of the counter is fetched, one is added, and then the answer is stored back into the variable counter.

TABLE 5.8

Set of Increment Instructions

Instruction	IMM	DIR	INDX	INDY	EXT	INH	Function	H	I	N	Z	V	C
INCA	—	—	—	—	—	X	(A) + 1 → A	—	—	X	X	X	—
INCB	—	—	—	—	—	X	(B) + 1 → B	—	—	X	X	X	—
INS	—	—	—	—	—	X	(S) + 1 → S	—	—	—	—	—	—
INX	—	—	—	—	—	X	(X) + 1 → X	—	—	—	X	—	—
INY	—	—	—	—	—	X	(Y) + 1 → Y	—	—	—	X	—	—
INC	—	—	X	X	X	—	(M) + 1 → M	—	—	X	X	X	—

Example 5.7

Assume that the initial contents of index register X and Y are $AF00 and $FC00, respectively. What will be the final contents of index register X and Y if 20 INX and INY instructions are executed?

SOLUTION

Each increment instruction will increment the value of the register by 1. So, 20 increments will increment the initial value by 14_{HEX}. Therefore, the final values of index register X and Y will be $AF14 and $FC14, respectively.

5.2.3 Subtraction

Each subtraction instruction in Table 5.9 performs a mathematical subtraction operation on the contents of an accumulator or a memory location. The result of each of these operations is stored back into the accumulator or memory location designated by the instruction. The instruction SBA subtracts the contents of AccB from the contents of AccA and writes back the result in AccA. The ALU uses the 2's complement method for subtraction. We have worked through many examples of the 2's complement in Chapter 1. Recall that the 2's complement of a number behaves like the negative of the original number. The ALU adds the 2's complement of the contents of AccB to the contents of AccA. The result affects the N, Z, and V flags in the same manner as the addition operation earlier. The C flag, however, is affected in a different way. After a subtraction operation is performed, the C flag is complemented so that it indicates the occurrence of a borrow. In a subtraction operation such as p-q, if p is smaller than q, then there will be a borrow. In this case the processor will set $C = 1$. On the other hand, if p is larger or equal to q, then there will be no borrow. In this case C will be cleared to zero. In short, the C flag is set if the absolute value of AccB is larger than the absolute value of the contents of AccA; otherwise, it is cleared.

TABLE 5.9

Set of Subtraction Instructions

Instruction	IMM	DIR	INDX	INDY	EXT	INH	Function	H	I	N	Z	V	C
SUBA	X	X	X	X	X	—	(A) – (M) → A	—	—	X	X	X	X
SUBB	X	X	X	X	X	—	(B) – (M) → B	—	—	X	X	X	X
SUBD	X	X	X	X	X	—	D – (M):(M+1) → D	—	—	X	X	X	X
SBA	—	—	—	—	—	X	(A) – (B) → A	—	—	X	X	X	X

Let us assume that AccA = 0000 1001$_2$ = 9$_{10}$ and AccB = 0000 0101$_2$ = 5$_{10}$. The SBA operation is shown here:

$$\begin{array}{r} \text{AccA} = 0000\ 1001_2 \\ \text{2's complement of AccB} = +\,1111\ 1011_2 \\ \hline \end{array}$$

Place result in AccA = 1 0000 0100$_2$ (note that carry has occurred)

Note that there is a carry of 1 in the result that will make C = 0. Here, the addition of the 2's complement of AccB and AccA produces a carry-out of bit 7. This will always happen when a number is subtracted from larger or equal number. The processor will make C = 0 to indicate that no borrow has occurred; that is, the processor complements the C flag to indicate the borrow status. If we reverse the numbers now and assume that AccB = 00001001$_2$ = 9$_{10}$ and AccA = 00000101$_2$ = 5$_{10}$, the SBA operation will occur as follows:

$$\begin{array}{r} \text{AccA} = 0000\ 0101_2 \\ \text{2's complement of AccB} = +\,1111\ 0111_2 \\ \hline \end{array}$$

Place result in AccA = 1111 1100$_2$ (note that no carry has occurred)

Note that there is no carry in the result, which will make C = 1.

In this case, there is no carry-out of bit 7. This will always happen when a number is subtracted from a smaller number. The processor will set C = 1 to indicate that a borrow has occurred. In the next example, we will see in detail how the instructions SUBA, SUBB, and SUBD work.

Example 5.8

With the saved data and various processor registers set to the values given here, answer the following for each of the instruction (I to III) listed.

(a) What is the addressing mode? How did you figure out?
(b) What is the effective address(es) of operand?
(c) What is the result?
(d) Describe the instruction in plain English.
(e) What is the resulting status of each of the flags in the CCR? How did you figure them out?

Data:

0040	11	22	22	33	22	22	D6	21
0150	CC	FF	B0	00	FF	00	00	00
AccA	$81							
AccB	$11							
X	$C600							
Y	$FF00							

```
S       $00EE
CCR:    $F3
```

	Address	Machine Code		Source Code	
(I)	0000	80	66	SUBA	#$66
(II)	0005	F0	01 52	SUBB	$0152
(III)	0015	93	44	SUBD	$44

SOLUTION

(I)
   (a) Looking at the # sign, we can easily figure out that we are dealing in IMM addressing mode.

   (b) The effective address is the address immediately following the opcode in memory. To get that we add 1 to 0000. Therefore, the effective address is $0001.

   (c) $81 (AccA) – $66 (operand) = $1B. This is stored in AccA.
Let us look into this subtraction operation in detail:

$$AccA = \quad 1000\ 0001_2$$
$$2\text{'s complement of }\$66 = +\ 1001\ 1010_2$$
$$\text{Place result in AccA} = 1\ 0001\ 1011_2$$

Note that carry has occurred which means that C = 0. The V will be set so V = 1.

   (d) Subtract contents of memory ($66) from the contents of AccA ($81). The result is stored in AccA.

   (e) Original value of CCR is $F3 = 1111\ 0011_2. Therefore, the sequence of SXHINZVC is 11110011. The result $1B (0001\ 1011_2) makes N = 0 (MSB of the result is zero), Z = 0 (result is not zero), V = 1 (sign overflow has occurred), and C = 0 (taking the complement of carry in the result). Therefore, the value of CCR changes from 1111\ 0011_2 ($F3) to 1111\ 0010_2 ($F2).

(II)
   (a) In the absence of # sign and in the presence of two bytes of the address, we conclude that it is an EXT addressing mode.

   (b) In EXT addressing mode, the effective address is $hhll. Therefore, the effective address is $0152.

   (c) $11 (AccB) – $B0 (from address $0152) = $61.
Let us look into the details of this subtraction:

$$AccB = \quad 0001\ 0001_2$$
$$2\text{'s complement of }\$B0 = +\ 0101\ 0000_2$$
$$\text{Place result in AccB} = \quad 0110\ 0001_2$$

Note that no carry has occurred which means that C = 1. The V will be clear so V = 0.

   (d) Subtract contents of memory ($B0) from the contents of AccB ($11). The result is stored in AccB.

   (e) Original value of CCR is $F3 = 1101\ 0011_2. Therefore, the sequence of SXHINZVC is 11010011. The result $61 (0110\ 0001_2) makes N = 0 (MSB of the result is zero), Z = 0 (result is not zero), V = 0 (sign overflow has not occurred), and C = 1 (taking the complement of carry in the result). Therefore, the value of CCR changes from 1111\ 0011_2 ($F3) to 1111\ 0001_2 ($F1).

(III) (a) The operand field is a single byte (44), and there is no # sign. Therefore, it is in DIR addressing mode.

(b) The effective address is equal to 0044 (putting 00 in front of 44), and the following memory location 0045 (adding 1 to 0044).

(c) The register D is a 16-bit register which is a combination of AccA and AccB. Here, we have AccA = $81 and AccB = $11. Therefore, D = $8111. Also, from the data provided in the problem statement, $22 is the content for each of the memory location $0044 and $0045 (the effective addresses from part (b)).

$$
\begin{aligned}
\text{The result} = (D) \quad &- (\$0044){:}(\$0045) \\
&= \$8111 - \$2222 \\
&= \$8111 + (2\text{'s complement of } \$2222) \\
&= 1000\ 0001\ 0001\ 0001_2 + 1101\ 1101\ 1101\ 1110_2 \\
&= 1\ 0101\ 1110\ 1110\ 1111_2 \\
&= \$5EEF.
\end{aligned}
$$

Note that carry has occurred which means that C = 0. The V will be set so V = 1.

(d) Subtract the content of memory location $0044 and $0045 (i.e., $2222) from the contents of register D ($8111). Store the result in register D.

(e) Original value of CCR is $F3 = 1111\ 0011_2. Therefore, the sequence of SXHINZVC is 11110011. Keeping in mind that the result is $$5EEF (or 0101 1110 1110 1111_2), we have N = 0 (MSB of the result is zero), Z = 0 (result is not zero), V = 1 (sign overflow has occurred), and C = 0 (taking the complement of carry in the result). Therefore, the value of CCR changes from 1111 0011_2 ($F3) to 1111 0010_2 ($F2).

Common Misconception: Students often assume that the C flag of CCR is affected in addition and subtraction instruction alike. As a matter of fact, after a subtract operation is performed, the C flag is complemented so that it indicates the occurrence of a borrow.

5.3.4 Negate and Decrement Instructions

The negate and decrement instructions (Table 5.10 and Table 5.11, respectively) are subset of subtraction instructions. The negate instructions change the sign of a value in AccA, in AccB, or in a memory location. Again, the 2's complement plays a key role in changing the sign on the 8-bit operand. The decrement instructions subtract one from the operand, and then overwrite the previous operand with this result.

TABLE 5.10

Set of Negate Instructions

Instruction	IMM	DIR	INDX	INDY	EXT	INH	Function	H	I	N	Z	V	C
NEGA	—	—	—	—	—	X	$00 – (A) → A	—	—	X	X	X	X
NEGB	—	—	—	—	—	X	$00 – (B) → B	—	—	X	X	X	X
NEG	—	—	X	X	X	—	$00 – (M) → M	—	—	X	X	X	X

TABLE 5.11

Set of Decrement Instructions

Instruction	IMM	DIR	INDX	INDY	EXT	INH	Function	H	I	N	Z	V	C
DECA	—	—	—	—	—	X	$(A) - 1 \rightarrow A$	—	—	X	X	X	—
DECB	—	—	—	—	—	X	$(B) - 1 \rightarrow B$	—	—	X	X	X	—
DES	—	—	—	—	—	X	$(S) - 1 \rightarrow S$	—	—	—	—	—	—
DEX	—	—	—	—	—	X	$(X) - 1 \rightarrow X$	—	—	—	X	—	—
DEY	—	—	—	—	—	X	$(Y) - 1 \rightarrow Y$	—	—	—	X	—	—
DEC	—	—	X	X	X	—	$(C) - 1 \rightarrow C$	—	—	X	X	X	—

Example 5.9

Assume that the initial contents of index register X and Y are $AF14 and $FC14, respectively. What will be the final contents of index register X and Y if 20 DEX and DEY instructions are executed?

SOLUTION

Each decrement instruction will decrement the value of the register by 1. So, 20 decrements will decrement the initial value by 14_{HEX}. Therefore, the final values of index register X and Y will be $AF00 and $FC00, respectively.

Section 5.3 Review Quiz

Increment and decrement instructions are considered as arithmetic instructions. (True/False)

5.4 Logic

Recall from Chapter 2 the basic logic gates and their operations. Table 5.12 presents nine logic instructions. The AND, OR, and XOR operation is performed in HC11 through ANDx, ORAx, and EORx instructions, respectively. Logic operations are performed between two operands, one of which is the contents of an accumulator. They take the contents of the accumulator and an operand from memory, perform a bit-bit-logic operation on them, and store the results in the accumulator. Bit-bit operation is, in a way, like drawing a timing diagram. Each bit position acts like a time epoch. The ANDA instruction is illustrated in Figure 5.4, where the bits of AccA are A_7 to A_0, bits of memory are M_7 to M_0, and bits of result are R_7 to R_0. Note that the result is stored in the same AccA. The flags affected by these instructions are N, Z, and V flags. If the result has a 1 in bit 7, the N flag will be set; otherwise, the N flag will be cleared. If the result is $0000\ 0000_2$, the Z flag will be set;

TABLE 5.12

Set of Logic Instructions

Instruction	IMM	DIR	INDX	INDY	EXT	INH	Function	H	I	N	Z	V	C
ANDA	X	X	X	X	X	—	$(A) \bullet (M) \to A$	—	—	X	X	0	—
ANDB	X	X	X	X	X	—	$(B) \bullet (M) \to B$	—	—	X	X	0	—
ORAA	X	X	X	X	X	—	$(A) + (M) \to A$	—	—	X	X	0	—
ORAB	X	X	X	X	X	—	$(B) + (M) \to B$	—	—	X	X	0	—
EORA	X	X	X	X	X	—	$(A) \oplus (M) \to A$	—	—	X	X	0	—
EORB	X	X	X	X	X	—	$(B) \oplus (M) \to B$	—	—	X	X	0	—
COMA	—	—	—	—	—	X	$!(A) \to A$	—	—	X	X	0	1
COMB	—	—	—	—	—	X	$!(B) \to B$	—	—	X	X	0	1
COM	—	—	X	X	X	—	$!(M) \to M$	—	—	X	X	0	1

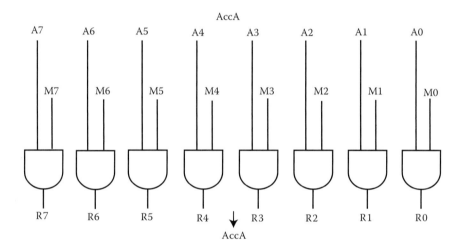

FIGURE 5.4
ANDA instruction.

otherwise, it will be cleared. The execution of any of these logical operations will always cause the V flag to be cleared. These three operations are illustrated in the following Table 5.13 for the same AccB and operand values. In all the three cases, the Z and V flags are clear. The V flag in case of ANDx and ORAx is set because the result has a 1 in bit 7. Since the bit 7 in EORx case is 0, the V flag is cleared.

Helpful Hint: ANDx, ORAx, and EORx work on each bit position separately and independently from the others.

Before we move on to logic instructions that perform inversions, we would like to discuss the concept of data masking. In general, data masking is the process of obscuring (masking) specific data elements within data stores. In our HC11 programming environment, the bits in an operand that need processing will be identified by a mask. The data mask basically blocks parts of

TABLE 5.13

Example of ANDB, ORAB, EORB Instructions

Bit	7	6	5	4	3	2	1	0
ACCB	1	0	1	0	1	0	1	0
OPERAND	1	1	0	1	0	0	1	0
AND RESULT	1	0	0	0	0	0	1	0
ORA RESULT	1	1	1	1	1	0	1	0
EOR RESULT	0	1	1	1	1	0	0	0

the operand to be processed. This may be required for protecting the data, or for filtering the data. We call a mask an 8-bit word that designates the locations of the bits within the operand to be processed. The bits that will be processed will be indicated as 1 in the mask. The bits to be ignored will be indicated as 0 in the mask. As seen in Table 5.13, when the logical AND operation is performed on AccB ($1010\ 1010_2$) and the operand ($1101\ 0010_2$), the result was $1000\ 0010_2$. The AccB can be called a mask such that the bits with 1 (i.e., b7, b5, b3, b1) are set to 1. This means that these bits of the operand are to be processes. The bits b6, b4, b2, b0 of AccB are set to zero. This means that these bits in the operand are unwanted and should be eliminated. We observe that bits b7, b5, b3, b1 of the operand remain the same in the result $1000\ 0010_2$. On the other hand, the bits b6, b4, b2, b0 of the operand become zero in the result $1000\ 0010_2$.

The other logic instructions are COMA, COMB, and COM. These instructions perform a logical NOT operation on an 8-bit operand. Recall from Chapter 2 that the digital logic NOT gate, the most basic of all the logical gates, is a single input device that "inverts" (complements) its input signal. COMA, COMB, and COM overwrite the contents of AccA, AccB, and contents of memory location, respectively, with the result of the NOT operation. The COMA instruction is illustrated in Figure 5.5, where the bits of AccA are A_7 to A_0, and the bits of result are R_7 to R_0. Note that the result is stored in the same AccA.

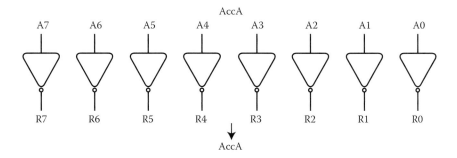

FIGURE 5.5
COMA instruction.

Example 5.10

Given the data following, for each of the following instructions, determine the result of the values of the N, Z, and V flags:

  (a) ANDB #$E3
  (b) ORAB #$C8
  (c) EORA #$36

Data:

```
0040   11   22   22   33   22   22   D6   21
0150   CC   FF   B0   00   FF   00   00   00
AccA   $AA
AccB   $37
X      $C600
Y      $FF00
S      $00EE
CCR:   $F3
```

SOLUTION

  (a) AccB AND $E3 = $37 AND $E3 = $23.

$37 = 0011 0111
$E3 = 1110 0011
─────────────────
$23 = 0010 0011

Flag status: Z = 0, result is not zero
             N = 0, bit 7 of the result is 0
             V = 0, always reset

  (b) AccB OR $C8 = $37 OR $C8 = $FF.

$37 = 0011 0111
$C8 = 1100 1000
─────────────────
$FF = 1111 1111

Flag status: Z = 0, result is not zero
             N = 1, bit 7 of the result is 1
             V = 0, always reset

  (c) AccA XOR $36 = $AA XOR $36 = $9C.

$AA = 1010 1010
$36 = 0011 0110
─────────────────
$9C = 1001 1100

Flag status: Z = 0, result is not zero
             N = 1, bit 7 of the result is 1
             V = 0, always reset

Section 5.4 Review Quiz

The instructions ANDB, ORAB, and EORB deal with AccA. (True/False)

5.5 Shifting and Rotating

Shift and rotate instructions include those that shift data left or right. There are eight logical shifts, three arithmetic shifts, and six rotate instructions. These are presented in Table 5.14 through 5.16.

Logical shifting causes a "0" to be shifted into one end of the data word, pushing all of the data bits over one position in the data word as shown in Figure 5.6. Figure 5.6 (a) illustrates the shifting of right. When shifting right, a "0" is shifted into the MSB of the data word. Also, the LSB is shifted into the C flag of the CCR. On the other hand, when shifting left, as illustrated in Figure 5.6 (b), a "0" is shifted into the LSB of the data word, and MSB is shifted into the C flag of the CCR. Figure 5.7 presents an example of LSR and LSL for an 8-bit data.

The arithmetic shift operation moves all the bits in an operand to the left or to the right by one bit position. The left or the right shift here is equivalent to a

TABLE 5.14

Set of Logical Shift Instructions

Instruction	IMM	DIR	INDX	INDY	EXT	INH	Function	H	I	N	Z	V	C
LSL	—	—	X	X	X	—	{U}(M)<<=1	—	—	X	X	X	X
LSLA	—	—	—	—	—	X	{U}(A)<<=1	—	—	X	X	X	X
LSLB	—	—	—	—	—	X	{U}(B)<<=1	—	—	X	X	X	X
LSLD	—	—	—	—	—	X	{U}(D)<<=1	—	—	X	X	X	X
LSR	—	—	X	X	X	—	{U}(M)>>=1	—	—	0	X	X	X
LSRA	—	—	—	—	—	X	{U}(A)>>=1	—	—	0	X	X	X
LSRB	—	—	—	—	—	X	{U}(B)>>=1	—	—	0	X	X	X
LSRD	—	—	—	—	—	X	{U}(D)>>=1	—	—	0	X	X	X

TABLE 5.15

Set of Arithmetic Shift Instructions

Instruction	IMM	DIR	INDX	INDY	EXT	INH	Function	H	I	N	Z	V	C
ASR	—	—	X	X	X	—	{I}(M)>>=1	—	—	X	X	X	X
ASRA	—	—	—	—	—	X	{I}(A)>>=1	—	—	X	X	X	X
ASRB	—	—	—	—	—	X	{I}(B)>>=1	—	—	X	X	X	X

TABLE 5.16

Set of Rotate Instructions

Instruction	IMM	DIR	INDX	INDY	EXT	INH	Function	H	I	N	Z	V	C
ROL	—	—	X	X	X	—	ROL(M) → M	—	—	X	X	X	X
ROLA	—	—	—	—	—	X	ROL(A) → A	—	—	X	X	X	X
ROLB	—	—	—	—	—	X	ROL(B) → B	—	—	X	X	X	X
ROR	—	—	X	X	X	—	ROR(M) → M	—	—	X	X	X	X
RORA	—	—	—	—	—	X	ROR(A) → A	—	—	X	X	X	X
RORB	—	—	—	—	—	X	ROR(B) → B	—	—	X	X	X	X

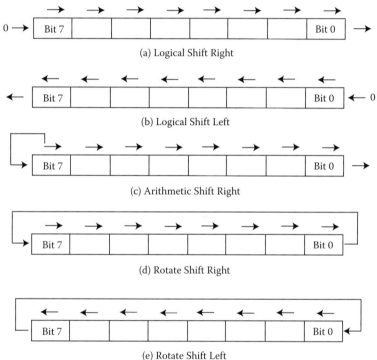

(a) Logical Shift Right

(b) Logical Shift Left

(c) Arithmetic Shift Right

(d) Rotate Shift Right

(e) Rotate Shift Left

FIGURE 5.6

Mechanism of shifting and rotating.

multiplication or division by 2 (respectively) upon the operand. Figure 5.6 (c) illustrates the arithmetic shifting of right. When performing a right arithmetic shift, a copy of the sign bit (MSB) is shifted, and the original sign bit remains in the MSB position of the data word. Also, the LSB is shifted into the C flag of the CCR. Figure 5.8 presents an example of ASR for an 8-bit data. The logical shift left and arithmetic shift left instructions are identical in function and share the same opcodes.

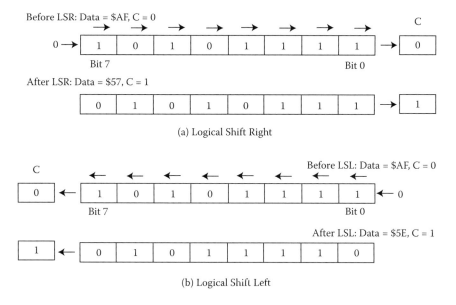

FIGURE 5.7
An example of logical shift instruction.

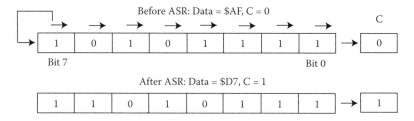

FIGURE 5.8
An example of arithmetic shift instruction.

Rotation operations are similar to the shift operations except that the bit shifted out of the high or low bit position (depending on the direction of the rotation) gets placed in the bit position vacated on the other side of the byte. The rotate left (ROL) instructions rotate a data word one bit to the left. The MSB is shifted into the C flag of the CCR, and the LSB is filled with the prior contents of the C flag. On the other hand, the rotate right (ROR) instructions rotate a data word one bit to the right. The LSB is shifted into the C flag of the CCR, and the MSB is filled with the prior contents of the C flag. Figure 5.9 presents an example of ROL and ROR instructions for an 8-bit data.

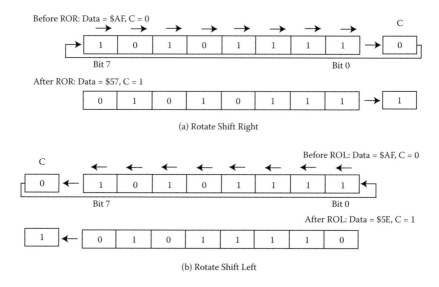

FIGURE 5.9
An example of rotate instruction.

Example 5.11

Assuming that the initial value of C is zero, perform ROR operations on $2F.

SOLUTION

$2F = 0010 1111$_2$. The ROR and ROL for $2F are shown in Figure 5.10.

Example 5.12

Assuming that the initial value of C is zero, perform ROL operations on $93.

SOLUTION

$93 = 1001 0011$_2$. The ROR and ROL for $93 are shown in Figure 5.11.

Section 5.5 Review Quiz

The ROL instruction rotates a data word one bit to the right. (True/False)

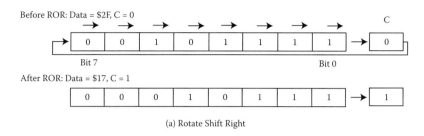

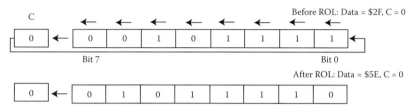

(a) Rotate Shift Right

(b) Rotate Shift Left

FIGURE 5.10
ROR and ROL on $2F.

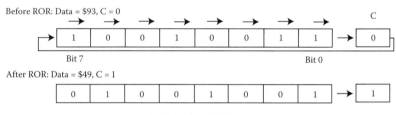

(a) Rotate Shift Right

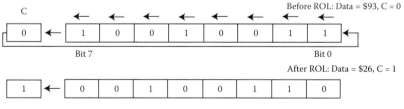

(b) Rotate Shift Left

FIGURE 5.11
ROR and ROL on $93.

TABLE 5.17

Set of Multiply-and-Divide Instructions

Instruction	IMM	DIR	INDX	INDY	EXT	INH	Function	H	I	N	Z	V	C
MUL	—	—	—	—	—	X	$(A) - (B) \rightarrow D$	—	—	—	—	—	4
FDIV	—	—	—	—	—	X	$D / X \rightarrow X$						
							$(D) \% (X) \rightarrow D$	—	—	—	X	X	X
IDIV	—	—	—	—	—	X	$(D) / (X) \rightarrow X$						
							$(D) \% (X) \rightarrow D$	—	—	—	X	0	X

5.6 Multiplication and Division

We now come back to these two arithmetic instructions presented on Table 5.17. The MUL instruction multiplies an 8-bit unsigned binary value in AccA by the 8-bit unsigned binary value in AccB. The result, which is the 16-bit unsigned product, is placed in the AccD. The C flag is set if bit-7 of the result, that is, the bit-7 of AccB, is set. On the other hand, for IDIV and FDIV one of the operands comes from the double accumulator and the other from index register X. The IDIV and FDIV are integer and fractional divide instructions, respectively. The IDIV instruction allows the processor to perform an unsigned integer divide of the 16-bit binary number in AccD (i.e., the numerator) by the 16-bit binary number in the index register X (i.e., the denominator). The result of the division instruction is executed; the C flag will be set if the denominator was $0000, or else the C flag will be cleared. The IDIV instruction will set the Z flag to 1 if the quotient is $0000. The FDIV instruction differs from the IDIV instruction in that the numerator, that is, AccD, is assumed to be less than the denominator, that is, index register X. The V flag is set if the original contents of the index register X are less than, or equal to, the contents of AccD.

Example 5.13

What operations are performed in the following code snippet? Rewrite with appropriate comments.

```
LDAB   #$0A
CLRA
LDX    #$000A
IDIV
```

Common Practice: In context of MUL, the C flag is helpful every time rounding of the most significant byte of the product is desired. Assuming multiplication of mixed numbers if the C = 1 after the MUL instruction is executed, it implies that the lower byte (AccB) of the result is greater than 0.5. In such case we may choose to round the most significant byte (AccA) of the product by executing the sequence: MUL ADCA #$00. This program sequence will add 1 to the most significant byte of the product (AccA) if the C flag is set due to the execution of the MUL instruction.

SOLUTION

First 10_{10} ($0A) is loaded in AccB. Next, the AccA is cleared. This makes the 16-bit register equal to $000A ($10_{10}$). The index register X is then loaded with 10_{10}($000A). The integer division is performed such that the numerator is set to D and denominator is set to X. The answer of the division operation is then stored in the index register X. Here the division is performed with numerator 10_{10} and denominator 10_{10}; therefore, the answer is 1_{10}.

```
LDAB    #$0A        ; Load AccA with $0A
CLRA                ; Clear AccA (D=AccA:AccB)
LDX     #$000A      ; Set denominator
IDIV                ; Divide D by X, answer is stored in X
```

Section 5.6 Review Quiz

The IDIV and FDIV are integer and fractional divide instructions, respectively. (True/False)

5.7 CCR (Flag) Manipulation

This group of instructions contains those used to set and clear some of the flags in CCR. Table 5.18 summarizes some of the status flag manipulation instructions. CLC and SEC clear or set the carry flag C in the CCR, respectively. The question arises, why will a programmer need to initialize the C flag to a desired state of 0 or 1? Perhaps shift or rotate instructions can provide us the best answer to this question. Since the shift and rotate instructions use the C flag of the CCR, it may be desired to initialize them in a program before using them.

Section 5.7 Review Quiz

The instruction that can clear the carry flag C in the CCR is _____. (CRC/SEC/CLV/SEV)

TABLE 5.18

Set of Status Flag Manipulation Instructions

Instruction	IMM	DIR	INDX	INDY	EXT	INH	Function	H	I	N	Z	V	C
CLC	—	—	—	—	—	X	$0 \rightarrow C$	—	—	—	—	—	0
SEC	—	—	—	—	—	X	$1 \rightarrow C$	—	—	—	—	—	1
CLV	—	—	—	—	—	X	$0 \rightarrow V$	—	—	—	—	0	—
SEV	—	—	—	—	—	X	$1 \rightarrow V$	—	—	—	—	1	—

5.8 Bit-Level Operations

We have seen the instructions that were related to bytes or 16-bit words. They evaluated data bytes for performing comparisons or making decisions for branching. The HC11 provides a set of instructions that have the ability to alter individual bits within data bytes or 16-bit words. These instructions evaluate the individual bits to perform comparisons or make decisions for branching. Recall from our discussion of logic instructions earlier in this chapter that a mask can be used to indicate the location of bits in an operand to be processed. Table 5.19 shows two-bit manipulation instructions. The BSET sets bits in contents of memory by using ORing the operand with the mask. All bits that are 1's in the mask are set in the memory location after BSET operation. The zero bits in the mask remain unchanged. For example, if the mask is $0000\ 1111_2$ and the operand is $1000\ 1001_2$, then BSET will set the bits from b0 to b3 and keep the bits from b4 to b7 unaffected. The result after the BSET operation will be $1000\ 1111_2$. Remember that we have done a bit by bit OR between the mask and the operand. On the other hand, BCLR clears bits in contents of memory by using ANDing the operand with the inverse of the mask. All bits that are 1's in the mask are reset in the memory location after BCLR operation. The zero bits in the mask remain unchanged. In our example of the mask as $0000\ 1111_2$ and the operand as $1000\ 1001_2$, the BCLR will clear the bits from b0 to b3 and keep the bits from b4 to b7 unaffected. The result after the BCLR operation will be $1000\ 0000_2$. Note that we have done a bit-by-bit AND between the inverse of the mask and the operand. Table 5.19 shows the bit manipulation instructions. The instruction used to set b0, b3, and b7 of $003A would look like:

```
BSET $3A $89
```

Bit tests are another set of bit-level instructions that perform bit-level comparisons. These instructions are used to test for zero conditions in a sequence of bits. Table 5.20 shows the bit-level comparisons instructions. BITA and BITB instructions perform logical AND on the contents of AccA and AccB, respectively, with a byte from a memory location. The result of AND operation updates the flags in the CCR. Table 5.21 shows bit-level conditional branch instructions. BRSET causes a relative mode branch if all the bits identified in

TABLE 5.19

Bit Manipulation Instructions

Instruction	IMM	DIR	INDX	INDY	EXT	INH	Function	H	I	N	Z	V	C
BSET	—	X	X	X	—	—	$(M)+MM \rightarrow M$	—	—	X	X	0	—
BCLR	—	X	X	X	—	—	$(M) \cdot MM' \rightarrow M$	—	—	X	X	0	—

TABLE 5.20

Bit-Level Comparison Instructions

Instruction	IMM	DIR	INDX	INDY	EXT	INH	Function	H	I	N	Z	V	C
BITA	X	X	X	X	X	—	A·(M)	—	—	X	X	0	—
BITB	X	X	X	X	X	—	B·(M)	—	—	X	X	0	—

TABLE 5.21

Bit-Level Conditional Branch Instructions

Instruction	IMM	DIR	INDX	INDY	EXT	INH	Function	H	I	N	Z	V	C
BRSET	—	X	X	X	—	—	IF (M')·MM = 0	—	—	—	—	—	—
							THEN P + SSRR = P	—	—	—	—	—	—
BRCLR	—	X	X	X	—	—	IF (M)·MM = 0	—	—	—	—	—	—
							THEN P + SSRR = P	—	—	—	—	—	—

the mask are set. An AND operation is performed between the mask and the inverse of the operand stored at the memory location. The result then decides if the branching should take place or not. If the result is zero, then branching takes place. The BRCLR causes a relative mode branch if all the bits identified in the mask are cleared. An AND operation is performed between the mask and the operand stored at the memory location. The result then decides if the branching should take place or not. If the result is zero, the branching takes place. Both BRSET and BRCLR neither permanently modify any data, nor do they have any effect on the CCR.

Section 5.8 Review Quiz

Give two examples of bit manipulation instructions.

5.9 Summary

1. An instruction is a single operation of a processor defined by an instruction set architecture.

2. Condition Code Register (CCR) is referred to as "C," "CCR," or "Status" register. It is an 8-bit status and control register. Out of the eight bits in CCR, five bits are H, N, Z, V, and C. These are called *status flags*. The remaining three bits are S, X, and I. These are called control bits.

3. There are several addressing modes in which instructions may be implemented, depending upon the instruction set architecture. Some of the addressing modes are inherent, immediate, extended, direct, indexed, and relative addressing modes.

4. Execution of instructions may affect CCR flags.

5. Moving data from one location to another is called *data movement*. The source and destination locations are determined by the addressing modes, and can be registers or memory. Load and store instructions are generally responsible for data movement.

6. Instructions that perform operations such as addition, subtraction, multiplication, and division are called *arithmetic instructions*.

7. Arithmetic shift moves all the bits in an operand to the left or to the right by one bit position. When shifting to the right, the MSB is shifted, and the original sign bit remains in the MSB position. When shifting to the left, the LSB is shifted into the C flag of the CCR.

8. Logical shifting causes "0" to be shifted into one end of the data word, pushing all of the data bits over one position in the data word.

9. Rotate operation is a type of shift operation where the bit that gets shifted out of the high or low bit position gets placed in the bit position vacated on the other side of the byte.

10. AND masking uses the AND operation to clear specific bits of a data word while not affecting the others.

11. OR masking uses the OR operation to set specific bits of a data word while not affecting the others.

12. The Instruction Set Table in the M68HC11E Series Programming Reference Guide is an important place to refer to while designing HC11 programs.

Glossary

AND Masking: A type of masking that uses the AND operation to clear specific bits of a data word while not affecting the others.

Arithmetic: Traditional operations such as addition, subtraction, multiplication, and division.

Arithmetic Shift: Moves all the bits in an operand to the left or to the right by one bit position. When shifting to the right, the MSB is shifted and the original sign bit remains in the MSB position. When shifting to the left, the LSB is shifted into the C flag of the CCR.

Carry Flag (C): A bit in the CCR that indicates that the result of an operation produced an answer greater than the number of available bits.

Clear: Equal to zero.

Comment: A comment is a programming language construct used to embed programmer-readable annotations in the source code of a computer program.

Condition Code Register (CCR): Referred to as "C", "CCR" or "Status" register, it is an 8-bit status and control register. Out of the eight bits in CCR, three bits are S, X, and I. These are called *control bits.*

Control Bits: Out of the eight bits in CCR, five bits are H, N, Z, V, and C. These are called *status flags.* The remaining three bits are S, X, and I. These are called *control bits.*

Data Movement: Moving data from one location to another. The source and destination locations are determined by the addressing modes, and can be registers or memory.

Direct Addressing (DIR): Two bytes, and typically takes only three clock cycles to execute.

Extended Addressing (EXT): Two-byte operand address following the opcode.

Half-Carry Flag (H): A bit in the CCR that indicates whether or not a carry is produced from bit position 3 to bit position 4 during the addition of two 8-bit binary numbers.

Immediate Addressing (IMM): The byte following the opcode is the operand.

Indexed Addressing (INDX, INDY): The operand address is obtained by adding the offset byte that follows the opcode to the contents of an index register.

Inherent Addressing (INH): Single-byte instruction that does not require an operand address.

Instruction: Commands that control the actions of the processor.

Interrupt Mask Flag (I): A bit in the CCR that can enable or disable interrupts by setting or clearing this flag.

Logic: Study of arguments.

Logical Shifting: Causes "0" to be shifted into one end of the data word, pushing all of the data bits over one position in the data word.

Negative Flag (N): A bit in the CCR that indicates that the result of a mathematical operation is negative.

Nonmaskable Interrupt Flag (X): A bit in the CCR that masks the XIRQ request when set. It is set by the hardware and cleared by the software, as well as set by unmaskable XIRQ.

Offset: A byte that follows the opcode for a conditional branch instruction such as BEQ. The offset is added to the PC to determine the address to which the processor will branch.

Opcode: Operation code, the unique multibit code given to identify each instruction to the microcontroller.

Operand: The parameters that follow the assembly language mnemonic to complete the specification of the instruction.

Operand Address: Address in memory where an operand is currently stored or is to be stored.

OR Masking: A type of masking where the OR operation is used to set specific bits of a data word while not affecting the others.

Overflow Flag (V): A bit in the CCR that indicates that the result of an operation has overflowed according to the processor's word representation, similar to the carry flag but for signed operations.

Rotate Operation: Shift operation where the bit that gets shifted out of the high or low bit position gets placed in the bit position vacated on the other side of the byte.

Set: Equal to one.

Status Flags: Out of the eight bits in CCR, five bits are H, N, Z, V, and C. These are called *status flags*.

Stop Disable Flag (S): A bit in the CCR that when set to 1 will prevent the STOP instruction from being executed.

Zero Flag (Z): A bit in the CCR that indicates that the result of a mathematical or logical operation was zero.

Answers to Section Review Quiz

5.2 True

5.3 True

5.4 False

5.5 False

5.6 True

5.7 CRC

5.8 BEST, BCLR

True/False Quiz

1. The HC11 instruction set has no equivalent instruction to add accumulator A to an index register.

2. No support is provided for the DIR addressing mode in Negate instructions.

3. No support is provided for the DIR addressing mode in decrement instructions.

4. AND masking uses the AND operation to clear specific bits of a data word while not affecting the others.

5. Zero flag of CCR plays an important role in the serial shifting of bits.

6. The logical shift left and arithmetic shift left instructions are identical in function and share the same opcodes.

7. The logical shift right and arithmetic shift right instructions are identical in function and share the same opcodes.

8. ROR and ROL are examples of arithmetic shift instructions.

9. The biggest disadvantage of HC11 is that it fails to provide a set of instructions that have the ability to alter individual bits within data bytes or 16-bit words.

10. An INX immediately followed by DEX keeps the final and initial values of index register X the same.

11. "LDAA #$FF" is an instruction in IMM addressing mode.

12. Two's complement representation allows the use of binary arithmetic operations on signed integers, yielding the correct 2's complement results.

13. CLRA and CLRB instructions clear AccA and AccB, respectively. only if the contents of AccA and AccB are not zero.

14. COMA, COMB, and COM instructions perform a logical NOT operation on an 8-bit operand.

15. If the C is equal to zero, there is no difference between logical shift and arithmetic shift.

Questions

QUESTION 5.1
What is the distinct advantage of clear instructions over load-and-store instructions when performing similar functions?

QUESTION 5.2
Why are the negate instructions grouped with subtract instructions?

QUESTION 5.3
What is the basic difference between IDIV and FDIV instructions?

QUESTION 5.4
How is the carry flag C affected in a different way in subtraction operations as compared to the addition operations?

QUESTION 5.5
Describe how logical AND and logical OR operations are used in data masking.

Problems

PROBLEM 5.1

With the stored data and CCR as given next, answer the following for each of the instructions (I to III) listed.

- (a) What is the addressing mode? How did you figure it out?
- (b) What is the effective address(es) of operand?
- (c) What is the operand?
- (d) Describe the instruction in plain English.
- (e) What is the resulting status of each of the flags in the CCR? How did you figure this out?

Data:

0030	88	FF	00	00	FF	FF	00	00
01C0	00	00	EE	FF	23	C3	D5	FF
CCR:	$D0							

	Memory Location	Machine Code			Source Code	
(I)	00FE	86	20		LDAA	#20
(II)	0000	D6	30		LDAB	$30
(III)	0002	FE	01	C0	LDX	$01C0

PROBLEM 5.2

With the saved data and various processor registers set to the values given as follows, answer the following for each of the instructions (I to III) listed.

- (a) What is the addressing mode? How did you figure it out?
- (b) What is the effective address(es) of operand?
- (c) What is the operand?
- (d) Describe the instruction in plain English.
- (e) What is the resulting status of each of the flags in the CCR? How did you figure them out?

Data:

0020	88	FF	00	00	FF	FF	80	00
0130	11	00	EE	00	00	00	80	8F
AccA	$8E							
AccB	$3D							
X	$FF00							
Y	$FF00							
S	$CCCC							
CCR:	$D1							

	Memory Location	Machine Code			Source Code	
(I)	00C0	B7	01	26	STAA	$0126
(II)	0000	D7	A0		STAB	$A0
(III)	0002	DD	0A		STS	$0A

PROBLEM 5.3

After all the operations are performed in the following code snippet, what is the value of the contents of AccA and AccB?

```
CLRB
CLRA
INCB
INCA
ABA
```

PROBLEM 5.4

Without using INCA instruction, how would you increment the value of AccA?

PROBLEM 5.5

Use HC11 exchange instructions to swap the contents of the index register X to the index register Y. Illustrate with an example.

PROBLEM 5.6

With the saved data and various processor registers set to the values given here, answer the following for each of the instructions (I to IV) listed.

(a) What is the addressing mode? How did you figure out?
(b) What is the effective address(es) of operand?
(c) What is the result?
(d) Describe the instructions in plain English.
(e) What is the resulting status of each of the flags in the CCR? How did you figure it out?

Data:

0140	11	22	22	33	44	44	D6	21
C120	CC	FF	C5	00	FF	00	00	00

AccA	$00
AccB	$00
X	$C600
Y	$FF00
S	$00EE
CCR:	$D3

	Address	Machine Code		Source Code	
(I)	00A0	8B	55	ADDA	#$55
(II)	00C5	FB	01	42	ADDB	$0142

PROBLEM 5.7

Assume that the initial content of index register X is $0C00. What will be the final contents of index register X if 20 INX instructions are executed after executing 12 DEX instructions?

PROBLEM 5.8

Assume that the initial content of index register Y is $FC14. What will be the final content of index register Y if 20 DEY instructions are executed followed by 25 INY instructions?

PROBLEM 5.9

Describe the operation performed by the instruction sequence following.

```
Instruction code    Assembly language
CE                   LDX #$C000
C0
00
18                   LDY #$C200
CE
C2
00
AB                   ADDA $32,X
32
18                   ADDB $50,Y
EB
50
A7                   STAA $00,X
00
18                   STAB $00,Y
E7
00
```

PROBLEM 5.10

Given the content of AccB equal to $73, determine the result of the values of the N, Z, and V flags for each of the following instructions

```
(a)         ANDB #$E3
(b)         ORAB #$C8
```

PROBLEM 5.11

Assuming that the initial value of C is zero, perform ROR and ROL operations on $EE.

PROBLEM 5.12

Assuming that the initial value of C is zero, perform arithmetic shift right operations on $2B.

PROBLEM 5.13

What is the value of the contents of the index register X after the following operations are performed? Explain each step in your solution.

```
CLRA
LDAB    #$0A
LDX     #$0002
IDIV
```

PROBLEM 5.14

For each of the following instructions, use the M68HC11E Series Programming Reference Guide to find the number of bytes it will occupy in memory, the name of each bytes (such as ii or dd, etc.), and the opcode.

(a) LDAB   #$22
(b) LDX $22
(c) STY $0FFF
(d) STAA $0000
(e) ABA
(f) ABY

PROBLEM 5.15

Write a code snippet to copy single byte of data using AccB from memory location $00F0 to $00C0.

6

Control Structures and Subroutines

Life is the sum of all your choices.

—Albert Camus (Nobel Prize Laureate)

OUTLINE

OBJECTIVES

Upon completion of this chapter you should be able to

1. Create short programs using various addressing modes.
2. Determine the effective address for an indexed mode instruction.
3. Create short programs using jump and branch instructions.
4. Determine the relative and destination address when employing branch instructions.
5. Appreciate the working of conditional branching.
6. Apply flowcharting as a design tool for HC11 programming.
7. Apply compare instructions to update the CCR.
8. Understand control flow and finite loops using IF-THEN-ELSE, and WHILE and UNTIL programming structures.
9. Understand the functionality of stack.
10. Write short programs that move data to and from the stack using push-and-pull instructions.
11. Identify the importance of subroutines.
12. Identify the role of instructions like JSR, BSR, and RTI in a subroutine.

13. Visualize the operation of nested subroutines.
14. Access subroutines in the BUFFALO monitor program.

Key Terms: Absolute Addressing, Branch, BUFFALO, Comment, Conditional Branch, Control Statement, Jump, Jump Table, Label, LIFO, Loops, Nested Subroutine, Offset, Program Counter (PC), Pull, Push, Relative Addressing, Sequential Execution, Stack, Stack Pointer, Subroutine, Utility Subroutine, Zero Filling

6.1 Introduction

In the last chapter, instructions in a program were executed one after the other in the order in which they were written. This is called *sequential execution*. In this chapter, we will discuss various assembly instructions that enable the programmer to move away from sequential execution. The programmer will now be able to specify that the next statement to be executed may be other than the next one in sequence. Then, a program is not limited to a linear sequence of instructions. This is called *transfer of control*.

During the process of execution, a program may bifurcate, repeat code, or take decisions. Recall that a flowchart is a graphical representation of an algorithm that contains certain special-purpose symbols, such as diamonds, ovals, and rectangles, etc., connected by arrows called *flow-lines*. One of the most important flowcharting symbols is the diamond symbol, also called the *decision symbol*. The diamond symbol indicates that a decision is to be made. It represents a situation where the next action item to be worked on may be other than the next one in sequence. A control statement in a program is equivalent to a diamond symbol. A control statement has the ability to change the computer's control from automatically reading the next line of code to reading a different one. Most of the programming languages provide control statements that serve to specify what has to be done by the program, when, and under which circumstances.

The main goal of this chapter is to understand the basic concepts of control structures and subroutines using HC11 instructions. A subroutine is a portion of code within a larger program that performs a specific task and is relatively independent of the remaining code. A subroutine is often created so that it can be started ("called") several times and/or from several places during a single execution of the program, and then branch back (*return*) to the next instruction after the "call," once the subroutine's task is done. The topics covered in this chapter are such that as the chapter progresses, the readers will be able to write simple programs that move, store, and transfer data using INH, IMM, DIR, and EXT addressing modes involving conditional flow using IF-THEN-ELSE control structure and program loops using

WHILE and UNTIL repetition structures. We will also introduce the concept of stack, and discuss its role in subroutines. Finally, we will list some of the frequently used BUFFALO utility subroutines.

6.2 Indexed Addressing Mode

Recall that a microcontroller operates by moving data from memory into its internal registers, processing them, and then copying it back into memory. These registers act as variables for the processor. The registers are used by the processor to perform computations. The general-purpose index registers are the X and Y registers. These are 16-bit registers and are most commonly used to address data in memory. Index registers are used to point at data that is located in memory. For example, in the add operation, the number getting "added in" to the sum might be indexed by the "X" register. In other words, the X register may be used to indicate the address of the data in memory. A code snippet that performs addition is given in the following text. Here, AccB is assumed to be a variable called SUM (see comments in the code). This variable is cleared in the first line of the code using CLRB instruction. Next, the index register X is initialized to a base address $00CE. In the third line of code, the offset is $01. The indexed addressing mode uses the X register as a pointer into memory. The value contained in the index register ($00CE) and the offset byte ($01) is added to specify the location of the desired memory byte or word ($00CF). The contents of the memory location $00CF ($00CE + $01) are added to the contents of AccB. Finally, the index register X is incremented to point to another memory location where the next data might be located.

```
        CLRB               ; Clear SUM
        LDX    #$00CE      ; Initialize base address
LOOP:   ADDB   $01,X       ; Add number to current SUM
        INX                ; Increment base address
```

The offset indicates the number of bytes to index forward in memory to the effective address. It is an offset from the current base address stored in one of the index registers. Since the offset is an unsigned 8-bit number, it is added to the contents of the 16-bit register by attaching 00 in front of the offset byte. In this way the effective address is determined. This process of attaching 00 in front of the offset byte is sometimes referred to as *zero filling*. Thus, the following equations can be formed in order to establish a relationship between effective address, address in the index register, and offset:

Effective Address = Address in the index register + offset

Suppose the X register currently has the value $B000. Then, the instruction "LDAA 0, X" will load AccA with the contents of location $B000 ($B000 + $00). The instruction "LDAA $19, X" will load AccA with the contents of location $B019. Here, $19 is the offset and X is the index register. The effective address is $B019, which is the addition of $B000 and $0019.

It is worth mentioning that the offset value is contained in one byte of data, and only positive or zero offsets are allowed. This means that only offsets in the range of 0_{10} to 255_{10} are possible. But why would a programmer use the indexed addressing mode, when the extended addressing mode will access the desired byte directly? The answer to this question lies in the fact that the indexed addressing mode is useful when a programmer is creating a program such that the program would need to repeatedly access locations from a particular region of memory. For example, the HC11 special register area begins at location $1000 and ends at location $103F. Suppose there were a series of instructions that accessed the registers located in this range. The programmer can set up the X register as a base pointer, and point it to the beginning of this range of memory. In other words, the programmer can load the X register with $1000 by "LDX #$1000." Now the program can use the two-byte indexed instructions to perform a series of loads, stores, etc., to the locations that fall in this range.

Common Practice: Experienced programmers often utilize the indexed addressing mode. The index register needs be loaded with the base address only once. This is usually done at the start of the program. Each indexed instruction saves a byte over the extended instruction, which in turn saves code space and execution time.

If an index register is set to point at the base of each data structure like arrays, the indexed operations can be used to access individual fields of that data element. To go to the subsequent data element, only the index base pointer needs to be altered, the offsets will then access the subsequent structure.

Example 6.1

Use the following equation and data to determine the effective address for each of the instructions from (I) to (IV):

$$\text{Effective Address} = \text{Address in the index register} + \text{offset}$$

Data:

0020	88	FF	00	00	FF	FF	00	00
0130	11	00	EE	00	00	00	8F	8F

AccA	$8E
AccB	$F0
X	$0010
Y	$B200
S	$00EE
CCR:	$D1

	Location	Machine Code		Source Code
(I)	0000	E6	D3	LDAB $D3, X
(II)	00B5	18	6F 13	CLR $13, Y
(III)	E000	18	E3 02	ADDD $02, Y
(IV)	C001	AE	10	LDS $10, X

SOLUTION

Observe the difference between the types of instructions from (1) to (IV). The instructions (I) and (II) require a 1-byte memory access, while (III) and (IV) need 2-byte memory access. Note that all the offset have to be zero filled before adding to the index register.

(I) X   = $0010
    Offset   = + $00D3

    Effective address   = $00E3

(II) Y = $B200
    Offset   = + $0013

    Effective address   = $B213

(III) Y = $B200
    Offset   = + $0002

    Effective address   = $B202 (M)
    The next address is M + 1 = $B203

(IV) Y = $0010
    Offset   = + $0010

    Effective address   = $0020 (M)
    The next address is M + 1 = $0021

Section 6.2 Review Quiz

In determining the effective address, the offset can be in the range from 0_{10} to 255_{10}. (True/False)

6.3 Jumping and Branching

Figure 6.1 illustrates an operation of copying a single byte of data from the memory location $01F0 to $01A0. Note that this is just one of the many

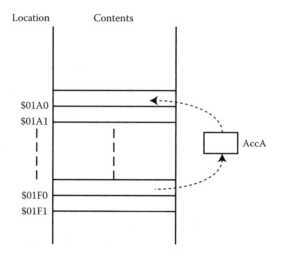

FIGURE 6.1
Illustrating a single byte copy in memory.

methods of copying data. Another way to copy may be to use AccB instead of AccA. This operation of copying a single byte of data from the memory location $01F0 to $01A0 can be coded with a load and a store instruction as given follows.

```
Address    Machine code    Source code    Comments
C000       B6  01  0F      LDAA  $01F0    ; ($01F0) → A
C003       B7  01  A0      STAA  $01A0    ; (A) → $01A0
```

In order to copy a byte from one memory location to another, the byte is loaded into a processor register, that is, AccA, and then stored to the new location. Through LDAA instruction, the processor reads the byte stored at location $01F0 and loads it into AccA. Then the instruction STAA allows the processor to take the value from AccA and write it out to memory location $01A0. This program can be described as a *straight-line program* due to the processor fetching the instructions from memory, starting at address C000, and continuing in sequence. This is accomplished by incrementing the PC after each fetch operation. Although straight-line programs are easy to create, they are limited to relatively simple programming tasks. What could be done if we wanted to repeat this load and store operation many times? To accomplish repetition, most programs use a technique called *looping*.

Common Practice: Students often use a load and a store instruction in pair for copying a single byte of data from one location to another.

Looping allows the processor to break out of the normal straight-line sequence, and jump or branch to a different area of program memory to continue executing instructions.

6.3.1 Jumping

The category of jumping and branching instructions contains instructions that cause the processor to change its execution sequence. For the next instruction, the JMP instruction causes the processor to "jump" to a specified memory address that is other than the next one in sequence.

Example 6.2

In the following code snippet, assuming that the processor has just completed execution of the ASLA instruction at 0000, what will the processor do in the next step? Draw a diagram to show this operation.

```
Address      Machine Code     Source Code
0000         48               ASLA
0001         7E               JMP $C15F
0002         C1
0003         5F
0004         -
0005         -
  •            •
  •            •
  •            •
C15F         B7               STAA $0300
C160         03
C161         00
```

SOLUTION

After executing ASLA, the processor will fetch the opcode for the JMP instruction at 0001. Once the processor recognizes that it is a JMP instruction, it knows that a two-byte address follows the opcode. It fetches its address, that is, $C15F in two steps. It then replaces the current contents of the PC, which is currently $0104 with $C15F. With the PC now holding the address $C15F, the processor will fetch its next opcode from this address. Thus, the processor jumps over the addresses $0104 through $C15E and goes to $C15F. Figure 6.2 illustrates the jump operation for the code above.

From the code snippet given in Example 6.2, we would like to comment on some of the observations we have made. First, we can observe that the JMP instruction is a three-byte instruction. The opcode byte is followed by the two-byte jump address that informs the processor of the location to jump to. This location is commonly referred to as the *destination address*. Jump instructions used in the source code are usually followed by a label. This lets the programmer refer to the location of an instruction using a label rather than by knowing its exact address. The same label is used to the left of the instruction that will be executed following the jump instruction. Labels must begin

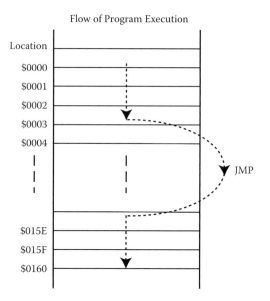

FIGURE 6.2
An example of JMP instruction.

in the first character position of a new line and may optionally end in a colon
(:). The code in the following is the same code from Example 6.1 except that
it uses a symbolic label "STORE." This label "STORE" identifies a place in
the memory that the assembler keeps track of. This location in the memory
is initially unknown to the programmer. The
colon (:) is not part of the label but acts as a
separator. Label symbols are limited to upper
and lower-case characters a–z and the digits
0–9. The three special characters that are also
allowed: the period "." and the dollar sign "$"
and the underscore "_".

Helpful Hint: The assembler only resolves
symbol and label names to eight characters. You can make them longer if you want,
but be careful. The names VARIABLE1 and
VARIABLE2 will be treated as identical by
the assembler.

```
Address         Machine Code        Source Code
0000            48                  ASLA
0001            7E                  JMP STORE
0002            C1
0003            5F
0004            –
0005            –
•               •
•               •
•               •
C15F            B7                  STORE    STAA $0300
C160            03
C161            00
```

6.3.2 Branching

An address "points to" a place in the memory where the processor wants to get or put a byte. In case of PC, the address in a JMP instruction is fetched out of the second and third bytes of the instruction itself and placed into PC. Jump instructions have to be provided with the absolute address of the next instruction. However, the branch instructions, a powerful set of instructions used to alter the execution sequence, involve calculation of destination address relative to the current PC. Let us compare and contrast relative addressing to absolute addressing.

6.3.2.1 Relative Addressing versus Absolute Addressing

We've just seen how the JMP instruction loads a $C15F into the PC. This is an example of absolute addressing because the address is right there in the instruction, it is absolute and unvarying. The address is just loaded directly into the PC. Branching is another way of changing the sequence of execution where a programmer can tell the processor something like "branch ahead seven lines" or "branch back three lines," instead of "jump to $C15F." For example, the BEQ instruction is a two-byte instruction such that the first byte is the opcode, and the second byte is a signed binary number called the *relative address offset* or *simple offset*. The offset is used to determine the destination address, that is, the address to which the processor will branch for its next instruction. The offset is not an address, but is a number that must be added to the PC to form the destination address. Let us look into the role of this offset in the following example.

Example 6.3

What does the BEQ instruction do in the code snippet given in the following text?

Address	Machine Code	Source Code
0200	B0	SUBA $02A2
0201	02	
0202	A2	
0203	27	BEQ #$09
0204	09	
0205	??	?
•	•	
•	•	
•	•	
020E	B7	STAA $0122
020F	01	
0210	22	

SOLUTION

When the above program starts executing, the machine code for "SUBA $02A2" is placed from memory location $0200 to $0202. The processor then fetches the opcode at 0203. When the processor identifies the BEQ instruction, it fetches the offset byte at $0204. The offset byte at $0204 is $09. Now the processor has to make a decision based on the result of the last operation. If the SUBA operation produced a result equal to zero, the processor will add the offset byte to the current contents of the PC. Note that the current content of the PC is $0205. With PC set to $0205, and offset equal to $09, the new value of PC is $0205 + $09 = $020E. Due to the new contents of the PC, the processor will branch to the new address $020E for its next instruction, and will continue from there. One the other hand, if the result of the SUBA operation is not equal to zero, the processor does not add the offset to PC. Instead, it will take its next instruction in sequence from address $0205. Figure 6.3 illustrates the program flow.

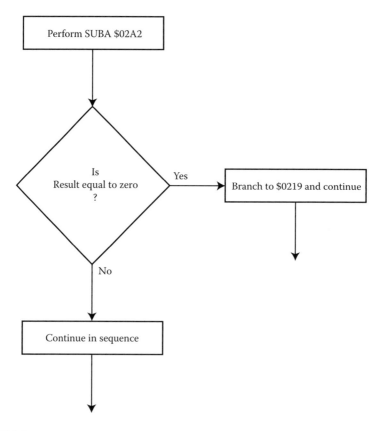

FIGURE 6.3
Program flow for Example 6.3.

The offset byte is a signed binary number. When its MSB is 0, the offset is a positive value ranging from 00_{16} to $7F_{16}$ (or $+127_{10}$). A positive offset, when added to the PC, can cause forward branching by as much as $7F_{16} = 127_{10}$ addresses. When its MSB is 1, the

Common Misconception: Students often think that branches may jump forward or backward by any number of bytes. Actually, branches may only jump forward or backward a maximum of about 128 bytes from the location of the branch instruction.

offset is a negative value ranging from FF_{16} (or -28_{10}) to $80_{16} = -128_{10}$. A negative offset, when added to the PC, can cause backward branching by as much as $80_{16} = -128_{10}$ addresses. Thus, the available branching range is -128 to $+128$. From the example we discussed above, we can write the following equation:

Destination Address = Address in the Program Counter + Sign-Extended Relative Address

We will illustrate more on the relation between offset and sign-extended relative address in Example 6.3.

6.3.2.2 Conditional and Unconditional Branching

When going out for an outdoor activity, you may ask yourself "Is it raining outside?" If the answer to this question is yes, then you will pick up an umbrella. On the other hand, if the answer is no, then you would walk out as usual. The action of "picking up an umbrella" based on the satisfaction of the condition "raining outside" is an example of branching. Conditional branch instructions are used to allow the processor to alter its execution sequence only if a specific condition is satisfied. This class of instructions gives the computing machine its "decision making" ability. The sequence following in the operation of conditional branch instructions is as follows:

1. The processor fetches the opcode and finds the condition to be checked.
2. The processor checks the specified condition. The conditions are presented in Table 6.1.
3. If the specified condition is satisfied, the contents of the PC are changed to the new address. This causes the processor to branch to the destination address.
4. If the condition is not satisfied, the processor takes its next instruction in sequence.

This process is depicted in a flowchart in Figure 6.4. Let us use this flowchart to see what happens if a conditional branch instruction such as BGE (Branch if Greater Than or Equal) is implemented. From the Table 6.1 we can see that the BGE stands for "branch if greater than or equal." In the Boolean

TABLE 6.1

Set of Branch Instructions

Mnemonic	Description	Boolean Condition Test
BCC	Branch if carry cleared	$? C = 0$
BCS	Branch if carry set	$? C = 1$
BEQ	Branch if equal to zero	$? Z = 1$
BGE	Branch if greater than or equal	$? N \oplus V = 0$
BGT	Branch if greater than	$? Z + (N \oplus V) = 0$
BHI	Branch if higher	$? C + Z = 0$
BHS	Branch if higher or same	$? C = 0$
BLE	Branch if less than or equal	$? Z + (N \oplus V) = 1$
BLO	Branch if lower	$? C = 1$
BLS	Branch if lower or same	$? C + Z = 1$
BLT	Branch if less than	$? N \oplus V = 1$
BMI	Branch if minus (negative)	$? N = 1$
BNE	Branch if not equal to zero	$? Z = 0$
BPL	Branch if plus (positive)	$? N = 0$
BRA	Branch always	Always passes
BRN	Branch never	Never passes
BVC	Branch if overflow cleared	$? V = 0$
BVS	Branch if overflow is set	$? V = 1$

condition test column, the condition to satisfy in order to branch to a destination address is given by the Boolean expression

$$? N \oplus V = 0$$

This Boolean expression stands for "Is (N XOR V) equal to 0?" The HC11 performs this test with logic gates, as shown in Figure 6.5. The condition is true only when N and V are identical.

Taking another instruction BHI from Table 6.1, we can see that the BHI stands for "branch if higher." In the Boolean condition test column the condition to satisfy in order to branch to a destination address is given by the Boolean expression

$$? C + Z = 0$$

This Boolean expression stands for "Is (C OR Z) equal to 0?" The HC11 performs this test with logic gates, as shown in Figure 6.6. The condition is true only when C and Z are both zero.

Common Practice: Many programmers use the instruction BRN during troubleshooting to replace another branch instruction as well as in timing loops to cause a time delay.

BRA and BRN are the only two unconditional branch instructions in HC11. While the BRA instruction always passes the branch test and will always branch, the BRN instruction never passes the branch test and will never branch.

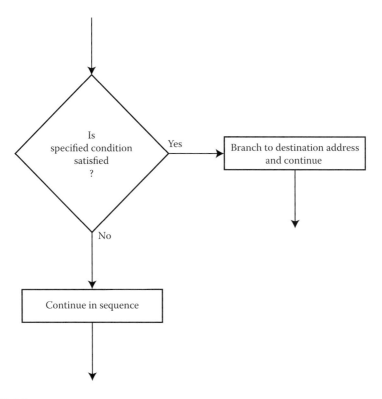

FIGURE 6.4
Flowchart illustrating conditional branch operation.

Example 6.4

Calculate the destination address for each of the following branch instructions:

	Address	Machine Code		Source Code	
(a)	0000	20	55	BRA	STORE
(b)	0000	20	90	BRA	CIRCLE
(c)	5C03	20	FE	BRA	$5C03
(d)	5003	20	80	BRA	DRUM

SOLUTION

We know that the destination address is equal to the address in PC plus the sign-extended relative address. We will use this relationship to get to the destination address.

(a) Since the instruction occupies two bytes of memory, the address of the next instruction (PC) will be two greater than the address of the BRA instruction. Thus, PC = $0000 + 2 = $0002.

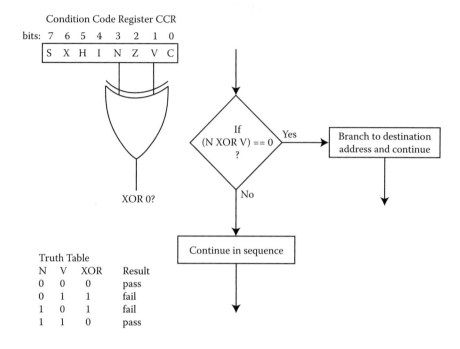

FIGURE 6.5
Conditional branch operation for BGE instruction.

Sign-Extended Relative Address = $0055. Note that we have attached "00"
to the offset byte because MSB in $55 is zero.

Therefore Destination Address = PC + Sign-Extended Relative Address
= $0002 + $0055
= $0057

(b) PC = $0000 + 2 = $0002.

Sign-Extended Relative Address = $FF90. Note that we have attached "FF"
to the offset byte because MSB in $90 is one.

Therefore Destination Address = PC + Sign-extended Relative Address
= $0002 + $FF90
= $FF92

(c) This will halt the execution of a program since it is an unconditional branch
instruction. When the opcode 20 is followed by the offset byte FE, the HC11
processor will branch backward to the address of the opcode for the BRA
instruction. Thus an endless loop is formed by continuous execution of the
BRA instruction.

(d) PC = $5003 + 2 = $5005.

Sign-Extended Relative Address = $FF80. Note that we have attached "FF"
to the offset byte because MSB in $80 is one.

Therefore Destination Address = PC + Sign-Extended Relative Address
= $5005 + $FF80
= $4F85

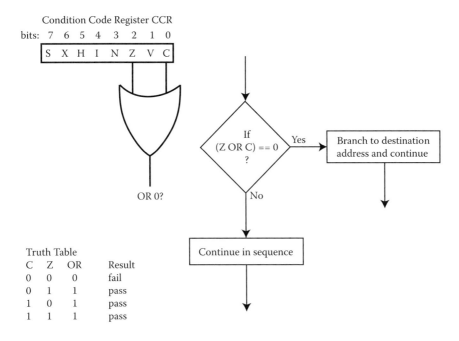

FIGURE 6.6
Conditional branch operation for BHI instruction.

Section 6.3 Review Quiz

Give two examples of unconditional branch instructions.

6.4 Compare Instructions

Various compare instructions are presented in Table 6.2. Compare instructions perform the subtraction operation on two operands, but do not store the result in any register. The result is used only to affect the flags in the CCR. These flags can then be checked using the appropriate conditional branch instruction to determine the relative values of the operands. In other words, the compare instructions provide information that can be tested by a subsequent branch instruction. For example, to branch if the contents of a register are less than an immediate or memory value, you would follow the compare instruction with a Branch if Carry Cleared (BCC) instruction. This is illustrated in Figure 6.7. The only compare instruction that is in INH addressing mode is the CBA. CBA compares the contents of AccA to the contents of AccB. The rest of the compare instructions are available in any of the addressing modes except the INH mode.

TABLE 6.2

Set of Compare Instructions

Instruction	IMM	DIR	INDX	INDY	EXT	INH	Function	H	I	N	Z	V	C
CBA	—	—	—	—	—	x	(a) – (b)	—	—	X	X	X	X
CMPA	X	X	X	X	X	—	(a) – (m)	—	—	X	X	X	X
CMPB	X	X	X	X	X	—	(b) – (m)	—	—	X	X	X	X
CPD	X	X	X	X	X	—	(d) – (m):(M+1)	—	—	X	X	X	X
CPX	X	X	X	X	X	—	(x) – (m):(M+1)	—	—	X	X	X	X
CPY	X	X	X	X	X	—	(y) – (m):(M+1)	—	—	X	X	X	X

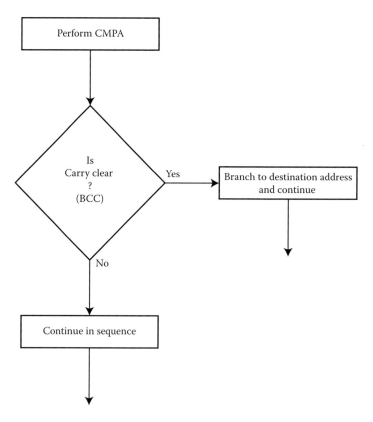

FIGURE 6.7
Illustrating CMPA followed by BCC instruction.

Example 6.5

Using AccA, compare $54 to $77. Which CCR flags are affected by this comparison?

SOLUTION

The syntax for comparison of $54 and $77 using AccA is given as follows:

```
Address    Machine Code     Source Code
0000        86   54         LDAA #$54
0002        81   77         CMPA #$77
```

The compare instruction will subtract $77 from $54. The results are not stored anywhere, but the CCR flags are updated.

$$
\begin{aligned}
\$54 &= 0101\ 0100_2 \\
\$77 &= 0111\ 0111_2 \\
2's\ complement\ of\ \$77 &= 1000\ 1001_2
\end{aligned}
$$

Subtracting $77 from $54 using 2's complement:

$$
\begin{array}{r}
0101\ 0100_2 \\
+\quad 1000\ 1001_2 \\
\hline
1101\ 1101 \quad = \$DD
\end{array}
$$

$DD is a negative number. Therefore, N is set (N = 1). Z is cleared (Z = 0) it is a nonzero number. There was no sign overflow (the sign of the result is correct), therefore, V is cleared (V = 0). A borrow was done during the subtraction, so C is set (C = 1).

Table 6.3 presents the set of compare-to-zero instructions. These instructions are often referred to as *test instructions*. They function identically to the compare instructions except that they never require that the second value be provided. The second value is assumed to be zero. The TST instruction examines the data in the specified memory location and adjusts the N and Z flags accordingly. TSTA and TSTB instructions perform the TST operation on data in AccA and AccB, respectively.

TABLE 6.3

Set of Compare To Zero Instructions

Instruction	IMM	DIR	INDX	INDY	EXT	INH	Function	H	I	N	Z	V	C
TST	—	—	X	X	X	—	(m) – $00	—	—	X	X	0	0
TSTA	—	—	—	—	—	X	(a) – $00	—	—	X	X	0	0
TSTB	—	—	—	—	—	X	(b) – $00	—	—	X	X	0	0

Example 6.6

In the following program, explain how branching would affect the CCR flags.
Draw the program flow.

```
Address Machine Code  Source Code   Comments
0000     7d  E0  00    tst $E000     ;Test data at address $E000
0003     27  B0        beq SUN       ;If zero, branch to SUN
0005     2b  C4        bmi MOON      ;If negative, branch to MOON
0007     3e            wai           ;Halt
```

SOLUTION

The TST instruction evaluates the data at address $E000 and makes N the same
as the MSB of the data. The Z flag is set (Z = 1) only if the data is zero. The BEQ
instruction will materialize into branching only if the data is zero (Z = 1). The
BMI instruction will result into branching only if the data is negative (N = 1).
Here, SUN and MOON are the labels. The program flow is shown in Figure 6.8.

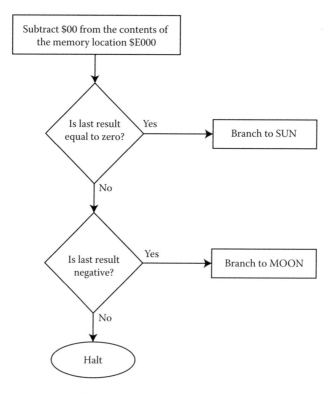

FIGURE 6.8
Program flow for Example 6.6.

Section 6.4 Review Quiz

The instruction CBA is an example of _____ addressing mode. (INH/DIR/REL)

6.5 Conditional Flow and Program Loops

By using different combinations of conditional branch instructions we can control the program execution flow as required. The IF-THEN-ELSE structure employs conditional branch instructions to establish whether a condition is true. This is the IF condition in the IF-THEN-ELSE structure. If the condition is found to be true (by passing the branch test), then the program will branch. This branching forms the THEN portion of the IF-THEN-ELSE structure. On the other hand, if the condition is found to be false (by failing the branch test), then the program will not branch. The program falls through to the next instruction. This falling through the next instruction is the ELSE portion of the IF-THEN-ELSE structure. The structure is shown in a flowchart in Figure 6.9.

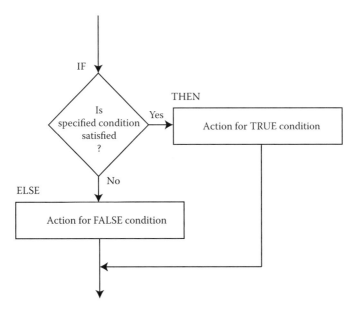

FIGURE 6.9
IF-THEN-ELSE flowchart.

Example 6.7

Describe the operation of the following code snippet. Create an IF-THEN-ELSE type of flowchart for this code.

```
TEST1:   CMPA    #$71
         BNE     LABEL1
         JMP     QUIT
```

SOLUTION

The first instruction has the label TEST1. Here, the contents of AccA are compared with $71 using CMPA. In the next instruction, a conditional branch instruction BNE is used to determine whether the result of the last operation performed (CMPA) is not equal to zero. This means that if the content of AccA is not equal to $71, then the conditional test will pass. In case the condition is satisfied, that is, (AccA) ≠ $71, the program execution will branch to LABEL1. On the other hand, if the condition is not met (i.e., AccA is equal to $71), the program execution will jump to the label QUIT. The flowchart is shown in Figure 6.10.

A repetition structure allows the programmer to specify that an action is to be repeated while some condition remains true. The following pseudocode statement describes the repetition that occurs during a shopping trip.

While there are more items on my shopping list
Purchase next item and cross it off my list

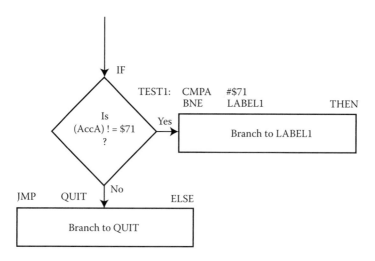

FIGURE 6.10
Flowchart for Example 6.7.

The condition, "there are more items on my shopping list" is binary in nature. It either is true or false. If it is true, then the action, "Purchase next item and cross it off my list" is performed. This action will be performed repeatedly while the condition remains true. The statement(s) contained in the while repetition structure constitute the body of the while. A WHILE-DO structure is a control flow that allows code to be executed repeatedly based on a given Boolean condition. If the structure and condition of the pseudocode is changed to the following, a new repetition structure comes into existence.

Purchase next item and cross it off my list

Until there are no more items on my shopping list

This is called the DO-UNTIL repetition structure. Notice how the purchase line has been put on the top and the condition has been changed to "no more items." Whereas the WHILE-DO repetition structure continues executing the body of the loop as long as the comparison test is true, the DO-UNTIL repetition structure executes the loop as long as the comparison test is false. The structure for both these types is shown in a flowchart in Figure 6.11.

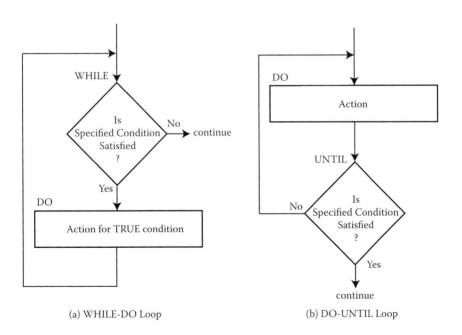

(a) WHILE-DO Loop   (b) DO-UNTIL Loop

FIGURE 6.11
WHILE-DO and DO-UNTIL flowcharts.

Example 6.8

Describe the operation performed in the following text? Assume that the keyboard data is available at memory location $C090.

```
Address        Instruction Code    Mnemonic
0000                F6             LDAB   $C090
0001                C0
0002                90
0003                C1             CMPB   #$FF
0004                FF
0005                27             BEQ    $0000
0006                F9
```

SOLUTION

The program first loads the keyboard data ($C090) into the AccB. Then, it compares it with $FF. If the contents of AccB is equal to $FF, the program branches back to the LDAB on the $0000 memory location. This can mean that if none of the keys are down, the program should branch back to reload the keyboard data into AccB. Otherwise, if AccB is not equal to $FF, the program falls through this decision block and continues further. The program flow is shown in Figure 6.12.

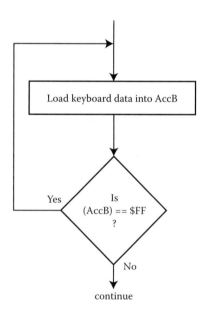

FIGURE 6.12
Flowchart for Example 6.8.

Section 6.5 Review Quiz

For a given problem, the DO-UNTIL and WHILE-DO program flows are identical in implementation. (True/False)

6.6 Stack Operation

In Chapter 3 we briefly mentioned a 16-bit internal register named stack pointer (SP) that contains the address of the last entry on the RAM stack. A stack is a special portion of RAM reserved for the temporary storage and retrieval of information—typically, the contents of the processor's internal registers. The stack is a special part of the memory, not because other memory operations can't reach it, but because it has its own special memory pointer—the SP—which works differently from the index registers or program counters in the way it operates.

Processor stacks are a simple and fast way of saving data during program execution. Stacks as shown in Figure 6.13 save data in a processor the same way you save papers on your desk. As you are working, the work piles up in front of you and you do the task that is at the top of the pile. A stack is known as Last In First Out (LIFO) memory. This should be pretty obvious because the first work item on the stack of paper will be last one you get to.

Unlike the PC, which marches forward through memory as it picks up instructions, the stack pointer marches backward in steps after each word is

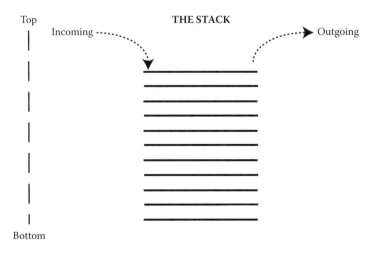

FIGURE 6.13
Stack data flow.

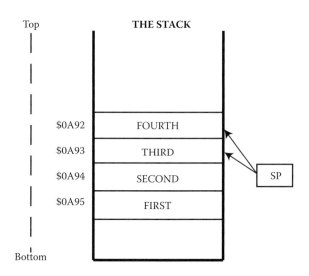

FIGURE 6.14
Stack operation.

pushed onto the stack. When a word is taken off the stack, that is, pulled or read, the SP marches forward. The stack thus operates like a shift register that can reverse directions. This is the reason such a register is called a LIFO. It is customary to initialize the stack at the top of user RAM. Then, as the stack grows, it moves downward toward location $0000. Refer to the Figure 6.14 for an example of the operation of stack and SP. Let us assume that we have four words that we want to store on the stack. Let us call these words as FIRST, SECOND, THIRD, and FOURTH. The word FIRST is stored at the memory location 0A95. Word SECOND is stored at memory location 0A94. Similarly, words THIRD and FOURTH are stored in memory locations 0A93 and 0A92, respectively. Words stored on the stack are read from the stack in the opposite order from that in which they were placed on the stack. Here, we are simply following the LIFO strategy. Thus, the word FOURTH must be read first, then the word SECOND, the word THIRD, and finally the FOURTH word. Once a word is read from the stack, its location on the stack becomes available for storage. Let us assume that the current value of SP is $0A92. If a stack READ operation is performed, the SP pointing to $0A92 is automatically incremented to $0A93, and the word stored there, that is, word THIRD is read by the processor.

The HC11 supports a number of special instructions designed to manage the data on the stack. The data in several different registers may be temporarily stored and retrieved by using these special instructions. A programmer can store data on the stack at any time by using the set of push instructions. The push instructions are also responsible for decrementing the stack pointer so that it continues to point to the next available location on the memory stack. An opposite operation can be performed using the pull instruction so that the stored stack data can be retrieved. Pull instructions first increment the stack pointer so that it points to the last byte written to the stack. The data pull is performed after the stack pointer is updated. Tables 6.4 and 6.5 present push-and-pull instructions. Push-and-pull instructions have no effect on the CCR.

Helpful Hint: The lower part of the stack has a higher address. The top part of the stack has a lower address.

TABLE 6.4

Set of Push Instructions

Instruction	IMM	DIR	INDX	INDY	EXT	INH	Function	H	I	N	Z	V	C
PSHA	—	—	—	—	—	X	$(A) \rightarrow M_S$	—	—	—	—	—	—
							$(S) - 1 \rightarrow S$						
PSHB	—	—	—	—	—	X	$(B) \rightarrow M_S$	—	—	—	—	—	—
							$(S) - 1 \rightarrow S$						
PSHX	—	—	—	—	—	X	$(X_L) \rightarrow M_S$	—	—	—	—	—	—
							$(S) - 1 \rightarrow S$						
							$(X_H) \rightarrow M_S$						
							$(S) - 1 \rightarrow S$						
PSHY	—	—	—	—	—	X	$(Y_L) \rightarrow M_S$	—	—	—	—	—	—
							$(S) - 1 \rightarrow S$						
							$(Y_H) \rightarrow M_S$						
							$(S) - 1 \rightarrow S$						

TABLE 6.5

Set of Pull Instructions

Instruction	IMM	DIR	INDX	INDY	EXT	INH	Function	H	I	N	Z	V	C
PULA	—	—	—	—	—	X	$(S) + 1 \rightarrow S$	—	—	—	—	—	—
							$M_S \rightarrow (A)$						
PULB	—	—	—	—	—	X	$(S) + 1 \rightarrow S$	—	—	—	—	—	—
							$M_S \rightarrow (B)$						
PULX	—	—	—	—	—	X	$(S) + 1 \rightarrow S$	—	—	—	—	—	—
							$M_S \rightarrow X_H$						
							$(S) + 1 \rightarrow S$						
							$M_S \rightarrow X_L$						
PULY	—	—	—	—	—	X	$(S) + 1 \rightarrow S$	—	—	—	—	—	—
							$M_S \rightarrow Y_H$						
							$(S) + 1 \rightarrow S$						
							$M_S \rightarrow Y_L$						

Example 6.9

With the saved data and various processor registers set to the values given here, state the effect each of the instruction (I to III) listed in the following has on the memory or registers.

Data:

0020	88	FF	00	00	FF	68	00	12
0130	11	00	EE	00	00	00	8F	8F

AccA	$8E
AccB	$F0
X	$FF00
Y	$FF00
S	$0025
CCR:	$D1

	Memory Location	Machine Code	Source Code
(I)	0000	36	PSHA
(II)	0000	32	PULB
(III)	0002	38	PULX

SOLUTION

(I)   (AccA) → $0025

The contents of AccA ($8E) are written to the location $0025.

$0024 → S

The SP is decremented to $0024 after the data is pushed.

(II) $0026 → S

The SP is incremented to $0026 first before data is pulled.

($0026) → AccB

The contents of the location $0026 ($00) are loaded to AccB.

(III) $0026 → S

The SP is incremented to $0026 first before the data is pulled.

($0026) → XL

The contents of the location $0026 ($00) are loaded to the low byte of index register X.

$0027 → S

SP is incremented again to $0027.

($0027) → XH

The contents of the location $0027 ($12) are loaded to the high byte of index register X.

Helpful Hint: It is important to remember that data goes on and comes off the stack in a particular order. If data is stored with a PSHA and then a PSHB (push AccA, push AccB), it must be restored with the sequence PULB, PULA (pull B register, pull A register)

Helpful Hint: During program execution, the stack changes size based on operations.

Helpful Hint: HC11 follows a postdecrement and preincrement convention.

Example 6.10

(a) Describe the operation that is actually being performed in the following sequence of code. Assume that the initial contents of AccA = $00, and AccB = $FF. SP is initialized to $0025.

```
PSHA
PSHB
PULA
PULB
```

(b) What advantage does stack provide in the following segment of code? Assume the initial value of AccA to be 10_{10}:

```
PSHA
<perform some operations>
PULA
DECA
BNE
```

SOLUTION

(a) The Table 6.6 shows the sequence of operation. Notice how the contents of AccA, AccB, and SP change at every step.

Helpful Hint: HC11 uses a 16-bit register as a stack pointer.

It can be observed from the table that the instructions simply swap the value of AccA and AccB by reversing the order of pulls.

(b) The advantage of using stack with PSHA and PULA instructions is that the programmer has not referenced memory for a loop counter.

TABLE 6.6

Swap Performed Using Push-and-Pull Instructions

Steps	(AccA)	(AccB)	(SP)	Comments
Initial	$00	$FF	$0025	Initial values
PSHA	$00	$FF	$0024	(A) → $0025
				The contents of AccA ($00) are written to the location $0025
				$0024 → S
				The SP is decremented to $0024 after the data is pushed
PSHB	$00	$FF	$0023	(B) → $0023
				The contents of AccB ($FF) are written to the location $0023
				$0023 → S
				The SP is decremented to $0023 after the data is pushed
PULA	$FF	$FF	$0024	$0024 → S
				The SP is incremented to $0024 first before data is pulled.
				($0024) → A
				The contents of the location $0024 ($FF) are loaded to AccA.
PULB	$FF	$00	$0025	$0025 → S
				The SP is incremented to $0025 first before data is pulled.
				($0025) → B
				The contents of the location $0025 ($00) are loaded to AccB.

Section 6.6 Review Quiz

LIFO stands for (a) Last In First Out, (b) Last In Found Out, (c) Last Input Fifth Output, (d) Later In Figure Out. (Choose one.)

6.7 Subroutines

As we develop bigger programs, we quickly find that there are program sections that are so useful that we would like to use them in different places. A subroutine is a program section structured in such a way that it can be called from anywhere in the program. Once it has been executed, the program continues to execute from wherever it was before. The idea is illustrated in Figure 6.15. At some point in the main program there is an instruction "Call ABC." When the program execution reaches this point, the control switches to Subroutine A, identified by its label. The subroutine must be terminated with a "return from subroutine" instruction. The program execution then continues from the instruction after the "Call ABC" instruction. A little later in the main program, another subroutine is called (Call XYZ), followed a little later again by another call to the first routine (Call ABC). The dashed arrows indicate the second call to the Subroutine A.

Generically, the term subroutine can be used to denote a piece of code, separate from the main body, fulfilling a discrete task. For example, a program might refer to time-delay subroutines. A self-contained program that can be used repeatedly as part of a main program is called a *subroutine*. Just like

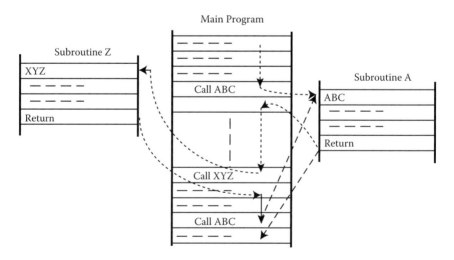

FIGURE 6.15
Subroutine calling.

a time-delay subroutine, each subroutine is associated with a specific task. A subroutine may contain one or more instructions to fulfill its objective. Various subroutines can be held in any order in memory because of their modular self-contained nature. The subroutines are executed in an order that is decided by the main program.

Let us look into how a subroutine works. A subroutine call is a special instruction that must be used to start the execution of a subroutine. The HC11 supports two subroutine call instructions: Jump-to-Subroutine (JSR) and Branch-to-Subroutine (BSR). When a subroutine is called from the main program, the contents of the PC are stored or pushed on the stack. This way the processor knows where to come back to after it has finished the subroutine. It then loads the subroutine start address into the PC. Program execution thus continues at the subroutine. Each subroutine must end with a special instruction that returns the flow to the calling program. This instruction is the Return-from-Subroutine (RTS). Therefore, when the subroutine is finished, the PC is loaded with the data held at the top of the stack. This data is the address of the instruction following the call instruction. Program execution then continues at this address. This way the control comes back to the main program.

Suppose a program often has the requirement to execute a delay, simply waiting half a second. Rather than repeatedly writing the code to perform the delay, it can be written just once, as a subroutine. Then, whenever the main code needs to execute the delay, it can just call the subroutine. Table 6.7 summarizes the instructions that manage the use of subroutines. A subroutine call from within another subroutine is called a *nested subroutine*. Figure 6.16 shows a program execution flow in a nested subroutine. The main program calls the Subroutine A, which then calls the Subroutine Z. After Subroutine Z is finished, the program flow moves back to Subroutine A. After Subroutine A is complete, the program execution returns back to the main program. In doing this, it must be remembered that every time a subroutine is called, one stack position is taken up, which becomes free again on the subroutine return. If we call a subroutine from within another subroutine, then two

TABLE 6.7

Set of Subroutine Instructions

Instruction	IMM	DIR	INDX	INDY	EXT	INH	Function	H	I	N	Z	V	C
BSR	X	—	—	—	—	—	$(P) \rightarrow M_s{:}M_{s-1}$, $S-2 \rightarrow S$, Eff. Address $\rightarrow P$	—	—	—	—	—	—
JSR	—	X	X	X	X	—	$(P) \rightarrow M_s{:}M_{s-1}$, $S-2 \rightarrow S$, Eff. Address $\rightarrow P$	—	—	—	—	—	—
RTS	—	—	—	—	—	X	$M_{s+1}{:}M_{s+21} \rightarrow P$, $S+2 \rightarrow S$,	—	—	—	—	—	—

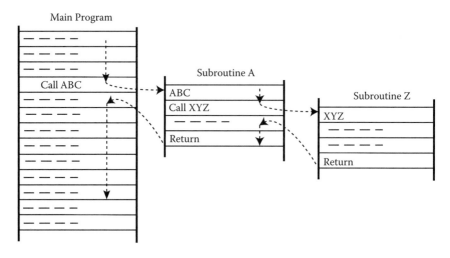

FIGURE 6.16
Program flow in a nested subroutine.

stack locations are used up, or three if there is another nested call. In the next chapter, we will explore more on subroutines where we will show parameter passing and nesting program examples.

Subroutines are a powerful programming tool. Judicious use of subroutines will often substantially reduce the cost of developing and maintaining a large program. Subroutines also improve a program's quality and reliability. It is important for the programmer to remember the function, starting address, and initializing of parameters when working with subroutines. Special attention must be given in tracking the resisters or memory locations changed by the subroutine.

Team Discussion: Discuss how program loops and subroutines can be used to give time delays of known and accurate duration.

Self-Learning: Do some Internet research and write a paragraph about "stack overflow."

Helpful Hint: Subroutine call and return instructions must always work in pairs.

Section 6.7 Review Quiz

RTS stands for (a) rest for stack, (b) round from subroutine, (c) return from stack, (d) return from subroutine. (Choose one)

6.8 BUFFALO Subroutine

Several utility subroutines are available along with BUFFALO for performing various tasks. These are prewritten blocks of code that may be called from different parts of a program. These subroutines are simple and are primarily related to serial input/output communications. A jump table (or branch table) for each of these subroutines has been set up in ROM. The implementation of these utility subroutines requires use of a subroutine (JSR) command to the appropriate jump table target address. However, to call a utility subroutine from the BUFFALO command line, the call command is used. A complete list of BUFFALO subroutines is provided in the BUFFALO documentation for HC11 Development Boards. Readers are encouraged to go through the document in order to get familiar with these subroutines. In the next chapter, we will explore some of the frequently used utility subroutines. The Table 6.8 presents frequently used BUFFALO utility subroutines. The column heading "Address" is the jump table target address used to make the corresponding call.

TABLE 6.8

Frequently Used BUFFALO Utility Subroutines

Address	Label	Description
$FF7C	WARMST	Go to the BUFFALO prompt ">", skip past "BUFFALO..." message, go to '>'
$FFA9	INIT	Initialize the I/O device
$FFB2	OUTLHL	Convert left nibble in A register to ASCII and output to serial communications port
$FFB5	OUTRHL	Convert right nibble in A register to ASCII and output to serial communications port
$FFB8	OUTA	Output the ASCII character in the A register
$FFBB	OUT1BY	Convert binary byte in memory that X register points at to ASCII and output to serial communications port. The X register is incremented by one
$FFBE	OUT1BS	Convert binary byte in memory that X register points at to ASCII characters followed by a space and output to serial communications port. The X register is incremented by one
$FFC1	OUT2BS	Convert binary byte in memory that X register points at to ASCII characters followed by a space and output to serial communications port. The X register is incremented by two
$FFC4	OUTCRL	Output a carriage return character followed by a line feed character
$FFC7	OUTSTR	Output a string of ASCII characters starting with that pointed to by the X register and terminated with the end of transmission character ($04)
$FFCA	OUTST0	The same as OUTSTRG except that a leading carriage return, line feed character pair are not produced
$FFCD	INCHAR	Wait for and echo an ASCII character, the received character is returned in the A register
$FFD0	VECINT	Initializes the Vector Jump Table in RAM with JMP instructions and default addresses

Example 6.11

Write a code snippet to enable a character input from the keyboard. The code shall then convert the character to uppercase. Assemble the program at $0100 in memory.

SOLUTION

```
Address    Instruction Code   Mnemonic
0100       BD FF CD           JSR $FFCD   ;get char from keyboard
0103       BD FF AO           JSR $FFA0   ;convert to uppercase
```

Common Practice: Programmers often use BUFFALO utility subroutines to make their programs more interactive and user friendly. Such user interactions can be in the form of accepting the input from the keyboard, or displaying the answers to the user.

We use BUFFALO subroutines for both the tasks. Using Table 6.8, we pick the two utility subroutines (viz., INCHAR and UPCASE) with the corresponding jump table target address $FFCD and $FFA0, respectively. The INCHAR will enable input from the keyboard (INCHAR), while UPCASE converts the character to uppercase.

Section 6.8 Review Quiz

Which BUFFALO utility subroutine would be used to output the ASCII character in the A register?

6.9 Summary

- Control statements have the ability to change the computer's control from automatically reading the next line of code to reading a different one.
- The indexed addressing uses a 16-bit base address in combination with an 8-bit unsigned offset.
- The indexed addressing mode allows stepping through the memory space from a common base address.
- Jumping and branching is done to change the execution sequence of the processor.
- A jump that occurs only if certain conditions are satisfied is called a conditional jump.
- Instructions that do not test any condition play a vital role in unconditional jumps. They either always cause a branch (e.g., BRA) or never cause a branch (e.g., BRN).

- One important thing to remember about branching instructions is that they use the *relative addressing mode,* which means that the destination of a branch is specified by a one-byte offset from the location of the branch instruction.
- Loops repeat a statement a certain number of times or while a condition is fulfilled.
- The IF-THEN-ELSE structure allows a process to occur only IF a condition is true.
- The WHILE structure allows a process only while some condition is true.
- The UNTIL structure allows the process to continue until some condition is true.
- A portion of RAM reserved for the temporary storage and retrieval of information is called a stack.
- A stack pointer (SP) is a 16-bit register that addresses the stack.
- Operation of a stack is a last in/first out (LIFO) manner.
- Data is pushed (saved) or pulled (read) in and from the stack.
- Pull instructions are complementary to push instructions.
- Subroutines are used to improve efficiency, reliability, and reusability of the program.
- Subroutine call and return instructions must always work in pairs.
- Nested subroutines typically form when the main program jumps to a subroutine that contains a jump to a subroutine, and so on.
- Utility subroutines are predefined subroutines available in BUFFALO monitor. A jump table target addressing is used in the main program to refer to these subroutines.

Glossary

Absolute Addressing: Addressing in which the instruction contains the address of the data to be operated on.

Branch: A jump that can be conditional or unconditional.

BUFFALO: BUFFALO stands for Bit User Fast Friendly Aid to Logical Operations. It is a monitor program that provides a controlled environment for the HC11 chip.

Comment: Personal notes in an assembly-language program that are not assembled. They refresh the programmer's memory at a later date.

Conditional Branch: A jump that occurs only if certain conditions are satisfied.

Control Statement: Ability to change the computer's control from automatically reading the next line of code to reading a different one.

Jump: A process of changing the execution sequence of the processor.

Jump Table: An efficient method of transferring program control (branching) to another part of a program using a table of branch instructions.

Label: A name given to an instruction in an assembly-language program. To just to this instruction, you can use the label rather than address. The assembler will work out the correct address of the label and will use this address in the machine-language program.

LIFO: Last in, first out. In a LIFO structured linear list, elements can be added or taken off from only one end. Therefore, the last element to be placed on the open end is also the first to be taken off the open end.

Loops: Repeat a statement a certain number of times or while a condition is fulfilled.

Nested Subroutine: The main program jumps to a subroutine that contains a jump to a subroutine, and so on.

Offset: A byte that follows the opcode for a conditional branch instruction such as BEQ. The offset is added to the PC to determine the address to which the processor will branch.

Program Counter (PC): A register that counts in binary. Its contents are the address of the next instruction to be fetched from the memory.

Pull: To read data from the stack.

Push: To save data in the stack.

Relative Addressing: Used in conditional branch instructions to determine the branching address by adding the offset to the program counter.

Sequential Execution: When the instructions in a program are executed one after the other in the order in which they were written.

Stack: A portion of RAM reserved for the temporary storage and retrieval of information, typically the contents of the processor's internal registers.

Stack Pointer: A 16-bit internal register that contains the address of the last entry on the RAM stack.

Subroutine: A program stored in higher memory that can be used repeatedly as part of a main program.

Unconditional Branch: Instructions that do not test any condition. They either always cause a branch (e.g., BRA) or never cause a branch (e.g., BRN).

Utility Subroutine: A predefined subroutine available in BUFFALO monitor.

Zero Filling: The process of filling all bit positions with zeros that are not occupied by data.

Answers to Section Review Quiz

  6.2 True

  6.3 BRA, BRN

  6.4 INH

  6.5 False

  6.6 (a)

  6.7 (d)

  6.8 OUTA

True/False Quiz

1. In the indexed mode, an offset follows the opcode in memory.

2. Pull instructions are complementary to push instructions.

3. The indexed addressing does not require the use of offset.

4. The indexed addressing mode allows stepping through the memory space from a common base address.

5. Jump instructions cause a change in the program flow to any location within the addressable range of the processor.

6. Destination Address = Address in the Program Counter + Sign-Extended Relative Address.

7. The instruction BEQ (branch if equal to zero) is complementary to branch if not equal to zero (BNE).

8. BRA and BRN are conditional branch instructions.

9. The branch test is a Boolean expression that must be evaluated before the branch can take place.

10. Compare instructions are a special set of instructions that do not affect status flags in the CCR.

11. The CBA instruction is responsible for comparing the data in AccA to the data in AccB.

12. A loop can never end with a branch or a jump instruction.

13. The NOP instruction stands for "No operation."

14. A subroutine is a portion of code within a larger program that performs a specific task and is relatively independent of the remaining code.

15. A stack can have any abstract data type as an element, but is characterized by only two fundamental operations: push and pop.

Questions

QUESTION 6.1

Every Monday when Mark starts his car to go to work, he checks to see if he has enough gas in the car. In case the gas is not enough, he drives to a nearby gas station. At the gas station, he checks if his car tires have enough air. If he finds that the air is not sufficient, he uses the air pump at the gas station. Draw a flowchart to represent this process of multiple conditional tests.

QUESTION 6.2

Why is a stack often referred to as a restricted data structure? What problems will arise if you change the stack push-pull operation from LIFO (last in/first out) to FIFO (first in/first out)?

QUESTION 6.3

Describe how branching instructions use the relative addressing mode. Is there a limit to the amount of jump (forward or backward, in term of bytes) that can be performed by the branching instructions?

QUESTION 6.4

Fill in the blanks:

(a) Branch if Equal to Zero (BEQ): Branch is made if Z flag is ___.
(b) Branch if Not Equal to Zero (BNE): Branch is made if Z flag is ___.
(c) Branch if Carry is Clear (BCC): Branch is made if C flag is ___, indicating that a carry did not result from the last operation.
(d) Branch if Carry is Set (BCS): Branch is made if C flag is ___, indicating carry occurred.
(e) Branch if Lower (BLO): Branch is made if the result of subtraction was less than ___.
(f) Branch if Greater Than or Equal (BGE): Branch is made if result of subtraction is greater than or equal to ___.
(g) When a number is placed on the stack (called a stack push), the number is stored in memory at the current address of the stack

pointer. Then the stack pointer is _____ to the next position in memory.

(h) When a number is taken off the stack (called a stack pull), the stack pointer is _____ to the last location stored, and then the number at that memory location is retrieved.

QUESTION 6.5

Refer to Table 6.7. What is the main difference in the way a user will experience the output of a program if the BUFFALO subroutine OUTST0 is used instead of OUTSTR?

QUESTION 6.6

Why should the stack be carefully accessed when writing programs with nested subroutines?

QUESTION 6.7

Why is it important to remember that data goes on and comes off the stack in a particular order?

Problems

PROBLEM 6.1

With the help of a diagram illustrate an operation of copying a single byte of data from the memory location $0101 to $0100. Use AccB as a temporary storage. Write an assembly code to accomplish this task.

PROBLEM 6.2

Use the following equation and the values of index register X and Y given here to determine the effective address for each of the instructions (I) and (II).

Effective Address = Address in the index register + offset

Initial contents of index registers:

X	$0010
Y	$B200

	Memory Location	Machine Code		Source Code		
(I)	0000	18	6F	01	CLR	$01,	Y
(II)	C001	ED	B0		STD	$B0,	X

PROBLEM 6.3

For each value of the content of AccA from (a) through (e), what would be the program flow for the following code snippet? QUIT and LABEL2 are located at memory location $0140 and $0118, respectively.

(a) $00, (b) $71, (c) $51, (d) 71_{10}, and (e) $FF

Address	Machine Code	Source Code
010A	81 71	CMPA #$71
010C	26 03	BNE LABEL1
010E	7E 01 40	JMP QUIT
0111	81 51	LABEL1: CMPA #$51
0113	26 03	BNE LABEL2
0115	7E 01 40	JMP QUIT

PROBLEM 6.4

Calculate the destination address for each of the following branch instructions:

	Address	Machine Code	Source Code
(a)	0C00	20 40	BRA PHONE
(b)	0000	20 FA	BRA SOLVE

PROBLEM 6.5

The initial value of the content of AccA is $AA. Write a single line code to compare $AA to $BB. Show the updated values of the affected CCR flags.

PROBLEM 6.6

Figure 6.17 shows a branch test for a conditional branch instruction. Which instruction is it?

PROBLEM 6.7

Draw a logical representation (using CCR, OR, and XOR gates) of the branch test for the conditional branch instruction BLE.

PROBLEM 6.8

Figure 6.18 shows a flowchart for a branch test of a conditional branch instruction. Which instruction is it?

PROBLEM 6.9

Write a fragment of code that will repeatedly decrement AccA until it is zero.

PROBLEM 6.10

Write a fragment of code such that AccB is first initialized to $00. Then AccB is repeatedly incremented until it is $0A.

Condition Code Register CCR

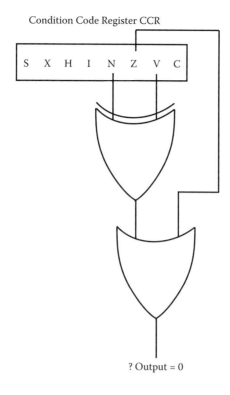

? Output = 0

FIGURE 6.17
A branch test.

PROBLEM 6.11

A programmer wrote the following code segment such that it loops only 10 times. He forgot to add a conditional branch test. Where do you think is the most appropriate location to place the conditional branch test. Which instruction would be used?

```
        CLRB
        LDY     #$0000
LOOP    CPY     #$000A
        ADDB    $00,Y
        INY
        BRA     LOOP
```

PROBLEM 6.12

In the Figure 6.19, a pull operation is performed. For this pull operation, the value of SP is first incremented to $0C33. What data will be read after this increment in the value of SP?

PROBLEM 6.13

What problems do you see in the program flow shown in Figure 6.20?

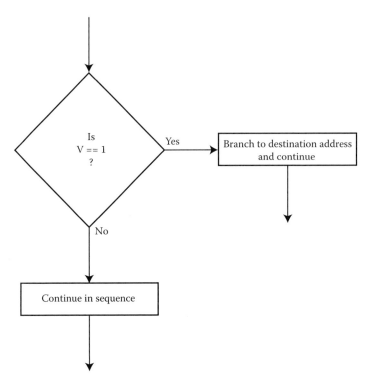

FIGURE 6.18
Flowchart for a branch test.

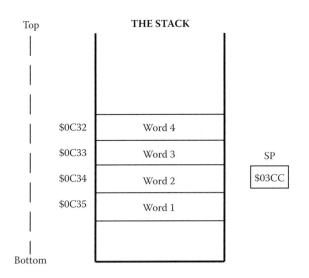

FIGURE 6.19
Stack and stack pointer.

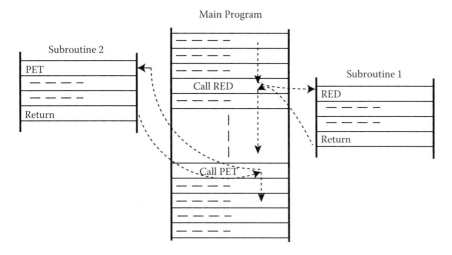

FIGURE 6.20
A problematic program flow.

PROBLEM 6.14

What function do the following instructions perform if executed one after the other? Convert this piece of code into a subroutine. Comment your code. List all of your assumptions.

```
TAB
MUL
```

PROBLEM 6.15

Identify the subroutine, calling instruction, and return instruction that has been used in the program given as follows. Identify the condition branch tests and compare instructions.

```
*  ****************************************************
*  Program Name: H2A.asm
*  Objective: Convert HEX into ASCII using a subroutine
*  Usage: Buffalo Call 140 command
*  Output: ASCII is saved at location OUT
*  ****************************************************
MAIN    EQU    $0140
DATA    EQU    $0000
OUT     EQU    $0010

        ORG    DATA
        FCB    $06

        ORG    MAIN
        LDX    #DATA
        LDY    #OUT
        LDAA   $00,X
```

```
        JSR   H2A
        STAA  $00,Y
        SWI

H2A     CMPA  #$0F
        BHI   SKIP
        CMPA  #$0A
        BLO   PASS
        ADDA  #$07
PASS    ADDA  #$30
SKIP    RTS
```

PROBLEM 6.16

List the problem(s) that would arise in the program given in Problem 6.15 if the changes given in the Table 6.9 are made.

PROBLEM 6.17

Identify BUFFALO utility subroutines from the program given here. What function does each of these subroutines perform?

```
ORG   MAIN
JSR   $FFCD
JSR   $FFB8
SWI
```

PROBLEM 6.18

The code given in Problem 6.17 is changed. The new code is given here:

```
ORG   MAIN
JSR   $FFB8
JSR   $FFCD
SWI
```

What would change in the behavior of the program experienced by the user?

TABLE 6.9

Changes Made to the Program in Problem 6.15

Change Number	Original Syntax		Modified Syntax			Changed Element
1	JSR	H2A	CALL	H2A		Changed call
2		BLO PASS		BLO	PASSAGE2011	Changed label
		ADDA #$07		ADDA	#$07	
	PASS	ADDA #$30	PASSAGE2011	ADDA	#$30	
3	LDAA $00,X		LOOP: LDAA $00,X			Added label

PROBLEM 6.19

Write a code fragment to compare the contents of AccA with 00110000_2. If the content of AccA is lower than 00110000_2, branch to NODGT. Again, compare the contents of AccA with 00111001_2. If the content of AccA is greater than 00111001_2, branch to NODGT.

PROBLEM 6.20

With the saved data and various processor registers set to the values given here, state the effect each of the instruction (I to III) listed in the following has on the memory or registers.

Data:

```
00A0    88    FF    00    00    FF    68    00    12
00B0    11    00    EE    00    00    00    8F    8F
AccA    $CC
AccB    $8E
X       $FF00
S       $00A4
```

	Memory Location	Machine Code	Source Code
(I)	0000	36	PSHA
(II)	0000	32	PULB
(III)	0002	38	PULX

7

Hello World!

Many things difficult to design prove easy to perform.

—Samuel Johnson (Author)

OUTLINE

OBJECTIVES

Upon completion of this chapter, you should be able to

1. Understand a typical assembly program development environment.
2. Write readable source code.
3. Utilize assembler directives, labels, and directives to enhance source code.
4. Describe the fields in the listing file.
5. Define and interpret the contents and structure of S-records.
6. Write a variety of decision-making statements.
7. Create new subroutines and implement BUFFALO subroutines.
8. Implement control flow and finite loops using IF-THEN-ELSE, and WHILE and UNTIL programming structures.
9. Understand the use of memory locations to access lists of values.
10. Understand the use of memory locations to move a block of data.
11. Design and develop conversion programs such as hexadecimal to corresponding ASCII value.

Key Terms: Assemble, Assembler Directive, Breakpoint, Checksum, Comments, Debugging, DOS, EVBU, Executable Code, Execute, Graphical User Interface (GUI), Hands-On, Hello World Program, Integrated Development Environment (IDE), Instruction, Linker, List File, MAKE, S-Record File, Single Stepping, Software Development Life Cycle Model, User Interface (UI).

7.1 Introduction

One of the simplest programs possible in most of the programming languages is a "Hello World!" program. This program prints out "Hello World!" on a display device such as a computer monitor. The program acts as an introduction to hands-on learning for most of the computer programming languages. It is often considered to be tradition among programmers for people attempting to learn a new programming language to write a "Hello World!" program as one of the first steps in learning that particular language. Most of these programs are very simple, especially those that rely heavily on a particular command line interpreter ("shell") to perform the actual output. In a microcontroller, the text may be sent to a liquid crystal display (LCD), or the message may be substituted by some other appropriate signal, such as an LED being turned on. However, in some programming languages, especially in some graphical user interface (GUI) contexts, the "Hello World!" program is surprisingly complex.

A "Hello World!" program has become the traditional first program that many people learn. In general, it is simple enough that people who have no previous experience with computer programming can easily understand it, especially with the guidance of a teacher or a written guide. Using this simple program as a basis, computer science principles or elements of a specific programming language can be explained to novice programmers. Experienced programmers learning new languages can also gain information about a given language's syntax and structure from a "Hello World!" program. Additionally, substantial amounts of time and effort can be involved in configuring a full programming tool-chain from scratch to the point where even small and simple programs can be compiled or assembled and run.

For this reason, many programmers run this program in order to ensure that a language's compiler or assembler, development environment, and run-time environment are correctly installed. A "Hello World!" in C looks like the following:

```c
#include <stdio.h>
int main()
{
    printf("Hello World!");
    return 0;
}
```

The main goal of this chapter is to provide hands-on assembly coding exposure so that the theoretical knowledge of instructions, decision making, loops, subroutines, etc., that has been acquired up till now can be utilized in executing microcontroller programs on the EVBU. Just after briefly introducing assembler functions, listing files, and "S" records, we will create our own "Hello World!" program and test it on the EVBU. This will be treated as the "kickoff" for our incredible programming journey that we are going to experience from here onwards.

7.2 Creating Source Code Files

Superior coding techniques and programming practices are hallmarks of a professional programmer. A good part of programming consists of making numerous small choices in an effort to solve a larger set of problems. This section addresses some fundamental assembly coding techniques and provides a collection of coding practices from which to learn. The coding techniques are primarily those that improve the readability and maintainability of code, whereas the programming practices are mostly performance enhancements. We have been learning programming practices in the previous chapters also under the heading "common practice" and "best practice."

To create an assembly program for the HC11, a programmer can use his or her favorite text editor. The standard suffix for assembly programs is .ASM (or .asm). This suffix is to be used to designate all of the source code files. Writing assembly code is analogous to the way you write source code in most other high-level languages. However, the important difference is that here the assembler assumes fixed fields in the source file. What does the fixed field mean? Figure 7.1 illustrates the fixed-field format. Anything starting in the first column is assumed to be a label; the first word not in column one

Helpful Hint: While creating an assembly program, if a line in the .ASM source code does not contain a label, you still have to tab or space over so that it is understood as an instruction mnemonic or assembler directive by the assembler.

is assumed to be an instruction mnemonic or assembler directive. We will introduce ourselves to the assembler directive shortly. The third field contains the operands, if any, and the fourth contains comments. Comments are ignored by the assembler. All fields are separated by one or more spaces or tab characters (or white space). A white space provides a significant degree of readability to assembly programs.

An assembler supports a special set of commands called directives. Assembler directives are instructions entered into the source code along with the assembly language. These directives do not get translated into object code but are used as special instructions to the assembler to perform some special functions. The assembler will recognize directives that assign

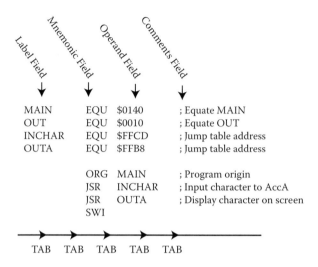

FIGURE 7.1
Fixed-field format for .ASM source code.

memory space, assign address to labels, format the pages of the source code, and so on. A directive is usually placed in the opcode field. If any labels or data are required by the directive, they are placed in the label or operand field as necessary. Some commonly used directives are listed in Table 7.1.

The Origin (ORG) statement is used by the programmer when it is necessary to place the program in a particular location in memory. As the assembler is translating the source code, it keeps an internal counter that keeps track of the address for the machine code. The counter is incremented automatically and sequentially by the assembler. If the programmer wishes to alter the locations where the machine code is going to be located, the ORG statement is used. We will be using ORG statement in all the HC11 programs throughout the text.

The Equate (EQU) directive is an assembler directive that is used to assign the data value or address in the operand field to the label field. EQU represents word substitutions for values. The EQU directive is valuable because it allows the programmer to write a source code in symbolic form and not be concerned with the numeric value needed. In some cases, the programmer may be developing a program without prior knowledge of the addresses or data that may be required by the hardware. The program may be written and debugged in symbolic form, and the actual data may be added at a later stage. Using the EQU instruction is also helpful when a data value is used several times in a program. For example, PORT A on the HC11, a port that has several input and output pins for external connection to the outside world, is mapped to memory address $1000. That

TABLE 7.1

Summary of Assembler Directives

Directive	Type	Syntax	Example	Description
BSZ	Data definition/ storage allocation	BSZ <n>	BSZ #$02	Block store zeros (reserves n numbers of memory bytes and fills them with zero)
EQU	Symbol definition	<label> EQU <value>	PORTB EQU $1004	Equate (value is replaced by the label)
FCB	Data definition/ storage allocation	FCB <n>	FCB $04	Form contract byte (a byte of data n is assigned to specific memory location)
FCC	Data definition/ storage allocation	FCC <'string'>	FCC 'Hello World!'	Form constant character (stores ASCII string into consecutive bytes of memory)
FDB	Data definition/ storage allocation	FDB <n>	FDB #$1234	Form double byte (store a double (two-byte) word)
FILL	Data definition/ storage allocation	FILL <val>, <n>	FILL #$FF, #$02	Fill memory (fill constant val in n memory locations)
ORG	Assembly control	ORG <address>	ORG $0000	Origin (sets program counter to origin)
RMB	Data definition/ storage allocation	<label> RMB <n>	BUFFER RMB 8	Reserve memory byte (reserve n number of memory location)
ZMB	Data definition/ storage allocation	ZMB <n>	ZERO ZMB 0 ZMB #$02	Zero memory byte (same as BSZ)

means that any reading from memory address $1000 or writing to memory address $1000 refers to PORTA. You could write $1000 throughout your code, but it would be easier if you just called it PORTA and referred to that. This can be done by using an EQU directive as "PORTA EQU $1000." An assembler will replace every instance of the PORTA with the value associated with it. So "LDAA PORTA" will be the same as "LDAA $1000." By using this technique, if it is found during debugging that the value in PORTA must be changed, it need only be changed at the EQU instruction rather than at each of the locations in the program. In Figure 7.1, MAIN is equated to $0140. Then the "ORG MAIN" assembler directive is used to start the main program from the memory location MAIN, which is $0140. We will learn more about using AS11 assembler directives in the examples to follow.

Example 7.1

Which of the HC11 assembler directives are used in the following program? Why has the programmer used the symbol # in the instructions LDX?

```
MAIN  EQU   $0140
DATA  EQU   $0000
END   EQU   $000A

      ORG   MAIN
      LDX   #DATA       ; Initialize source base-address
LOOP: LDAB  $00,X       ; Get a number from source list
```

SOLUTION

The Equate directive is used to substitute MAIN, DATA, and END labels for values $0140, $0000, and $000A respectably. The Origin directive is used to start the program from MAIN, which is $0140. If the programmer would have used LDX DATA, it would have been an error. In the absence of #, it would have meant "load memory location of $100A to the index register X," which does not make any sense. The programmer has correctly used the # in order to load the value $000A to the index register X.

Best Practice: All things being equal, the more Equates you use, the easier your programs will be to debug and modify.

It is worth mentioning that expressions can be used at multiple places throughout the code. An expression is a combination of symbols, constants, algebraic or logical operators, and parentheses. Expressions may consist of symbols or constants joined together by one of the following operators: +- * / % & | ^. These operators are listed in Table 7.2. They are evaluated by the assembler before it makes an assignment. A simple example of expression is BASE + 1. If the label BASE equals the base address $0000, the BASE + 1 will be equal to $0001.

TABLE 7.2

Operators Defined for Expressions

Operator	Description
+	Integer addition
−	Integer subtraction
*	Integer multiplication
/	Integer division
%	Integer remainder
&	Bitwise AND
|	Bitwise OR
^	Bitwise XOR

TABLE 7.3

The AS11 Assembler Data Types

Numeric Data Type	Preceded By
Decimal	Numbers are considered decimal unless they are preceded by a special symbol
Hex	$
Binary	%
Octal	@
ASCII character	Single quote (')

The AS11 assembler supports five numeric data types. These are decimal, hex, binary, octal, and ASCII characters. Table 7.3 lists these AS11 assembler data types.

If a number is not preceded by a special symbol, it is considered as a decimal. If the user wants the assembler to interpret numbers as hex, binary, or octal values, the numbers must be preceded by the "$", "%", or "@" symbol, respectively. ASCII characters will be interpreted as the ASCII character codes when they are preceded by the single quote (') character. Examples of these data types are shown as follows:

Common Practice: Although ASCII characters will be interpreted as the ASCII character codes when they are preceded by only a single quote (') character, experienced programmers prefer to provide a trailing single quote also.

Group Discussion: Which data types are read most easily by novice programmers?

Common Misconception: Students often tend to avoid using binary types in assembly programming thinking that it would be difficult to read. However, binary type is more readable in place of hex when specific bits are being emphasized.

```
LDAA #200      ; Decimal 200 will be converted by the assembler
LDAA #%10110100; Binary 10110100 will be converted by the assembler
LDAA #$FF       ; Hex FF will be used by the assembler
LDAA #@503     ; Octal 503 will be converted by the assembler
```

Example 7.2

Comment on the label END in the code fragment given as follows:

```
MAIN    EQU    $0140      ; Equate MAIN
DATA    EQU    $0B00      ; Equate DATA
END     EQU    DATA+$0A   ; Equate END
```

SOLUTION

In the second line, $0B00 is substituted by the label DATA. Then, in the third line, sum of DATA to $0A is substituted by label END. This means that END is equal to $0B00 + $0A = $0B0A.

Going back to the fixed fields in the source code illustrated in Figure 7.1, note that each line of code can be commented to the right of the instruction. The AS11 assembler assumes any characters that follow the last byte in the operand field to be a comment. These are known as end-of-line comments and usually begin with a semicolon character (;). Programmers insert comments to document programs and improve program readability. Comments do not cause the program to perform any action when the program is run. Comments are ignored by the assembler or compiler and do not cause any machine code to be generated. A comment that generally begins with an asterisk character (*) and occupies an entire line or several lines of the source file is called full-line comment. The asterisk must appear in the left-most position of the line. This indicates to the assembler that this line contains only a comment. Blank lines form a subset of full-line comments. The following is a fragment of code with in-line and full-line comments beginning with semicolon (;) and asterisk (*) characters, respectively.

Best Practice: A good programming practice to follow is to begin every program with a comment describing the purpose of the program. Comments also help other people read and understand your program.

```
*  ****************************************************
*  Program Name: Copy.asm
*  Objective: Copy Block of Memory
*  Usage: Buffalo Call 140 command
*  Output: Block of memory is copied
*  ****************************************************
MAIN    EQU   $0140        ; Equate MAIN
DATA    EQU   $0000        ; Equate DATA
END     EQU   DATA+$0A     ; Equate END
NDATA   EQU   $0010        ; Equate NDATA
        ORG   DATA         ; Data origin
*  Populate an array with ten numbers
ARRAY   FCB   $02,$04,$06,$08,$0A,$0C,$0E,$10,$12,$14
        ORG   MAIN         ; Program origin
```

Recall that labels were briefly mentioned in Chapter 6. A label lets the programmer refer to the location of an instruction using a word rather than by knowing its exact address. The following code is the same code from Example 6.1 except that it uses a symbolic label "STORE." This label "STORE" identifies a location in the memory that the assembler keeps track of. This location is initially not known to the programmer.

The same label is used to the left of the instruction that will be executed following the jump instruction. Labels must begin in the first character position of a new line and may optionally end in a colon (:). The colon is not part of the label but acts as a separator. Label symbols are limited to upper- and lower-case characters a–z and the digits 0–9. The three special characters that are also allowed: the period ".", the dollar sign "$", and the underscore "_".

```
Address           Machine Code       Source Code
0000              48                 ASLA
0001              7E                 JMP STORE
0002              C1
0003              5F
0004              -
0005              -
  •                 •
  •                 •
  •                 •
015F              B7          STORE  STAA $0300
0160              03
0161              00
```

The label field starts in the left-most column of each line of the source file. If the first character is the asterisk "*", the entire line is treated as a comment. If the first character of a line is a space, the assembler assumes that there is no label defined for the line of code.

Helpful Hint: The assembler only resolves symbol and label names to eight characters. You can make them longer if you want, but be careful. The names VARIABLE1 and VARIABLE2 will be treated as identical by the assembler.

Since now we know the basic structure of simple and short programs using instructions set, assembler directives, etc., we can begin to demonstrate the complete microcontroller system development cycle. One way to look at the total development of a microcontroller-based system is to classify it into three phases: software design, hardware design, and program diagnostic design. Note that there are other phases such as project planning, requirements analysis, etc., but just for simplicity we will ignore them for now. In such a system development cycle, a systems programmer will be assigned the task of writing the application software, a logic designer will be assigned the task of designing the hardware, and

Best Practice: For readability, align all your instructions.

Best Practice: In software engineering, the Software Development Life Cycle Model concept underpins many kinds of software development methodologies. These methodologies form the framework for planning and controlling the software development process. The Software Development Life Cycle Model must be tailored during project planning to meet the specific requirements of the project.

typically, both designers will be assigned the task of developing diagnostics to test the system. For small systems, one engineer may do all three phases, while on large systems, several engineers may be assigned to each phase. Figure 7.2 shows a flowchart for the total development for a system. In order to save time, software and hardware development may occur simultaneously.

7.2.1 Writing the "Hello World!" Program

Let us write a program that just does what a "Hello World!" program is supposed to do. By going through a complete development cycle, we will get

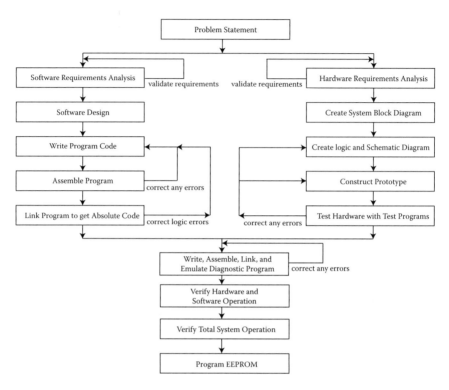

FIGURE 7.2
Microcontroller system development flowchart.

hands-on experience of what it takes to develop and execute an assembly program on an EVBU. The following is a program that outputs the string "Hello World!" to the terminal window. The first few lines are the comments that describe the program name, objective of the program, its usage, and the location where the output is to be expected. MAIN, DATA, and EXIT are the substitute labels for $0140, $0100, and $FF7C. OUTCRLF and OUTSTRG are the BUFFALO utility subroutines that will be used in the program. $FFC4 and $FFC7 are jump table addresses for OUTCRLF and OUTSTRG, respectively. The data starts from the $0100 memory address. This is indicated by the programmer by the ORG DATA statement. This address is up to the programmer to select from an appropriate available range of memory locations. Recall from Chapter 3 that the BUFFALO monitor program on the EVBU uses a substantial piece of the RAM for temporary storage. This storage can be used for calculations, variables, data, the user program, the interrupt vector jump table, or work in progress. Figure 3.24 shows the layout of the scratchpad memory.

The line containing FCC in the Hello.asm program will assign the message "Hello World!" to memory by the FCC directive. FCC converts each character of the message to the corresponding ASCII character codes. The label "Messg" is the base-address for this ASCII character code string. Since the utility subroutine OUTSTRG outputs the ASCII character codes until end-of-transmission ($04) is found, we use FCB to place $04 at the end of the string "Hello World!"

Now let us look into the main program. The program starts from the location $0140. This is denoted by the ORG MAIN statement. The label "Start" is just to let the readers know that the main program starts from that point. The program execution will remain the same even if the label "Start" is completely removed from the source code. Since the OUTSTRG subroutine outputs the ASCII character codes with base-address in the index register X, we first load the index register X with "Messg." Next we call the OUTSTRG subroutine followed by a call to the subroutine OUTCRLF. The OUTSTRG subroutine outputs the message to the terminal and stops at $04. Next, the OUTRLF subroutine writes a newline to screen. The program will end with SWI instruction.

```
* ***************************************************
* Program Name: Hello.asm
* Objective: An introductory program "Hello World!"
* Usage: Buffalo CALL 140 command
* Output: On Screen
* ***************************************************
MAIN       EQU    $0140             ; Equate MAIN to $0140
DATA       EQU    $0100             ; Equate DATA to $0100
OUTCRLF    EQU    $FFC4             ; Jump table address
OUTSTRG    EQU    $FFC7             ; Jump table address

           ORG    DATA              ; Data origin
Messg      FCC    'Hello World!'    ; Message to display
           FCB    $04               ; EOT Character - Delimiter

           ORG    MAIN              ; Program origin
Start      LDX    #Messg            ; Message pointer
           JSR    OUTSTRG           ; Send message
           JSR    OUTCRLF           ; Send CR,LF
           SWI                      ; Program exit
```

Section 7.2 Review Quiz

(a) Unlike source code, program comments cannot be tested. (True/False)

(b) Incorrect and obsolete comments cannot mislead a programmer. (True/False)

(c) _____ do not cause the program to perform any action when the program is run. (Directives/Comments/Labels)

7.3 Assembling Programs

From this point onward, we will cover essential topics in order for the readers to get started with the specifics of the CME11E9-EVBU development process. It is highly recommended that the readers read this chapter and try to create, assemble, and execute the programs (given as examples) on EVBU to get hands-on experience. Additionally, while developing design and code, it is extremely important to keep the "BUFFALO Manual," the "M68HC11E9-EVBU Development Board Manual," and the "Instruction Set available in 68HC11 Microcontroller Reference Manual" handy. In order to use EVBU in the software development mode, you will need to install the HC11 support CD. Refer to the Installation Guide for step-by-step installation instructions and troubleshooting information. If the installation is properly performed as directed by the BUFFALO manual, a Buffalo Monitor prompt (> symbol similar to the one that follows) will appear in the Terminal window.

```
BUFFALO 3.4 (ext) - Bit User Fast Friendly Aid to Logical Operation
>
```

Generally, a program is written in .ASM file using a text editor. Figure 7.3 shows the HELLO.ASM program being written in Microsoft's WordPad. After a program is written, it has to be assembled by the assembler so that

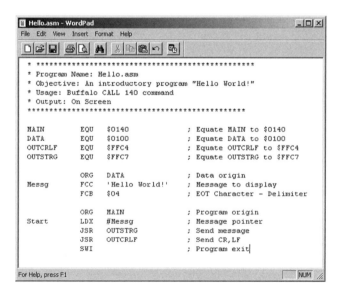

FIGURE 7.3
Creating source code for the "Hello World!" program.

a machine language object program is created from an assembly language source program. You can assemble your source code using command-line tools under a DOS prompt by typing

```
AS11 HELLO.ASM -l cre s >HELLO.LST
```

Many assemblers allow several command-line options, so using a MAKE utility or batch file can be used to assemble programs. For example, utility software or Integrated Development Environment (IDE) provided with your board from a manufacturer other than Motorola can contain a simple interface to the free assembler. It may contain a menu with a "Build" option that may prompt for the file to be assembled.

Helpful Hint: MAKE is a utility that automatically builds executable programs and libraries from source code by reading files called Make-files. These Make-files specify how to derive the target program.

The assembler will check for syntax errors and print error messages to help in the correction of errors. If there are no errors in the Hello.asm source code, two output files will be created:

1. HELLO.S19: a Motorola S-Record (hex) format file that can be programmed into memory
2. HELLO.LST: a common listing file (or list file) that shows the relationship between source and output

Helpful Hint: There are eight different S record types defined. Only S1 and S9 records are supported by the EVBU monitor program.

A Motorola S-Record (hex) format file usually has a .MOT, .HEX, or .S19 file extension and is in a format that can be read by the programming utilities to be programmed into the HC11 board. If the source code did not assemble correctly, a .S19 file will not be generated (or it will be empty). The S19 files are intended to facilitate simpler data transfer to the memory after assembly. Figure 7.4 shows the HELLO.S19 file. S records are character strings that consist of up to five fields. These fields are summarized in Table 7.4. The S9 record acts as a termination record for a block of S1 records. The address of the first instruction to be executed after the download is complete is optional in the address field of an S9 record. There is no data field in the S9 records.

Helpful Hint: In the S-record format, the last field is for the checksum. The checksum is the sum of all characters, starting with the record length through the data field.

```
S110010048656C6C6F20576F726C642104AD
S10D0140CE0100BDFFC7BDFFC43FA0
S9030000FC
```

FIGURE 7.4
The HELLO.S19 file.

TABLE 7.4

S-Record Format for AS11

Fields in Order	Standard Size	Description
Type	2 characters	Must be "S1" or "S9" for AS11 records for the HC11 EVBU
Record length	2-digit hex value	Sum of the number of bytes in the address field and the number of bytes in the data field plus one for the checksum
Address	4 characters	Character codes for a 16-bit address
Code and data	Actual program codes	Hex value in the machine code
Checksum	2 characters	1's complement of the two-digit hex checksum value

The list file is a text file that shows two things:

1. The source code instructions and their respective translation into hex
2. Any error messages that were found during assembly

The assembler output is a detailed listing that shows the programmer exactly what machine instructions were created and where they will be placed in memory, along with the original source code. Each line of the listing file starts with a line number. Next to the line number is the memory location that the machine code will start for that instruction. Next to the address is the actual machine code. To the right of the machine code is the original source code. The HELLO.LST file is shown in Figure 7.5 with various fields. The list file may be sent to a disk file for use in debugging, or it may be directed to the printer. If the program being assembled has errors, they will be displayed, and no output will be generated; otherwise, the listing file will be displayed.

Team Discussion: Map each field from Table 7.4 to the HELLO.S19 file given in Figure 7.4. You may refer to the HELLO.LST file given in Figure 7.5.

Helpful Hint: For error messages, look at the assembler output listing file generated during assembly.

Common Practice: Many times the free assembler is an old DOS tool that does not recognize long file names. For that reason, most programmers do not use long file names (> 8 characters).

The linker can now take the object code generated by the assembler and create the final absolute code that will be executed on the target system. The emulation phase will take the absolute code and load it into the development system RAM. From here, the program may be debugged using breakpoints or single-stepping. A breakpoint in software development is an intentional stopping or pausing place in a program, put in place for debugging purposes. It is also referred to as a pause. More generally, a breakpoint is a means of acquiring knowledge about a program during its execution. Single-stepping is used when precise control over instruction execution is required. As each instruction is executed, control is passed back to the debugger, which disassembles the next instruction to be executed.

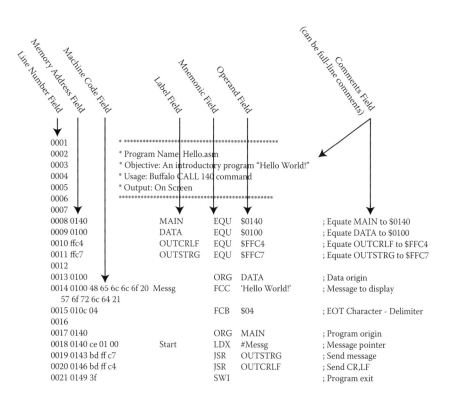

FIGURE 7.5
The HELLO.LST file.

After successfully assembling the HELLO.ASM source code and creating a HELLO.S19 S-Record file, we can "upload" the S-Record file to the development board for a test run. Since our HELLO.ASM has been created to run from RAM, we will use the BUFFALO Monitor to test our code without programming it. After verifying that the EVBU board is connected and operating properly, we will type "CALL 140" at the BUFFALO monitor prompt. This command tells BUFFALO to execute the program at address $140, which is the start of our test program. A sample execution of the "Hello World!" program on EVBU by a user is shown in Figure 7.6.

Best Practice: An assembly program should contain the following elements in this sequence: Equates section, ORG data section, ORG main program section, subroutines, interrupt service routines.

Section 7.3 Review Quiz

A Motorola S-Record (hex) format file has usually a .MOT, .HEX, or .S19 file extension. (True/False)

FIGURE 7.6
Running of the "Hello World!" program on EVBU.

7.4 Ten Useful Programs

Example 7.3

Write a program such that it inputs a character from the keyboard, then displays that character back to the terminal window.

SOLUTION

This is a simple program that can be written by using just two BUFFALO utility subroutines: INCHAR and OUTA. Recall from Chapter 6 that frequently used BUFFALO utility subroutines were summarized in Table 6.7. From that table, we can get the jump table address for a particular subroutine. The utility subroutines INCHAR and OUTA have the jump table addresses $FFCD and $FFB8, respectively. These addresses are set in the program using Equate (EQU) directives. The program starts from $0140 using the Origin (ORG) directive (Figure 7.7). The starting point of the program is set by the programmer. The program first calls the utility subroutine INCHAR. INCHAR takes the ASCII character code from the terminal and stores it in the AccA. This subroutine does not return until a valid character code is received, which means that the program keeps waiting for the character to be typed by the user on the keyboard. Once the character is typed, the main program returns from the INCHAR subroutine and calls the OUTA subroutine. The OUTA subroutine simply outputs the ASCII character code in the AccA to terminal port. The SWI instruction finally ends the program. Figure 7.8 shows a sample execution of this program.

```
0001                   * **************************************************
0002                   * Program Name: IOChar.asm
0003                   * Objective: Inputs a char from user, outputs to the screen
0004                   * Usage: Buffalo Call 140 command
0005                   * Output: On screen
0006                   * **************************************************
0007 0140       MAIN      EQU    $0140            ; Equate MAIN
0008 ffcd       INCHAR    EQU    $FFCD            ; Jump table address
0009 ffb8       OUTA      EQU    $FFB8            ; Jump table address
0010
0011 0140                 ORG    MAIN             ; Program origin
0012 0140 bd ff cd        JSR    INCHAR           ; Input character to AccA
0013 0143 bd ff b8        JSR    OUTA             ; Display character on screen
0014 0146 3f              SWI
0015
```

FIGURE 7.7
IOChar.LST file.

```
>CALL 140
aa
P-0146 Y-FFFF X-010C A-61 B-FF C-D0 S-0041
```

FIGURE 7.8
Sample execution to output a user-provided character.

```
*  **************************************************
* Program Name: IOChar.asm
* Objective: Inputs a char from user, outputs to the screen
* Usage: Buffalo Call 140 command
* Output: On screen
***************************************************
MAIN     EQU    $0140    ; Equate MAIN
INCHAR   EQU    $FFCD    ; Jump table address
OUTA     EQU    $FFB8    ; Jump table address

         ORG    MAIN     ; Program origin
         JSR    INCHAR   ; Input character to AccA
         JSR    OUTA     ; Display character on screen
         SWI
```

Helpful Hint: The program is terminated on the HC11 EVBU with an SWI instruction.

Example 7.4

Write a program to find the average of 10 numbers that are stored in sequential memory locations. Use the UNTIL method to accomplish this task. Draw the corresponding flowchart.

SOLUTION

```
*  **************************************************
* Program Name: DoAvg.asm
* Objective: Find Average of a list using DO loop
* Usage: Buffalo Call 140 command
* Output: End of the list ($000A and $000B)
***************************************************
MAIN  EQU  $0140      ; Equate MAIN
DATA  EQU  $0000      ; Equate DATA
END   EQU  DATA+$0A   ; Equate END

      ORG  DATA       ; Data origin
* Populate an array with ten numbers
ARRAY FCB  $02,$04,$06,$08,$0A,$0C,$0E,$10,$12,$14

      ORG  MAIN       ; Program origin
      CLRB            ; Clear SUM
      LDY  #DATA      ; Initialize base address
```

```
LOOP   ADDB $00,Y        ; Add number to current sum
       INY               ; Increment base address
       CPY  #END         ; Reached End?
       BNE  LOOP         ; If not, do again
       CLRA              ; Clear AccA (D=AccA:AccB)
       LDX  #$000A       ; Set denominator
       IDIV              ; Divide D by X
       STX  $00,Y        ; Store AVG at the list end
       SWI
```

This program first adds all the numbers (there are 10 numbers) stored in sequential memory locations and then divides the sum by 10. Since the UNTIL method is to be used here, a loop will be employed such that the condition test will be evaluated at the end of the loop. The flow of the program is shown in Figure 7.9.

First of all, we reset (or clear) the variable sum. We will use AccB as that sum variable. You may ask why we are not using AccA. Each number that is retrieved

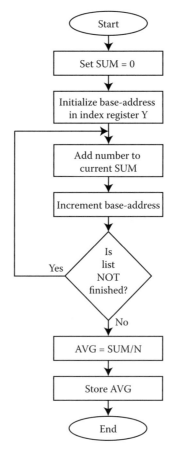

FIGURE 7.9
Flowchart to find the average using the UNTIL method.

from memory will be added to the current total stored in the variable sum, that is, AccB. Later in the program, in order to determine the average of the list, we will use the divide operation. This division operation is the reason why we are not using AccA right now as the variable sum. You will appreciate this decision when we get closer to the divide operation.

Next, we initialize the base address or the reference to the list. This program uses the index register Y to hold the base-address of the list. Again, since the divide operation (performed later in the program at the end of the summation) employs index register X for its denominator, we will not use index register X for now. Index register Y is initially loaded with $0000. The $0000 is the address of the first number in the list or the list's base address. The next steps to follow are the steps that take place in a loop. The summation starts by retrieving the first number from the list and adding it to the current sum. Remember that the current sum is stored in AccB. Initially, AccB is set to zero. The summation is done using ADDB. The "ADDB $00,Y"instruction will always offset $00 bytes from the base-address. The offset always remains constant, but since the base-address can change in the X register, this instruction can access a different memory location each time it is executed in the loop. To utilize this concept, the base-address is incremented next, and it is checked if the list is finished. This is done by comparing the value of the index register Y to the address following the list. The program uses a CPY instruction to compare the value in the index register Y to the address following the list. If the value in the index register Y is smaller than the address $000A, the Z flag is cleared for the BNE instruction, and the program loops back and continues its addition of list elements. When the address equals $000A, the Z flag is set, and the BNE conditional test fails. Thus, the program execution comes out of the loop to perform the divide operation.

Let us look closely to the divide operation performed using the instruction IDIV. IDIV basically divides register D by register X. We know that register D is a combination of Accumulator A and Accumulator B in order, that is, D = AccA:AccB. We currently have the sum of the numbers in the list stored in AccB. Therefore, we will have to clear AccA. This is done using the CLRA instruction. Now AccA is set to $00 and AccB contains the sum of the number in the list. This makes the register D (D = AccA:AccB) equal to the sum of the list. Next we load decimal number 10 or its equivalent in hexadecimal, that is, $000A, to the denominator, which is the index register X. Since we have our numerator and denominator ready, the next step is to perform the division. IDIV performs the division of register D by register X. The average of the list is saved in the register X. The statement "STX $00, Y" then stores the average of the list at the next memory location, and the program is complete. You would have appreciated by now the decision to not use AccA and the index register X initially in the summation portion of the program.

Note that this program also sets the list of the ten numbers by using the Form Constant Byte (FCB) directive. FCB causes a byte of data to be assigned to specific memory locations. Since successive bytes can be separated by commas, we have used $02, $04, $06 … to populate out the list of numbers starting from memory location $0000. We also gave the label ARRAY to this list but never used it in the program. The way

Helpful Hint: Often, an infinite loop is unintentionally created by a programming error in a condition-controlled loop, wherein the loop condition uses variables that never change within the loop.

Best Practice: In writing good programs, remember to design the program such that it adapts easily to different sets of data, that is, length and size of data.

```
>MD 0000

0000 02 04 06 08 0A 0C 0E 10 12 14 FF FF FF FF FF FF
0010 FF FF FF FF FF FF FF FF FF FF FF FF FF FF FF FF
0020 FF FF FF FF FF FF FF FF FF FF FF FF FF FF FF FF
0030 FF FF E4 E4 E4 6D E3 D0 FF 61 01 0C FF FF 01 47
0040 FA 29 FF E4 E4 6D E3 D4 00 E4 6D E3 E4 E4 6D E3
0050 D4 00 E4 6D E3 D4 00 57 07 20 E5 02 E8 10 E1 AA
0060 01 46 FF FF 01 0C 61 FF D0 00 41 4D 44 20 30 30
0070 30 30 0D 0D FF FF FF FF FF FF FF FF FF FF FF FF
0080 FF FF FF FF FF FF FF FF FF FF FF FF FF FF 4D 44
>CALL 140

P-0158 Y-000A X-000B A-00 B-00 C-D0 S-0041
>MD 0000

0000 02 04 06 08 0A 0C 0E 10 12 14 00 0B FF FF FF FF
0010 FF FF FF FF FF FF FF FF FF FF FF FF FF FF FF FF
0020 FF FF FF FF FF FF FF FF FF FF FF FF FF FF FF FF
0030 FF FF E4 E4 E4 6D E3 D0 00 00 00 0B 00 0A 01 59
0040 FA 29 FF E4 E4 6D E3 D4 00 E4 6D E3 E4 E4 6D E3
0050 D4 00 E4 6D E3 D4 00 57 07 20 E5 02 E8 10 E1 AA
0060 01 58 00 0A 00 0B 00 00 D0 00 41 4D 44 20 30 30
0070 30 30 0D 0D FF FF FF FF FF FF FF FF FF FF FF FF
0080 FF FF FF FF FF FF FF FF FF FF FF FF FF FF 4D 44
```

FIGURE 7.10
Sample execution to find the average using the UNTIL method.

we have started this list from the address $0000 is by putting FCB next to the ORG DATA statement. ORG DATA is the origin of data section in the program. As can be seen in Figure 7.10, the list is populated from address $0000 onward. The average is stored at the end of the list.

Best Practice: In writing good programs, remember to separate the instructions and data in memory, that is, the program section and the data section.

Example 7.5

Write a program to find the average of 10 numbers that are stored in sequential memory locations. Use the WHILE method to accomplish this task. Draw the corresponding flowchart.

SOLUTION

```
* ****************************************************
* Program Name: WhileAvg.asm
* Objective: Find Average of a list using WHILE loop
* Usage: Buffalo Call 140 command
* Output: End of the list ($000A and $000B)
* ****************************************************
```

```
MAIN  EQU  $0140       ; Equate MAIN
DATA  EQU  $0000       ; Equate DATA
END   EQU  DATA+$0A    ; Equate END


      ORG  DATA        ; Data origin
* Populate an array with ten numbers
ARRAY FCB  $02,$04,$06,$08,$0A,$0C,$0E,$10,$12,$14


      ORG  MAIN        ; Program origin
      CLRB             ; Clear SUM
      LDY  #DATA       ; Initialize base address
LOOP  CPY  #END        ; Reached End?
      BEQ  AVG         ; If so, perform average
      ADDB $00,Y       ; Else, add number to current sum
      INY              ; Increment base address
      BRA  LOOP        ; keep looping
AVG   CLRA             ; Clear AccA (D=AccA:AccB)
      LDX  #$000A      ; Set denominator
      IDIV             ; Divide D by X
      STX  $00,Y       ; Store AVG at the list end
      SWI
```

This program first adds all the numbers (there are 10 numbers) stored in sequential memory locations and then divides the sum by 10. Since the WHILE method is to be used here, a loop will be employed such that the condition test will be evaluated at the beginning of the loop. The flow of the program is shown in Figure 7.11.

Recall from Chapter 6 that the WHILE-DO repetition structure continues executing the body of the loop as long as the comparison test is true, and the DO-UNTIL repetition structure executes the loop as long as the comparison test is false. The comparison test can also be reversed depending upon the kind of tests and variable values. In this program, we first add all the numbers stored in sequential memory locations and then divide that sum by 10 similar to the previous example. However, since the WHILE method is to be used here, a loop will be employed such that the reversed condition test will be performed at the beginning of the loop. The difference between the previous example and this one is that, instead of asking "if the list is NOT finished?" at the end of the loop, we will ask the question "if the list is finished?" at the beginning of the loop. In the previous example, we used the BNE instruction, but here the BEQ instruction will be used. BEQ basically tests if the current value in the index register Y is equal to the address of the end of the list. In case the list is finished (i.e., Z flag in CCR is set), the program will jump out of the loop to find the average by using the same IDIV instruction as in the previous example. This jump is accomplished by using the label AVG in the program. On the other hand, if the list is still remaining (i.e., Z flag in CCR is clear), then we will just add the current number to the current sum variable. Recall that we used AccB in the previous example as a sum variable and that it was set to zero initially. The summation was done using ADDB. We will use the same strategy here also.

The "ADDB $00,Y" instruction will always offset $00 bytes from the base-address. It is worth mentioning again that the offset always remains constant,

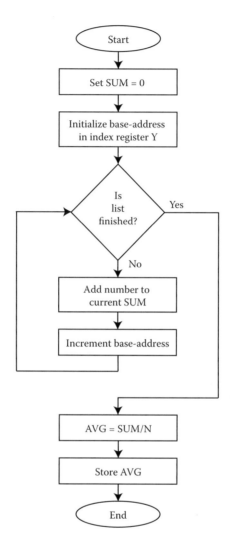

FIGURE 7.11
Flowchart to find the average using the WHILE method.

but since the base-address can change in the index register X, this instruction can access a different memory location each time it is executed in the loop. The base-address is incremented next using the INY instruction. After the base-address is incremented, the program goes back to the beginning of the loop using the instruction BRA. Recall from Chapter 6 that there are two unconditional branch instructions: BRA (Branch Always) and BRN (Branch Never). The BRA instruction always passes the branch test and will always branch. At this point of time, the program is again at the label LOOP. Again, as before, the same question is asked: Is the list finished? If the list is finished, then the program jumps to the label AVG to compute the average using the steps shown in the previous example.

```
>MD 0000

0000 02 04 06 08 0A 0C 0E 10 12 14 FF FF FF FF FF FF
0010 FF FF FF FF FF FF FF FF FF FF FF FF FF FF FF FF
0020 FF FF FF FF FF FF FF FF FF FF FF FF FF FF FF FF
0030 FF FF E4 E4 E4 6D E3 D0 00 00 00 0B 00 0A 01 59
0040 FA 29 FF E4 E4 6D E3 D4 00 E4 6D E3 E4 E4 6D E3
0050 D4 00 E4 6D E3 D4 00 57 07 20 E5 02 E8 10 E1 AA
0060 01 58 00 0A 00 0B 00 00 D0 00 41 4D 44 20 30 30
0070 30 30 0D 0D 31 38 38 43 30 30 30 41 32 36 46 35
0080 34 46 43 45 30 30 30 41 30 32 43 44 45 46 4D 44
>CALL 140

P-015A Y-000A X-000B A-00 B-00 C-D0 S-0041
>MD 0000

0000 02 04 06 08 0A 0C 0E 10 12 14 00 0B FF FF FF FF
0010 FF FF FF FF FF FF FF FF FF FF FF FF FF FF FF FF
0020 FF FF FF FF FF FF FF FF FF FF FF FF FF FF FF FF
0030 FF FF E4 E4 E4 6D E3 D0 00 00 00 0B 00 0A 01 5B
0040 FA 29 FF E4 E4 6D E3 D4 00 E4 6D E3 E4 E4 6D E3
0050 D4 00 E4 6D E3 D4 00 57 07 20 E5 02 E8 10 E1 AA
0060 01 5A 00 0A 00 0B 00 00 D0 00 41 4D 44 20 30 30
0070 30 30 0D 0D 31 38 38 43 30 30 30 41 32 36 46 35
0080 34 46 43 45 30 30 30 41 30 32 43 44 45 46 4D 44
```

FIGURE 7.12
Sample execution to find the average using the WHILE method.

If the list is not finished, then the program adds the current number to the current sum variable.

As observed in Figure 7.12, the list is populated from address $0000 onward. The average is stored at the end of the list. Since the lists in this example and the previous example are identical in size and values, the same average is computed although two different program flows were employed.

Another way to implement the WHILE structure can be as follows:

```
1. get a value for n
2. set the value of sum to 0
3. while n > 0 do
      a. set the value of sum to sum + n
      b. set the value of n to n - 1
4. end of the loop
5. Perform avg = sum/n
6. return the value of avg
```

Helpful Hint: A good program must not modify itself. The program should be correct and complete with respect to the solution design for the problem to be solved.

Readers are encouraged to write a program using the foregoing pseudocode.

Example 7.6

Write a program to find the smallest of 10 numbers that are stored in sequential memory locations. Draw the corresponding flowchart.

SOLUTION

Assume that you have a reference that is set to a very large number. With that reference in hand, you can simply walk through a list and, whenever you find an element smaller than your reference, you remember (or store) its value. When you will come out of the list, your reference variable will contain the smallest number of the list. This is a simple way of narrating one of the "smallest number" algorithms. A pseudocode for such an algorithm where the reference is not set to a very large number but set to the first number of the list follows:

```
1. get a value for n
2. get a value for L₁, … , Lₙ
3. set the value of smallest to L₁
4. set the value of i to 2
5. while i <= n do
        a. if Lᵢ < smallest then
             i.  set the value of smallest to Lᵢ
        b. set the value of i to i + 1
6. end of the loop
7. return the value of smallest
```

In the current example, we set the list of the 10 numbers by using the Form Constant Byte (FCB) directive similar to the previous two examples. This list is located such that its base-address is $0000. This is accomplished by putting the FCB next to the ORG DATA statement. The ORG DATA is the origin of the data section in the program because DATA is set by the Equate (EQU) directive to be $0000. The ORG MAIN provides the origin of the program from $0140 (alias as MAIN) memory location.

```
*****************************************************
* Program Name: Smallest.asm
* Objective: Find smallest number from a list
* Usage: Buffalo Call 140 command
* Output: End of the list ($000A)
*****************************************************
MAIN   EQU  $0140        ; Equate MAIN
DATA   EQU  $0000        ; Equate DATA
END    EQU  DATA+$0A     ; Equate END

       ORG  DATA         ; Data origin
* Populate an array with ten numbers
ARRAY FCB   $0C,$0E,$10,$12,$14,$02,$04,$06,$08,$0A

       ORG  MAIN         ; Program origin
       LDAA #$FF         ; Set reference
       LDX  #DATA        ; Initialize base address
LOOP:  CMPA $00,X        ; Compare number to reference
       BLO  SKIP         ; If reference is lower, don't replace
       LDAA $00,X        ; Else, replace the reference
```

```
SKIP   INX              ; Next item in list
       CPX  #END        ; Reached End?
       BNE  LOOP        ; If not, keep looping
       STAA $00,X       ; Store smallest at the list end
       SWI
```

One after another, each number stored in the list will be compared to the reference. If the value of the reference is lower than that number of the list, the loop will continue to the next value in the list. Otherwise, if the value of the reference is higher or equal to the number of the list, the reference is replaced by that number of the list. This algorithm ensures that, at any given time, the reference maintains the smallest value of the portion of the list traversed. Therefore, by the time we finish traversing our list, the final reference will contain the smallest value from the entire list.

We first set a reference (AccA) to a very large value ($FF). The $FF (or 1111 1111$_2$) is the maximum value that AccA can hold. As in the previous examples, we start by initializing the reference (AccA = $FF), base-address (register X = $0000). The instruction CMPA is used to compare the number from the list to the reference. As soon as the compare is complete, the flags in the CCR are updated with the appropriate CCR flag values. To find out if the reference is smaller, the BLO instruction is used, which performs this test and branches if the reference is smaller than the number from the list that is currently being compared.

In case of branching (done through the SKIP label), the base address is incremented next, and it is checked if the list is finished. This is done by comparing the value of the index register X to the address following the list. The program uses a CPX instruction to compare the value in the X register to the address following the list. If the value of the index register X is smaller than the address $000A, the Z flag is cleared for the BNE instruction, and the program loops back and continues its comparison of reference to the list elements. When the address equals $000A, the Z flag is set, and the BNE conditional test fails. Thus, the program execution comes out of the loop and stores the reference at the end of the list. On the other hand, if the reference is NOT smaller than the number of the list under comparison, the branch test fails and the program will not branch. The program will fall through to the next instruction where the reference is replaced by that number using the statement "LDAA $00 X." In this way the reference maintains its status of keeping the lowest value of the list traversed up to that point of time. The program flow is shown in Figure 7.13.

As observed in Figure 7.14, the list is populated from address $0000 onward. The smallest number is stored at the end of the list. The program to find the greatest number will be left for the students as an end-of-chapter problem.

Self-Learning: A selection algorithm is an algorithm for finding the *k*th smallest number in a list. Perform Internet research and find out more about a selection algorithm.

Example 7.7

Using a subroutine, write a program to determine the square of a number stored at the memory location $0000.

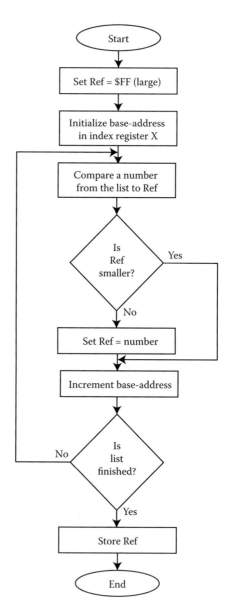

FIGURE 7.13
Program flow to find the smallest number.

```
>MD 0000

0000  0C 0E 10 12 14 02 04 06 08 0A FF FF FF FF FF FF
0010  FF FF FF FF FF FF FF FF FF FF FF FF FF FF FF FF
0020  FF FF FF FF FF FF FF FF FF FF FF FF FF FF FF FF
0030  FF FF E4 E4 E4 6D E3 D0 00 00 00 0B 00 0A 01 5B
0040  FA 29 FF E4 E4 E4 6D E3 D4 00 E4 6D E3 E4 E4 6D E3
0050  D4 00 E4 6D E3 D4 00 57 07 20 E5 02 E8 10 E1 AA
0060  01 5A 00 0A 00 0B 00 00 D0 00 41 4D 44 20 30 30
0070  30 30 0D 0D 31 38 38 43 30 30 30 41 32 36 46 35
0080  34 46 43 45 30 30 30 41 30 32 43 44 45 46 4D 44
>CALL 140

P-0153 Y-000A X-000A A-02 B-00 C-D0 S-0041
>MD 0000

0000  0C 0E 10 12 14 02 04 06 08 0A 02 FF FF FF FF FF
0010  FF FF FF FF FF FF FF FF FF FF FF FF FF FF FF FF
0020  FF FF FF FF FF FF FF FF FF FF FF FF FF FF FF FF
0030  FF FF E4 E4 E4 6D E3 D0 00 02 00 0A 00 0A 01 54
0040  FA 29 FF E4 E4 E4 6D E3 D4 00 E4 6D E3 E4 E4 6D E3
0050  D4 00 E4 6D E3 D4 00 57 07 20 E5 02 E8 10 E1 AA
0060  01 53 00 0A 00 0A 02 00 D0 00 41 4D 44 20 30 30
0070  30 30 0D 0D 31 38 38 43 30 30 30 41 32 36 46 35
0080  34 46 43 45 30 30 30 41 30 32 43 44 45 46 4D 44
```

FIGURE 7.14
Sample execution to find the smallest number.

SOLUTION

Since by now we know quite well about the assembler directives, we will focus more on the code components of the program. We first look into finding out how the square can be accomplished. The square of a number is a multiplication of that number by itself. The MUL instruction multiplies the contents of AccA to AccB and stores it in register D. Therefore, if we can create a duplicate of the value of AccA in AccB, then, just by simply using the MUL instruction, we can get the square of a number stored in AccA. So, how do we duplicate the value of AccA to AccB? The answer to this question can be found in the instruction TAB. The instruction transfers the contents of AccA into AccB. This copying of the contents of AccA into AccB does not affect the contents of AccA. Since the subroutine has to return back to the calling program, an RTS will be needed at the end of the subroutine.

```
*  ***************************************************
*  Program Name: Square.asm
*  Objective: Find Square using a subroutine
*  Usage: Buffalo Call 140 command
*  Output: Square is saved at location OUT
*  ***************************************************
```

```
MAIN   EQU   $0140      ; Equate MAIN
DATA   EQU   $0000      ; Equate DATA
OUT    EQU   DATA+$0A   ; Equate OUT

       ORG   DATA       ; Data origin
* Store a number
NUM    FCB   $02

       ORG   MAIN       ; Program origin
       LDX   #DATA      ; Initialize source base-address
       LDAA  $00,X      ; Get a number
       JSR   SQUR       ; Call subroutine SQUR
       STD   OUT        ; Save the answer
       SWI

SQUR   TAB              ; Duplicate value (A) to B
       MUL              ; Square value, (A) X (B) to D
       RTS              ; Return to calling program
```

The main program just loads the number to AccA and calls the subroutine. When the program is back from the subroutine, accessing register D fetches the result of the MUL instruction. We have used STD to save the contents of register D to the required memory location. The program flow is shown in Figure 7.15.

Team Discussion: Can the program given in Example 7.7 be used to find the square of the number $FF?

As observed in Figure 7.16, the number is located at the address $0000. The square is stored next to that number, that is, $0001. The program to find the square of each of the 10 numbers that are stored in sequential memory locations is our next endeavor.

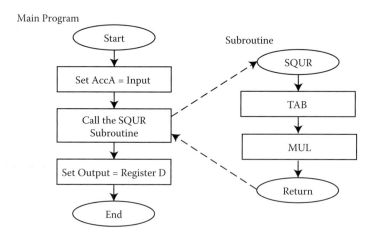

FIGURE 7.15
Program flow to find the square of a number.

```
>MD 0000

0000 02 FF FF FF FF FF FF FF FF FF FF FF FF FF FF FF
0010 FF FF FF FF FF FF FF FF FF FF FF FF FF FF FF FF
0020 FF FF FF FF FF FF FF FF FF FF FF FF FF FF FF FF
0030 FF FF E4 E4 E4 6D E3 D0 04 00 00 00 00 0A 01 4B
0040 FA 29 FF E4 E4 6D E3 D4 00 E4 6D E3 E4 E4 6D E3
0050 D4 00 E4 6D E3 D4 00 57 07 20 E5 02 E8 10 E1 AA
0060 01 4A 00 0A 00 00 00 04 D0 00 41 4D 44 20 30 30
0070 30 30 0D 30 46 43 0A 0D 53 31 31 31 30 31 34 30
0080 43 45 30 30 30 30 41 36 30 30 42 44 30 31 4D 44
>CALL 140

P-014A Y-000A X-0000 A-00 B-04 C-D0 S-0041
>MD 0000

0000 02 FF FF FF FF FF FF FF FF FF 00 04 FF FF FF FF
0010 FF FF FF FF FF FF FF FF FF FF FF FF FF FF FF FF
0020 FF FF FF FF FF FF FF FF FF FF FF FF FF FF FF FF
0030 FF FF E4 E4 E4 6D E3 D0 04 00 00 00 00 0A 01 4B
0040 FA 29 FF E4 E4 6D E3 D4 00 E4 6D E3 E4 E4 6D E3
0050 D4 00 E4 6D E3 D4 00 57 07 20 E5 02 E8 10 E1 AA
0060 01 4A 00 0A 00 00 00 04 D0 00 41 4D 44 20 30 30
0070 30 30 0D 0D 46 43 0A 0D 53 31 31 31 30 31 34 30
0080 43 45 30 30 30 30 41 36 30 30 42 44 30 31 4D 44
```

FIGURE 7.16
Sample execution to find the square of a number.

Example 7.8

Using the subroutine in Example 7.7, write a program to find the square of each of the 10 numbers that are stored in sequential memory locations. Note that the numbers are 8 bits, but the squares are 16-bit values.

SOLUTION

We will use the subroutine SQUR developed in the previous example. Also, by now we know how to navigate in a list using the instructions LDX, LDY, INX, INY, etc.

```
* ************************************************
* Program Name: SqSeries.asm
* Objective: Find Square of a series using a subroutine
* Usage: Buffalo Call 140 command
* Output: Square is saved at location OUT
**************************************************
MAIN   EQU  $0140      ; Equate MAIN
DATA   EQU  $0000      ; Equate DATA
END    EQU  DATA+$0A   ; Equate END
OUT    EQU  $0010      ; Equate OUT
```

```
        ORG   DATA      ; Data origin
* Populate an array with ten numbers
NUM    FCB   $00,$01,$02,$03,$04,$05,$06,$07,$08,$09


        ORG   MAIN      ; Program origin
        LDX   #DATA     ; Initialize source base-address
        LDY   #OUT      ; Initialize destination base-address
LOOP   LDAA  $00,X     ; Get a number
        JSR   SQUR      ; Call subroutine SQUR
        STD   $00,Y     ; Save the answer
        INX             ; Increment pointer to source list
        INY             ; Increment pointer to destination list
        INY             ; Increment pointer to destination list again
        CPX   #END      ; Reached End?
        BNE   LOOP      ; If not, keep looping
        SWI


SQUR   TAB             ; duplicate value (A) into B
        MUL             ; square value, (A) x (B) into D
        RTS             ; return to calling program
```

The key in this program is to acquire the number to be loaded in the AccA before calling the SQUR subroutine. This is performed by using our usual LDX and INX instructions. Also, the test for the end of the list is something we very well know by now. But the unique side of this problem, as mentioned in the problem statement, is that the input numbers are 8 bits but the output squares are 16-bit values. Therefore, for each 8-bit number (input), we will reserve a 16-bit (output) space for storage. Since the navigation on the input side uses INX to increment the register X by 1, we will use INY two times on the output side to increment the register Y by 2. In Figure 7.17, for each INX, we have placed two INY. This ensures that the square values have appropriate space for storage. Also, in Figure 7.18, observe how the squares of the numbers in the input list (8-bits elements) starting from $0000 are saved in the output list (16-bit elements) starting from $0010. The 00 element in the input list corresponds to the 0000 element in the output list. Similarly, 01 corresponds to 0001.

Helpful Hint: Always ensure that your subroutines end with the instruction RTS.

Common Practice: In navigating through a list, the most common instructions that are used include LDX, LDY, INX, INY, etc.

Example 7.9

In the field of Digital Image Processing, a Laplace filter is a filter that seeks out points in the signal stream where the digital signal of an image passes through a pre-set "0" value and marks this out as a potential edge point. Because the signal has crossed through the point of zero, it is called a zero-crossing. We are interested in writing a program that counts the number of times a signal crosses zero. For the sake of simplicity, we assume that the signal values are stored in an array. Create a program flow and describe what kind of branch tests would be performed to accomplish this task.

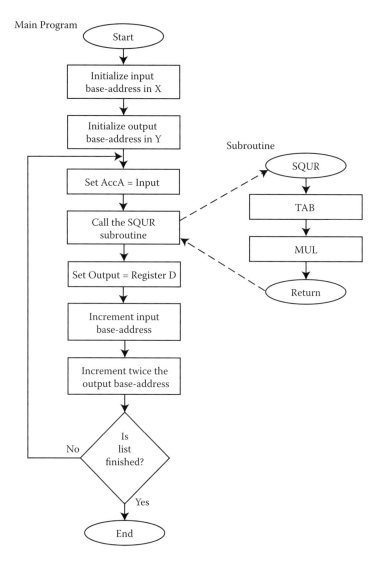

FIGURE 7.17
Program flow to find the square of numbers in a list.

```
>MD 0000

0000 00 01 02 03 04 05 06 07 08 09 FF FF FF FF FF FF
0010 FF FF FF FF FF FF FF FF FF FF FF FF FF FF FF FF
0020 FF FF FF FF FF FF FF FF FF FF FF FF FF FF FF FF
0030 FF FF E4 E4 E4 6D E3 D0 04 00 00 00 00 0A 01 4B
0040 FA 29 FF E4 E4 6D E3 D4 00 E4 6D E3 E4 E4 6D E3
0050 D4 00 E4 6D E3 D4 00 57 07 20 E5 02 E8 10 E1 AA
0060 01 4A 00 0A 00 00 00 04 D0 00 41 4D 44 20 30 30
0070 30 30 0D 0D 46 43 0A 0D 53 31 31 31 30 31 34 30
0080 43 45 30 30 30 30 41 36 30 30 42 44 30 31 4D 44
>CALL 140

P-0159 Y-0024 X-000A A-00 B-51 C-D4 S-0041
>MD 0000

0000 00 01 02 03 04 05 06 07 08 09 FF FF FF FF FF FF
0010 00 00 00 01 00 04 00 09 00 10 00 19 00 24 00 31
0020 00 40 00 51 FF FF FF FF FF FF FF FF FF FF FF FF
0030 FF FF E4 E4 E4 6D E3 D4 51 00 00 0A 00 24 01 5A
0040 FA 29 FF E4 E4 6D E3 D4 00 E4 6D E3 E4 E4 6D E3
0050 D4 00 E4 6D E3 D4 00 57 07 20 E5 02 E8 10 E1 AA
0060 01 59 00 24 00 0A 00 51 D4 00 41 4D 44 20 30 30
0070 30 30 0D 0D 46 43 0A 0D 53 31 31 31 30 31 34 30
0080 43 45 30 30 30 30 41 36 30 30 42 44 30 31 4D 44
```

FIGURE 7.18
Sample execution to find the square of numbers in a list.

SOLUTION

We will use AccB as a variable that would count the number of occurrences of zeros in the list. Let us call this a counter.

```
********************************************************
* Program Name: Zeros.asm
* Objective: Find the number of zeros in a list
* Usage: Buffalo Call 140 command
* Output: End of the list ($000A)
********************************************************
MAIN   EQU   $0140        ; Equate MAIN
DATA   EQU   $0000        ; Equate DATA
END    EQU   DATA+$0A     ; Equate END


       ORG   DATA         ; Data origin
* Populate an array with ten numbers
ARRAY FCB   $02,$04,$00,$08,$00,$0C,$0E,$00,$12,$14


       ORG   MAIN         ; Program origin
       CLRB               ; Clear counter
       LDX   #DATA        ; Initialize base address
```

```
LOOP   LDAA   $00,X        ; Get a number from the list
       BNE    SKIP         ; If not zero, don't count
       INCB                ; Increment counter
SKIP   INX                 ; Increment base address
       CPX    #END         ; Reached End?
       BNE    LOOP         ; If not, keep looping
       STAB   $00,X        ; Store counter at the list end
       SWI
```

We first set this counter to zero and initialize the base-address. We then load a number from the list to the AccA by using the LDAA. This operation updates the CCR flags to indicate the sign of the number so that the test can be immediately executed following the load. To find out if the contents of AccA is zero, a BNE instruction can be used that can perform this test and can branch if the contents of AccA is not zero.

In the case of branching (done through the SKIP label), the base-address is incremented next. The test to see if the list is finished is now done. This is performed by comparing the value of the index register X with the last memory location of the list. The program uses a CPX instruction to compare the value in the index register X to the address following the list. If the value in the X register is smaller than the address $000A, the Z flag is cleared for the BNE instruction, and the program loops back and continues its comparison of reference to the list elements. When the address equals $000A, the Z flag is set, and the BNE conditional test fails. Thus, the program execution comes out of the loop and stores the counter to the end of the list. On the other hand, if the contents of AccA is zero, the branch test fails and the program will not branch. The program will fall through to the next instruction where the counter is incremented to register that a zero number has appeared in the list. In this way, the counter maintains its status of keeping the number of zeros in the list traversed up to that point of time. The program flow is shown in Figure 7.19.

As observed in Figure 7.20, the list is populated from address $0000 onward. The number of zero found in the list is stored at the end of the list. The program to find the number of positive and negative numbers will be left for the students as end-of-chapter problems.

Team Discussion: Is the program given in Example 7.9 following the UNTIL-DO structure or the DO-WHILE structure? What is the key in identifying the looping structure being used in a program?

Example 7.10

In signal processing, the reverse effect is useful in creating special effects where a waveform may be reversed from right to left so it plays backward. We are interested in reversing a block of data stored at some memory location. Write a simple program to move a list of 10 numbers stored from $0000 to $0009 to memory location $0019 to $0010.

SOLUTION

By now we have seen how to initialize the base-address and how to check if the list has reached its end. We also have seen many instances of loading and storing of data from one location to another.

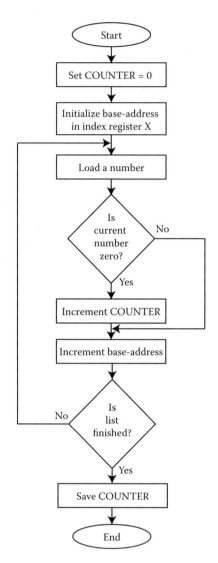

FIGURE 7.19
Program flow for a zero-crossing algorithm.

```
* ****************************************************
* Program Name: RevCopy.asm
* Objective: Copy block of memory in reverse order
* Usage: Buffalo Call 140 command
* Output: Block of memory is copied in reverse order
* ****************************************************
MAIN   EQU   $0140      ; Equate MAIN
DATA   EQU   $0000      ; Equate DATA
```

```
>MD 0000

0000 02 04 00 08 00 0C 0E 00 12 14 FF FF FF FF FF FF
0010 FF FF FF FF FF FF FF FF FF FF FF FF FF FF FF FF
0020 FF FF FF FF FF FF FF FF FF FF FF FF FF FF FF FF
0030 FF FF E4 E4 E4 6D E3 D4 51 00 00 0A 00 24 01 5A
0040 FA 29 FF E4 E4 6D E3 D4 00 E4 6D E3 E4 E4 6D E3
0050 D4 00 E4 6D E3 D4 00 57 07 20 E5 02 E8 10 E1 AA
0060 01 59 00 24 00 0A 00 51 D4 00 41 4D 44 20 30 30
0070 30 30 0D 0D 46 43 0A 0D 53 31 31 31 30 31 34 30
0080 43 45 30 30 30 30 41 36 30 30 42 44 30 31 4D 44
>CALL 140

P-0151 Y-0024 X-000A A-14 B-03 C-D0 S-0041
>MD 0000

0000 02 04 00 08 00 0C 0E 00 12 14 03 FF FF FF FF FF
0010 FF FF FF FF FF FF FF FF FF FF FF FF FF FF FF FF
0020 FF FF FF FF FF FF FF FF FF FF FF FF FF FF FF FF
0030 FF FF E4 E4 E4 6D E3 D4 D0 03 14 00 0A 00 24 01 52
0040 FA 29 FF E4 E4 6D E3 D4 00 E4 6D E3 E4 E4 6D E3
0050 D4 00 E4 6D E3 D4 00 57 07 20 E5 02 E8 10 E1 AA
0060 01 51 00 24 00 0A 14 03 D0 00 41 4D 44 20 30 30
0070 30 30 0D 0D 46 43 0A 0D 53 31 31 31 30 31 34 30
0080 43 45 30 30 30 30 41 36 30 30 42 44 30 31 4D 44
```

FIGURE 7.20

Sample execution for a zero-crossing algorithm.

```
END    EQU   DATA+$0A    ; Equate END
NDATA  EQU   $0010       ; Equate NDATA
NEND   EQU   NDATA+$09   ; Equate NEND


       ORG   DATA        ; Data origin
* Populate an array with ten numbers
ARRAY  FCB   $02,$04,$06,$08,$0A,$0C,$0E,$10,$12,$14


       ORG   MAIN        ; Program origin
       LDX   #DATA       ; Initialize source base-address
       LDY   #NEND       ; Initialize destination base-address
LOOP   LDAB  $00,X       ; Get a number from source list
       STAB  $00,Y       ; Put the number in destination list
       INX               ; Increment pointer of source list
       DEY               ; Decrement pointer of destination list
       CPX   #END        ; Reached End?
       BNE   LOOP        ; If not, keep looping
       SWI
```

The program flow for reverse block copy is shown in Figure 7.21. The key point in this example is that the initialization of the base-address for the source list is $0000, while the initialization of the base-address for the destination list is $0019. The source list needs its pointer (index register X) to be incremented every time a copy occurs, whereas the destination list requires its pointer (index register Y)

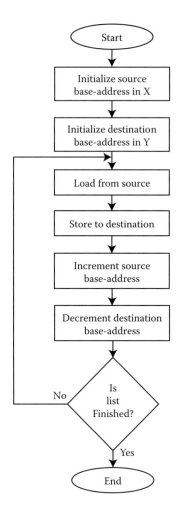

FIGURE 7.21
Program flow for reverse copy.

to be decremented for each such copy. Hence we use DEY to decrement the Y register.

With 10 iterations, the program terminates such that the X and Y registers will contain $000A and $000F, as shown in Figure 7.22.

Team Discussion: If the program in Example 7.10 is written using AccA instead of AccB, will it produce the same results? How can this program be written using other HC11 instructions like TAB etc?

Example 7.11

The Fibonacci series

0, 1, 1, 2, 3, 5, 8, 13, 21, ...

```
>MD 0000

0000 02 04 06 08 0A 0C 0E 10 12 14 FF FF FF FF FF FF
0010 FF FF FF FF FF FF FF FF FF FF FF FF FF FF FF FF
0020 FF FF FF FF FF FF FF FF FF FF FF FF FF FF FF FF
0030 FF FF E4 E4 E4 6D E3 D0 03 14 00 0A 00 24 01 52
0040 FA 29 FF E4 E4 6D E3 D4 00 E4 6D E3 E4 E4 6D E3
0050 D4 00 E4 6D E3 D4 00 57 07 20 E5 02 E8 10 E1 AA
0060 01 51 00 24 00 0A 14 03 D0 00 41 4D 44 20 30 30
0070 30 30 0D 0D 46 43 0A 0D 53 31 31 31 30 31 34 30
0080 43 45 30 30 30 30 41 36 30 30 42 44 30 31 4D 44
>CALL 140

P-0154 Y-000F X-000A A-14 B-14 C-D4 S-0041
>MD 0000

0000 02 04 06 08 0A 0C 0E 10 12 14 FF FF FF FF FF FF
0010 14 12 10 0E 0C 0A 08 06 04 02 FF FF FF FF FF FF
0020 FF FF FF FF FF FF FF FF FF FF FF FF FF FF FF FF
0030 FF FF E4 E4 E4 6D E3 D4 14 14 00 0A 00 0F 01 55
0040 FA 29 FF E4 E4 6D E3 D4 00 E4 6D E3 E4 E4 6D E3
0050 D4 00 E4 6D E3 D4 00 57 07 20 E5 02 E8 10 E1 AA
0060 01 54 00 0F 00 0A 14 14 D4 00 41 4D 44 20 30 30
0070 30 30 0D 0D 46 43 0A 0D 53 31 31 31 30 31 34 30
0080 43 45 30 30 30 30 41 36 30 30 42 44 30 31 4D 44
```

FIGURE 7.22
Sample execution of reverse copy.

occurs in nature and, in particular, describes a form of spiral. The ratio of successive Fibonacci numbers converges on a constant value of 1.618…. This number, too, repeatedly occurs in nature and has been called the golden ratio or the golden mean. Write a program to create the Fibonacci series with the first Fibonacci element at address $0000 and the tenth element of the series at $0009.

SOLUTION

The Fibonacci series begins with 0 and 1 and has the property that each subsequent Fibonacci number is the sum of the previous two Fibonacci numbers. Ancient Greek mathematicians first studied what we now call the golden ratio because of its frequent appearance in geometry. Humans tend to find the golden mean aesthetically pleasing. Architects often design windows, rooms, and buildings whose length and width are in the ratio of the golden mean. Some studies of the Acropolis, including the Parthenon, conclude that many of its proportions approximate the golden ratio. Postcards, playing cards, posters, wide-screen televisions, photographs, and light-switch plates are often designed with a golden mean length-to-width radio.

The Fibonacci series can be generated using both iteration and recursion. They both are based on a control structure: iteration uses a repetition structure; recursion uses a selection structure. Both iteration and recursion involve repetition: Iteration explicitly uses a repetition structure; recursion achieves repetition through repeated subroutine calls. Although this series can be generated by recursion, but because recursion is expensive for the processor and memory space assignment

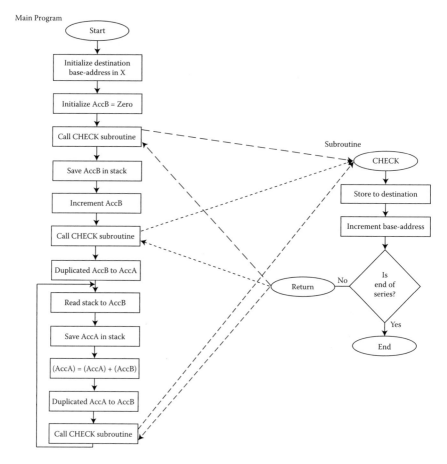

FIGURE 7.23
Program flow for the Fibonacci series.

due to the overhead of subroutine calls, we will use an iterative approach to create the first 10 elements of this series. The program flow is shown in Figure 7.23.

```
* ****************************************************
* Program Name: Fibonacci.asm
* Objective: Create a Fibonacci Series
* Usage: Buffalo Call 140 command
* Output: Saved at location OUT onwards
* ****************************************************
MAIN    EQU    $0140        ; Equate MAIN
OUT     EQU    $0000        ; Equate DATA
END     EQU    DATA+$0A     ; Equate END

        ORG    MAIN         ; Program origin
        LDX    #OUT         ; Initialize destination base-address
        CLRB                ; Initialize AccB
```

```
        JSR    CHECK        ; First element
        PSHB                ; Save AccB in Stack
        INCB                ; Increment AccB
        JSR    CHECK        ; Second element
        TBA                 ; Duplicate value (AccB) to AccA
FIB     PULB                ; Get the stack value to AccB
        PSHA                ; Save AccA in Stack
        ABA                 ; Add AccA and AccB, store in AccA
        TAB                 ; Duplicate value (AccA) to AccB
        JSR    CHECK        ; Next element
        BRA    FIB          ; Keep looping
QUIT    SWI

CHECK   STAB   $00,X        ; Save Fibonacci element
        INX                 ; Increment for next position
        CPX    #END         ; Reached End?
        BEQ    QUIT         ; If end, quit
        RTS                 ; Return to calling program
```

Similar to the previous programs, we first initialize the register X with the base-address and clear the AccB. To make our code reusable, we create a subroutine called CHECK. The purpose of this subroutine is to do three things:

1. Store the current Fibonacci element.
2. Increment the pointer for the next position.
3. Check if the program should finish the series at that element.

If the program finds out that 10 Fibonacci elements have been saved, it will quit. So, when we call this subroutine for the first time, it will save the first Fibonacci element. Next, we save the contents of AccB, which is the current Fibonacci element, to the stack using the PSHB instruction. This is followed by incrementing AccB and calling the subroutine CHECK again. The second Fibonacci element is saved now. Next, we duplicate the contents of AccB in AccA. In order to duplicate the contents of AccB in AccA, we use the TBA instruction. In this way, we have our current Fibonacci element now in AccA. Remember that the previous Fibonacci element is still saved in the stack. The next few steps combine to form a loop such that the loop starts with a label FIB. To use the previous element, we now get back the data from the stack using the PULB instruction. The data (previous Fibonacci element) in the stack are now stored in AccB. To save the current Fibonacci element in the stack, we will use PSHA. This saves the contents of AccA in the stack. When the contents of AccA (current Fibonacci element) and AccB (previous Fibonacci element) are added together using ABA instruction, the summation is saved in AccA. Note that the summation has replaced the previous data stored in AccA, which actually was the current Fibonacci element. But since we have already stored it in the stack, we do not have to worry about its retrieval. Next, we want to save the new element of the Fibonacci series that has been obtained by summing the last two elements. This is done by calling the subroutine CHECK. But before saving the new element of the Fibonacci series, we must save it to AccB. Why AccB? AccB is chosen because it is the means to store the Fibonacci element to the next memory location. The duplication is done

```
>MD 0000

0000 FF FF FF FF FF FF FF FF FF FF FF FF FF FF FF FF
0010 FF FF FF FF FF FF FF FF FF FF FF FF FF FF FF FF
0020 FF FF FF FF FF FF FF FF FF FF FF FF FF FF FF FF
0030 FF FF E4 E4 E4 6D E3 D4 14 14 00 0A 00 0F 01 55
0040 FA 29 FF E4 E4 6D E3 D4 00 E4 6D E3 E4 E4 6D E3
0050 D4 00 E4 6D E3 D4 00 57 07 20 E5 02 E8 10 E1 AA
0060 01 54 00 0F 00 0A 14 14 D4 00 41 4D 44 20 30 30
0070 30 30 0D 30 30 38 38 43 30 30 30 41 32 37 46 37
0080 33 39 42 39 0A 53 30 33 30 30 30 30 46 43 4D 44
>CALL 140

P-0156 Y-000F X-000A A-22 B-22 C-F4 S-003E
>MD 0000

0000 00 01 01 02 03 05 08 0D 15 22 FF FF FF FF FF FF
0010 FF FF FF FF FF FF FF FF FF FF FF FF FF FF FF FF
0020 FF FF FF FF FF FF FF FF FF FF FF FF FF FF FF FF
0030 FF FF E4 E4 F4 22 22 00 0A 00 0F 01 57 01 54 15
0040 FA 29 FF E4 E4 6D E3 D4 00 E4 6D E3 E4 E4 6D E3
0050 D4 00 E4 6D E3 D4 00 57 07 20 E5 02 E8 10 E1 AA
0060 01 56 00 0F 00 0A 22 22 F4 00 3E 4D 44 20 30 30
0070 30 30 0D 0D 30 38 38 43 30 30 30 41 32 37 46 37
0080 33 39 42 39 0A 53 30 33 30 30 30 30 46 43 4D 44
```

FIGURE 7.24
Sample execution for the Fibonacci series.

by TAB instruction. When the program returns from the subroutine CHECK, a BRA instruction is used to send it back to the beginning of the loop. A sample program execution to generate the first few elements of the Fibonacci series is shown in Figure 7.24.

Helpful Hint: Notice that Fibonacci tend to become large quickly. Therefore, creat only the first few elements of this series i your program.

Example 7.12

The following is a program that takes a character as an input from the keyboard. It then checks to see if it belongs to A to Z or a to z alphabets. The program outputs the case type to the screen. So, if the input is 'P' or 'r,' the output on the screen is 'Upper-case' or 'Lower-case,' respectively. You are required to provide appropriate comments to this program.

```
MAIN      EQU    $0140
DATA      EQU    $0000
OUT       EQU    $0010
INCHAR    EQU    $FFCD
OUTA      EQU    $FFB8
OUTSTRG   EQU    $FF

          ORG    DATA
MSGL      FCC    'Lower-case:'
          FCB    $04
```

```
MSGU      FCC    'Upper-case:'
          FCB    $04


          ORG    MAIN
          JSR    INCHAR
          JSR    IFLCASE
          JSR    IFUCASE
          SWI


IFUCASE   CMPA   #$5A
          BHI    SKIP
          CMPA   #$41
          BLO    SKIP
          LDX    #MSGU
          JSR    OUTSTRG
SKIP      RTS


IFLCASE   CMPA   #$7A
          BHI    PASS
          CMPA   #$61
          BLO    SKIP
          LDX    #MSGL
          JSR    OUTSTRG
PASS      RTS
```

SOLUTION

As per our earlier discussion regarding the comments, the program follows with in-line and full line comments. A sample execution of program in Example 7.12 is shown in Figure 7.25.

Common Practice: Appropriate program comments are extremely important. Program comments have long been used as a common practice for improving interprogrammer communication and code readability by explicitly specifying the programmers' intentions and assumptions.

```
* ***********************************************************
* Program Name: IOAlpha.asm
* Objective: Inputs a char from user, checks if it is alpha.
*        Outputs its case type to screen.
* Usage: Buffalo Call 140 command
* Output: On screen
* ***********************************************************
MAIN      EQU    $0140          ; Equate MAIN
DATA      EQU    $0000          ; Equate DATA
OUT       EQU    $0010          ; Equate OUT
INCHAR    EQU    $FFCD          ; Jump table address
OUTA      EQU    $FFB8          ; Jump table address
OUTSTRG   EQU    $FFC7          : Jump table address

          ORG    DATA           ; Data origin
MSGL      FCC    'Lower-case:'  ; Message to display
          FCB    $04            ; EOT Character - Delimiter
MSGU      FCC    'Upper-case:'  ;Message to display
          FCB    $04            ; EOT Character - Delimiter
```

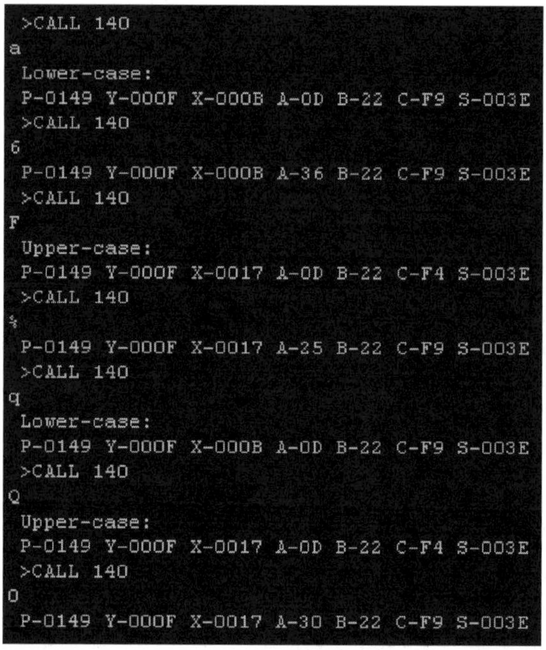

```
>CALL 140
a
 Lower-case:
 P-0149 Y-000F X-000B A-0D B-22 C-F9 S-003E
 >CALL 140
 6
 P-0149 Y-000F X-000B A-36 B-22 C-F9 S-003E
 >CALL 140
F
 Upper-case:
 P-0149 Y-000F X-0017 A-0D B-22 C-F4 S-003E
 >CALL 140
 ;
 P-0149 Y-000F X-0017 A-25 B-22 C-F9 S-003E
 >CALL 140
q
 Lower-case:
 P-0149 Y-000F X-000B A-0D B-22 C-F9 S-003E
 >CALL 140
Q
 Upper-case:
 P-0149 Y-000F X-0017 A-0D B-22 C-F4 S-003E
 >CALL 140
O
 P-0149 Y-000F X-0017 A-30 B-22 C-F9 S-003E
```

FIGURE 7.25
Program execution for Example 7.12.

```
            ORG    MAIN           ; Program origin
            JSR    INCHAR         ; Input character to AccA
            JSR    IFLCASE        ; Call subroutine H2A
            JSR    IFUCASE        ; Call subroutine H2A
            SWI                   ; End of program

IFUCASE     CMPA   #$5A           ; Compare AccA with #$5A?
            BHI    SKIP           ; If greater, then return
            CMPA   #$41           ; Compare AccA with #$41?
            BLO    SKIP           ; If smaller, then return
            LDX    #MSGU          ; Display message MSGU
            JSR    OUTSTRG        ; Call OUTSTRG
SKIP        RTS                   ; Return to calling program

IFLCASE     CMPA   #$7A           ; Compare AccA with #$7A?
            BHI    PASS           ; If greater, then return
            CMPA   #$61           ; Compare AccA with #$61?
            BLO    SKIP           ; If smaller, then return
            LDX    #MSGL          ; Display message MSGL
            JSR    OUTSTRG        ; Call OUTSTRG
PASS        RTS                   ; Return to calling program
```

Section 7.4 Review Quiz

What do the following instructions stand for? (a) RTS, (b) BRA, (c) INX

7.5 Summary

1. A "Hello World!" program is a simple program that prints out "Hello World!" on a display device such as a computer monitor.
2. To create an assembly program for the HC11, a programmer can use his or her favorite text editor.
3. The standard suffix for assembly programs is .ASM (or .asm).
4. The assembler in HC11 assumes fixed fields in the source file.
5. Anything starting in the first column is assumed to be a label; the first word not in column one is assumed to be an instruction mnemonic or assembler directive. The third field contains the operands, if any, and the fourth contains comments.
6. All fields are separated by one or more spaces or tab characters (or white space).
7. Assembler directives are instructions entered into the source code along with the assembly language. These directives do not get translated into object code but are used as special instructions to the assembler to perform some special functions.
8. The Origin (ORG) statement is used by the programmer when it is necessary to place the program in a particular location in memory.
9. The Equate (EQU) directive is an assembler directive that is used to assign the data value or address in the operand field to the label field.
10. Expressions can be used at multiple places throughout the code.
11. Programmers insert comments to document programs and improve program readability.
12. The assembler only resolves symbol and label names to eight characters.
13. FCC converts each character of the message to the corresponding ASCII character codes.
14. BUFFALO utility subroutines are quite frequently used in HC11 programs. For example, the OUTSTRG subroutine outputs the message to the terminal and stops at $04.
15. During assembly, if there are no errors in the source code, two output files will be created: (1) a Motorola S-Record (hex) format file that can be programmed into memory, (2) a common listing file (or list file) that shows the relationship between source and output.
16. The list file is a text file that shows two things: (1) the source code instructions and their respective translation into hex, (2) any error messages that were found during assembly.
17. After successfully assembling the source code and creating an S-Record file, the S-Record file is "uploaded" to the development board for a test run.

Glossary

Assemble: The processes of creating object code done by an assembler by translating assembly instruction mnemonics into opcodes and by resolving symbolic names for memory locations and other entities.

Assembler Directive: An assembler directive is a message to the assembler that informs the assembler information it requires recognizing in order to execute the assembly process.

Breakpoint: A place in a program where the processor will suspend execution and branch to a special routine that allows the operation to examine the processor registers.

Checksum: A checksum is a fixed-size datum computed from an arbitrary block of digital data for the purpose of detecting accidental errors that may have been introduced during its transmission or storage.

Comments: The descriptive statements intermixed with source code that are used to document the logic of a program.

Debugging: Procedures for checking a program for logic and syntax errors.

DOS: Disk Operating System.

EVBU: A Motorola development system board that allows the user to test and debug software to be used in a 68HC11-based application.

Executable Code: Software (in machine language) in a form that can be run in the computer.

Execute: To run a program, which causes the computer to carry out its instructions.

Graphical User Interface (GUI): Often pronounced gooey, it is a type of user interface that allows users to interact with programs in more ways than typing such as computers.

Hands-on: Involving practical experience of equipment, etc.

Hello World Program: A "Hello World" program is a computer program that prints out "Hello World!" on a display device.

Integrated Development Environment (IDE): A software application that provides comprehensive facilities to computer programmers for software development.

Instruction: Commands that control the actions of the processor.

Linker: A utility program that links a compiled or assembled program to a particular environment.

List File: A file that typically shows the relationship between source and output.

MAKE: MAKE is a utility that automatically builds executable programs and libraries from source code by reading files called Make-files. These Make-files specify how to derive the target program.

S-Record File: Hex format file that can be programmed into memory.

Single Stepping: A debugging procedure whereby the processor executes an instruction and halts until it gets a command to continue to the next instruction. In this way, the register and memory contents can be examined by the operator after each execution.

Software Development Life Cycle Model: Description of phases of the software cycle and the order in which those phases are executed.

User Interface (UI): A place where interaction between humans and machines occurs.

Answers to Section Review Quiz

7.2 (a) True, (b) False, (c) Comments

7.3 True

7.4 (a) Return to subroutine, (b) branch always, (c) increment register X

True/False Quiz

1. Configuring a full programming tool-chain from scratch to the point where programs can be compiled or assembled and run is always a trivial task.

2. HC11 assembly programs typically have the extension of .ASM (or .asm).

3. A white space has no value in assembly programs and must always be avoided in source code development.

4. Comments are ignored by the assembler.

5. An assembler recognizes directives that assign memory space, assign address to labels, format the pages of the source code, and so on.

6. The ORG assembly directive indicates the origin of the output of the program.

7. EQU represent word substitutions for values.

8. Comments do not cause the program to perform any action when the program is run.

9. FCC converts each character of the message to the corresponding BCD codes.

10. The utility subroutine OUTSTRG outputs the ASCII character codes until end-of-transmission ($04) is found.

11. The OUTRLF utility subroutine writes a newline to screen.
12. BUFFALO stands for Bit User Fast Friendly Aid to Logical Operation.
13. Even if the source code did not assemble correctly, a nonempty .S19 file will be generated.
14. The program is terminated on the HC11 EVBU with an SWI instruction.
15. A subroutine name/label can have any number of characters.

Questions

QUESTION 7.1

What is an assembler? What is the significance of using ORG as an assembler directive in HC11 source codes?

QUESTION 7.2

List some advantage of commenting in source code.

QUESTION 7.3

What is the basic difference between a pseudocode and an algorithm?

QUESTION 7.4

You assembled your source code using an assembler. Upon executing your valid program, it did not behave as intended. What could be wrong in your program?

QUESTION 7.5

What is an infinite loop?

Problems

PROBLEM 7.1

Find the errors in the following piece of code with assembler directives:

```
MAIN        EQUATE    $0140       ; Equate MAIN
DATA        EQU       $EFCX       ; Equate DATA
OUT         EQU       $0010       : Equate OUT
OUTSTRING   EQU       $FFC7       ; Jump table address
```

PROBLEM 7.2

If the following subroutine is used in a program, what would be some of the errors that the assembler would catch immediately?

```
SQUARE_SUBROUTINE    TAB     * duplicate value (A) to B
               MUL     * square value, (A) x (B) to D
               RTS     * return to calling program
```

PROBLEM 7.3

If the source code of Example 7.11 (Fibonacci series) is saved as Fibonacci.asm and assembled, what error would you get? How would you get rid of this error?

PROBLEM 7.4

The following is the main program from Example 7.6 (smallest number in a list). The programmer thought that the LDX and CPX instructions did not need a # symbol before DATA and END, respectively. What would be the problem that would result from his action?

```
      ORG    MAIN    ; Program origin
      LDAA   #$FF    ; Set reference
      LDX    DATA    ; Initialize base address
LOOP: CMPA   $00,X   ; Compare number to reference
      BLO    SKIP    ; If reference is lower, don't replace
      LDAA   $00,X   ; Else, replace the reference
SKIP  INX            ; Next item in list
      CPX    END     ; Reached End?
      BNE    LOOP    ; If not, keep looping
      STAA   $00,X   ; Store smallest at the list end
      SWI
```

PROBLEM 7.5

How would you modify Example 7.3 to output three times the character to the terminal window that was input from the keyboard? In other words, if the input is "r," the output should be "rrr."

PROBLEM 7.6

What address does the label END represent in the following assembler directive section of the program?

```
MAIN    EQU    $0140     ; Equate MAIN
DATA    EQU    $0000     ; Equate DATA
END     EQU    DATA+$0A  ; Equate END
```

PROBLEM 7.7

Find the errors in the following data section of the program.

```
ORG    $0000
MSGL         FCB    'North Pole'
             FCC    $04
MSGU         FCB    'South Pole'
             FCC    $04
```

PROBLEM 7.8

Write a program to find the largest of 10 numbers that are stored in sequential memory locations. Draw the corresponding flowchart.

PROBLEM 7.9

Write a program to count the number negatives in a list of 10 numbers that are stored in sequential memory locations. Draw the corresponding flowchart, execute the program on EVBU, and show the results.

PROBLEM 7.10

Write a program to count the positives in a list of 10 numbers that are stored in sequential memory locations. Draw the corresponding flowchart, execute the program on EVBU, and show the results.

PROBLEM 7.11

Write a program to copy a data block from one memory location to another. The source memory location is from $0000 to $0009, and the destination memory location is from $0010 to $0019. Draw the corresponding flowchart, execute the program on EVBU, and show the results.

PROBLEM 7.12

Write a program to copy only the elements that are located at the even-numbered index of a data block from one memory location to another. Note that only the second, fourth, sixth …. elements have to be copied. The source memory location is from $0000 to $0009, and the destination memory location is from $0010 to $0019.

PROBLEM 7.13

Write a program to copy a block of memory from a source memory location to a destination memory location. Any element with value zero has to be replaced by $FF in this process of copying. The source memory location is from $0000 to $0009, and the destination memory location is from $0010 to $0019.

PROBLEM 7.14

Write a subroutine to check if the data in the AccA is a single digit. If it is a single digit, then the subroutine should check if it is a digit from 0_{16} to 9_{16} or alphabet from A_{16} to F_{16}. If that is also true, then the subroutine should convert it into its ASCII equivalent.

PROBLEM 7.15

Use the subroutine developed in Problem 7.14 to convert a number into its ASCII equivalent such that the number is stored in the memory location $0000. Use the BUFFALO utility subroutine OUTA in your program.

PROBLEM 7.16

Use the subroutine developed in Problem 7.14 to convert 10 numbers in a list into their ASCII equivalent such that the source list is stored from the memory location $0000 to $0009, and the destination list will be stored from the memory location $0010 to $0019. Run the program and show the results.

PROBLEM 7.17

What are some of the assembler directives that are incorrect or missing in the program given below? Execute your program after correcting them and show your results.

```
*  ***************************************************
*  Program Name: NegCopy.asm
*  Objective: Copy Block of Memory, replace all negatives with $00
*  Usage: Buffalo Call 140 command
*  Output: Block of memory is copied with negatives replaced by $00
*  ***************************************************
DATA    EQUATE    $0000        ; Equate DATA
END     EQU       DATA+$0A     ; Equate END
NDATA   EQU       $0010        ; Equate NDATA


*  Populate an array with ten numbers
ARRAY   FCB    $FF,$FF,$FF,$08,$0A,$0C,$0E,$FF,$FF,$FF

        ORIGIN    MAIN         ; Program origin
        LDX       #DATA        ; Initialize source base-address
        LDY       #NDATA       ; Initialize destination base-address
LOOP    LDAB      $00,X        ; Get a number from source list
        BPL       COPY         ; If positive, then copy
        LDAA      #$00         ; Else, load AccA with $00
        STAA      $00,Y        ; Put $00 in destination list
        BRA       PASS         ; Branch to PASS
COPY    STAB      $00,Y        ; Put the number in destination list
PASS    INX                    ; Increment pointer of source list
        INY                    ; Increment pointer of destination list
        CPX       #END         ; Reached End?
        BNE       LOOP         ; If not, keep looping
        SWI
```

PROBLEM 7.18

Find out the main objective of the following program. Insert appropriate comments as much as you can.

```
MAIN    EQU    $0140
DATA    EQU    $0000
END     EQU    DATA+$0A
NDATA   EQU    $0010
```

```
        ORG    DATA
        FCB    $00,$F9,$F7,$00,$0A,$0C,$0E,$00,$FA,$01

        ORG    MAIN
        LDX    #DATA
        LDY    #NDATA
XYZ:    LDAB   $00,X
        BMI    PQR
        LDAA   #$FF
        STAA   $00,Y
        BRA    LMN
PQR:    STAB   $00,Y
LMN:    INX
        INY
        CPX    #END
        BNE    XYZ
        SWI
```

PROBLEM 7.19

Find out the function of the following program. Insert appropriate comments as much as you can.

```
MAIN   EQU    $0140
DATA   EQU    $0000
END    EQU    DATA+$0A
NDATA  EQU    $0010
       ORG    DATA
ARRAY  FCB    $02,$04,$06,$08,$0A,$0C,$0E,$10,$12,$14

       ORG    MAIN
       LDX    #DATA
       LDY    #NDATA
LOOP:  LDAB   $00,X
       STAB   $00,Y
       INX
       INX
       INY
       INY
       CPX    #END
       BNE    LOOP
       SWI
```

PROBLEM 7.20

What is the objective of the following program? Insert appropriate comments as much as you can.

```
MAIN     EQU    $0140
OUT      EQU    $0010
INCHAR   EQU    $FFCD
OUTA     EQU    $FFB8
```

```
            ORG    MAIN
            JSR    INCHAR
            JSR    IFNUM
            SWI
IFNUM       CMPA   #$39
            BHI    SKIP

            CMPA   #$30
            BLO    SKIP
            JSR    OUTA
SKIP        RTS
```

8

Input/Output (I/O) Ports

"We have two ears and one mouth so that we can listen twice as much as we speak."

—Epictetus (Greek Stoic philosopher)

OUTLINE

OBJECTIVES

Upon completion of this chapter, you should be able to

1. Understand the basic concept of programmed input/output ports.
2. Understand the pin assignments of Port A to Port E.
3. Write code to output data onto the Port B pins.
4. Understand the concept of the data direction register.
5. Understand the operation and pin assignments of Port C.
6. Program the data direction of the Port C bits.
7. Input and output data using Port C.
8. Program the direction of the Port D pins and use them for digital I/O.
9. Receive parallel data at the HC11 from an external device via Port E.
10. Data transfer using simple strobed mode.
11. Input data from an external system using full-input handshaking.
12. Output data to an external system using full-output handshaking.

Key Terms: Bidirectional, Data Direction Register, Data Direction Control Bit, External World, Input/Output (I/O), Outside World, Peripherals, Ports, Protocol, Handshaking, Simple Strobed Mode, Full-Input Handshaking Mode, Full-Output Handshaking Mode, Unidirectional.

8.1 Introduction

A few decades ago, there were mainframes, each shared by many people. In the last decade or so, we witnessed the personal computing era, that is, persons and machines interacting through desktops. With the popularity of mobile, hand-held, and smart devices, we have already entered the third wave of computing called Ubiquitous Computing. We rely on these equipments and devices in our lives to such an extent that one cannot imagine the world without these multifunctional gadgets. Described as "the calm technology that recedes into the background of our lives," ubiquitous computing relies on the use of tiny devices embedded in everyday objects and environments, collecting and delivering information and communicating wirelessly, and intelligently, between themselves. Such virtually imperceptible sensor networks could cooperatively monitor physical or environmental conditions such as temperature, sound, vibration, pressure, motion, or pollutants. In addition to one or more sensors, each node in such a sensor network is typically equipped with a radio transceiver or other wireless communications device, a small microcontroller, and an energy source, usually a battery. The microcontroller remains at the center point in such applications because of its connections to sensors and controls. The microcontroller collects data from the sensors and output signals to the controls to manage a machine process or the task of a device. These peripheral devices are coupled to the microcontroller via ports. Recall from Chapter 3 that the microcontroller core has internal data and address buses. Buses are just like motorways, or interstate freeways, carrying large amounts of traffic in both directions to a variety of different destinations. The microcontroller needs to be provided with a way of allowing that data flow to connect with the outside world, so that it can read in external digital values or output other values. In terms of motorways, it needs the equivalent of motorway junctions, where data can enter or leave the bus at designated times and locations. Since there are many different ways that data can be input and output, a wide range of forms exists for such junctions. The most general purpose of these is the parallel input/output (I/O) port. These ports are one of the microcontroller's most essential peripherals.

8.2 Data Transfer Mode

There are three ways of transferring data between the microcontroller and the outside world:

1. Programmed input/output
2. Interrupt-driven input/output
3. Direct memory access (DMA)

A microcontroller executes a program to communicate with an external device via a register called the input/output (I/O) ports for programmed I/O. The main goal of this chapter is to explore programmed I/O in detail. Other types of I/O will not be the focus in this chapter. Interrupt-driven I/O and direct memory access will be covered in detail in the later chapters. However, just to give readers an idea of what these two data-transferring methods are based on, we will provide few pointers here. An external device belonging to the outside world requests the microcontroller to transfer data by activating a signal on the microcontroller's interrupt line during interrupt I/O. In response, the microcontroller executes a program called the interrupt-service routine to carry out the function desired by the external device, again by way of one or more I/O ports. We will study interrupts in detail in the next chapter. Data transfer between the microcontroller's memory and an external device occurs without microcontroller involvement with direct memory access

Table 8.1 presents a summary of programmed I/O ports of HC11. There are five I/O ports: Port A, Port B, Port C, Port D, and Port E. These I/O ports are of two types. For one type, each bit in the port can be individually configured as either input or output. PORTC and PORTD are such I/O ports. For the other type, all or some bits in the port are either all parallel input or output bits. PORTB is an output port, whereas PORTE is an input port. PORTD supports 6 bits of general-purpose digital I/O. The rest of the ports support 8 bits of general-purpose digital I/O. The last row in Table 8.1 points to 11 dedicated input pins, 11 dedicated output pins, and 16 bidirectional pins in HC11. Many features from the Motorola/Freescale peripheral interface adapters (PIAs) and Asynchronous Communications Interface Adapter (ACIA) peripheral chips have been integrated into these five ports.

Self-Learning: Readers are encouraged to read Section 7 of the HC11 Reference Manual and Section 6 of the Technical Data Manual.

Section 8.2 Review Quiz

What are the three ways of transferring data between the microcontroller and the outside world?

TABLE 8.1

Summary of HC11 I/O Ports

Port	Input Pins	Output Pins	Bidirectional Pins	Total Number of Pins	Broad Area of Usage	Address	RESET
A	3	3	2	8	Timer event	PORTA: $1000	
						PACTL: $1026	00000000
B	0	8	0	8	Output only	PORTB: $1004	
							00000000
C	0	—	8	8	Programmable I/O	PORTC: $1003	
						PORTCL:$1005	
						DDRC: $1007	UUUUUUUU
D	0	—	6	6	Serial communication	PORTD: $1008	
						DDRD: $1009	00000000
E	8	—	0	8	Analog capture	PORTE: $100A	UUUUUUUU
TOTAL	11	11	16	38			

8.3 Port A

The first port in Table 8.1 is Port A. It is an 8-bit I/O port. As shown in Figure 8.1, the eight general-purpose I/O bits are labeled PA7–PA0. The bits can be grouped according to their function as:

1. Dedicated outputs (PA6, PA5 and PA4)
2. Dedicated inputs (PA2, PA1 and PA0)
3. Programmable input or output (PA7 and PA3)

The dedicated output bits are capable of sampling a signal when the micro-controller executes an input instruction. Similarly, the dedicated input bits hold the output data for an indefinite time—until the program changes it. On the other hand, the programmable bits are controlled by two data direction control bits. Here, DDRA7 and DDRA3 are the data direction control bits. A data direction control bit is a bit whose value decides if the corresponding bit of the port register will act as input or output. The DDRA7 controls the data direction of Port A b7 (PA7), and DDRA3 controls the data direction of PORTA b3 (PA3). The Pulse Accumulator Control (PACTL) register located at the memory location $1026 hosts both the DDRA3 and DDRA7 control bits. If DDRAx bit = 1, the corresponding I/O pin acts as output. Otherwise, if DDRAx bit = 0, the corresponding I/O pin acts as input. For example, if DDRA7 is set to 0 and DDRA3 is set to 1, then PA7 and PA3 will act as input and output, respectively. After reset, PORTA becomes 00000000_2, that is,

Port A

PORTA: $ 1000	b7	b6	b5	b4	b3	b2	b1	b0
	PA7	PA6	PA5	PA4	PA3	PA2	PA1	PA0
Digital I/O	In/Out	Out	Out	Out	In/Out	In	In	In
RESET	0	0	0	0	0	0	0	0

Data direction control bits DDRA7 and DDRA3

PACTL: $ 1026	b7	b6	b5	b4	b3	b2	b1	b0
	DDRA7				DDRA3			
RESET	0	0	0	0	0	0	0	0

FIGURE 8.1
Detials of Port A.

cleared. Consequently, PA3 and PA7 become input pins. We will study more about Port A in Chapter 11.

Example 8.1

Comment on the PORTA bit configuration if both the data direction control bits (PA7 and PA3) are set to 1.

SOLUTION

Common Practice: Usually for the data direction control bits, 1 denotes an output and 0 denotes an input for the corresponding I/O pin.

Helpful Hint: The Port A pins are alternatively used as the on-chip timing system.

Since both the data direction control bits (PA7 and PA3) are set to 1, the corresponding I/O bits in PORTA (i.e., PA7 and PA3) will act as output bits. We know that PA6, PA5, and PA4 are dedicated output bits, and PA2, PA1, and PA0 are dedicated input bits. Therefore, PA7 to PA3 will be output bits, whereas PA2 to PA1 will be input bits.

Section 8.3 Review Quiz

How many bidirectional pins are there in Port A?

8.4 Port B

Port B is an output port. It is one of the simplest of all the ports. Contrary to an input port such as Port E, capable of sampling a signal when the microcontroller executes an input instruction, Port B holds output data for an indefinite time—until the program changes it. This output port is therefore a latch or register that is capable of clocking data from the data bus whenever the microcontroller executes an output instruction.

Port B is composed of an 8-bit data register (PORTB) and eight Port B pins. Remember when we say PORTB, we mean the 8-bit data register. As mentioned earlier, this port outputs the digital data contained in PORTB to the output pins in parallel form. Figure 8.2 shows the internal details of Port B. The 64-byte register block holds PORTB. PORTB is accessed via read and write operations to the $1004 memory location. Pay attention to the details of this mechanism as shown in Figure 8.2. When a write instruction is performed by the processor on the PORTB, the data is output to the Port B pins. On the other hand, a read of PORTB by the processor returns the last data written to PORTB. This read has no effect on

Helpful Hint: In single-chip mode, the function of the Port B pins is limited to digital I/O. However, in expanded mode, the Port B pins are used as the upper byte of the 16-bit external address bus.

Helpful Hint: Since Port B is an output port, there are no data direction control bits associated with it.

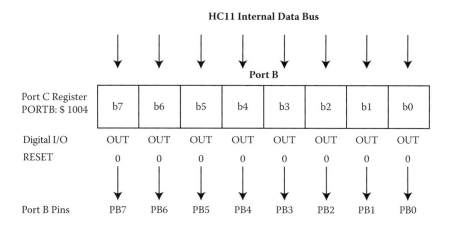

FIGURE 8.2
Details of Port B.

the contents of PORTB as well as on the data latched on the Port B pins. The Port B pins are named PB7–PB0. The PORTB and the Port B pins are cleared after reset.

Example 8.2

Write a fragment of code that will load data from the memory locations $0030 and output the result on the Port B pins.

SOLUTION

The problem statement states that the input data is located at a memory location that starts with $00. Therefore, we can use direct mode to load the data. The result can be directly stored to PORTB by writing to memory location $1004. Once the data is written to PORTB, the data will be latched on the Port B output pins.

```
LDAA    $03        ; Get data from ($0030) to AccA
STAA    $1004      ; Store AccA to ($1004)
```

Example 8.3

Use assembler directive EQU for Port B, and rewrite the fragment of code in Example 8.3.

SOLUTION

We will use the EQU directive as "PORTB EQU $1004". The assembler will replace every instance of PORTB with the value associated with it.

Best Practice: All things being equal, the more Equates you use, the easier your programs will be to debug and modify.

```
PORTB EQU    $1004       ; Substitute PORTB for $1004
      LDAA   $03         ; Get data from ($0030) to AccA
      STAA   PORTB       ; Store AccA to PORTB
```

Example 8.4

The fragment of code for Example 8.3 is written as follows:

```
PORTB EQU    $1004       ; Substitute PORTB for $1004
      LDAA   $03         ; Get data from ($0030) to AccA
      STAA   #PORTB      ; Store AccA to PORTB
```

Identify the syntax error in the code above.

SOLUTION

Since the programmer intends to store the contents of AccA at the memory location $1004, he or she must not use the # symbol. The # symbol makes no sense in this context of storing the contents of AccA.

Section 8.4 Review Quiz

How many output pins are there in Port B?

8.5 Port C

Port C is a programmable I/O port. It is one of the most useful of all the ports. Contrary to an input port such as Port E or an output port such as Port B, Port C is built such that it can output the digital data contained in the Port C register (PORTC) to the output pins or input the data from the Port C pins to the Port C register in parallel. Because of its ability to move data in both the directions, it is called a general-purpose bidirectional port.

Port C is composed of an 8-bit data register (PORTC), an 8-bit data direction register (DDRC), and eight Port C pins. Remember when we say PORTC, we mean the 8-bit data register. As mentioned earlier, this port inputs and outputs the digital data in parallel form. Figure 8.3 shows the internal details of Port C.

The Port C pins can be configured as an input or output pins by another register called the command, or data direction register (DDRC). We have seen data direction control bits in the context of Port A. Figure 8.4 illustrates the DDRC register, which is an 8-bit register in the context of Port C. PORTC contains the actual input or output data. The DDRC register is an output register and can be used to configure the bits in Port C as inputs or outputs.

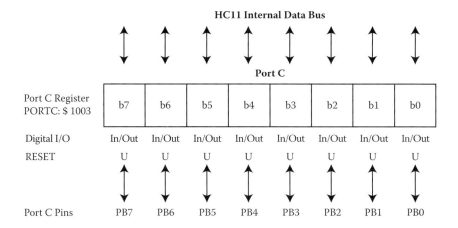

FIGURE 8.3
Details of Port C.

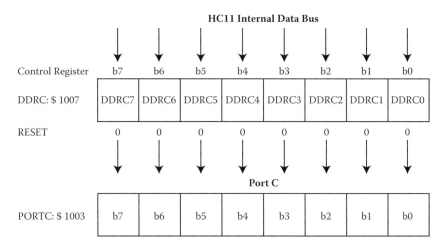

FIGURE 8.4
Details of data direction register (DDRC).

Each bit in PORTC can be set up as an input or output by writing a 0 or a 1 in the corresponding bit of the DDRC. A bidirectional buffer (one input buffer and one output buffer) is connected at each bit of PORTC. A '1' written to a particular bit in DDRC enables the output buffer, while a '0' enables the input buffer connected at the corresponding bit of the PORTC. As an example, if an 8-bit DDRC register contains $34 (i.e., 34_{HEX}), then the corresponding port is defined as shown in Figure 8.5. Since $34 (i.e., 00110100_2) is sent as an output into the DDRC register, bits 0, 1, 3, 6, and 7 of the port are set up as inputs, and bits 2, 4, and 5 of the port are defined as outputs. The microcontroller

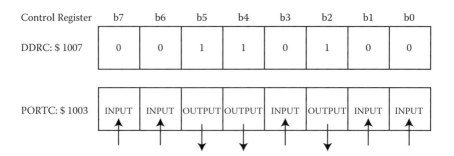

FIGURE 8.5
An example of Port C configuration.

can then send outputs to external devices, such as LEDs, connected to bits 2, 4, and 5 through a proper interface. Similarly, the microcontroller can input the status of external devices, such as switches, through bits 0, 1, 3, 6, and 7.

The 64-byte register block holds the PORTC and DDRC registers. These registers are accessed via read and write operations to the $1003 and $1007 memory locations, respectively. Pay attention to the details of the mechanism for setting the input and output as shown in Figure 8.3. When a write instruction is performed by the processor on PORTC, the data is output to the Port C pins. Note that bits that have been programmed for input operation remain unaffected by a write operation. On the other hand, a read of the PORTC returns the data present on the Port C pins that have been programmed as inputs. The last data written to the output bits of the register will be read back. This read has no effect on the contents of PORTC as well as on the data latched on the Port C pins. Port C pins are named PC7–PC0. PORTC is in an undefined state after reset. The DDRC register is cleared after reset.

Example 8.5

Configure Port C such that:

(a) All pins act as inputs
(b) All pins act as outputs
(c) All even and odd bits act as inputs and outputs, respectively

SOLUTION

Each bit in PORTC can be set up as an input or output by writing a 0 or a 1 in the corresponding bit of the DDRC.

(a) For all the pins to act as inputs, all the bits of DDRC must be LOW (0). Therefore, DDRC should be equal to 00000000_2.
(b) For all the pins to act as outputs, all the bits of DDRC must be HIGH (1). Therefore, DDRC should be equal to 11111111_2.

(c) For all the even bits to act as inputs, all the even bits of DDRC must be LOW (0). We will set b0, b2, ..., b6 to 0. Similarly, for all the odd bits to act as outputs, all the odd bits of DDRC must be HIGH (1). We will set b1, b3, ..., b7 to 1.Therefore, DDRC should be equal to 10101010_2.

Example 8.6

How will Port C behave if the DDRC is set to the following?

(a) 240_{10}
(b) CC_{HEX}
(c) $0000\ 1111_2$

SOLUTION

Since the bits of the DDRC decide the corresponding bits of the PORTC to act either as input or output, we will pay attention to the 0 or 1 bit values of DDRC register. We know that 0 represents an input and 1 represents an output.

(a) First, we convert the decimal number 240 into its binary equivalent. We get $240_{10} = 1111\ 0000_2$. Therefore, the bits b7 through b4 (upper nibble) of PORTC will act as output bits, and bits b3 through b0 (lower nibble) will act as input bits.
(b) Again, we convert the hexadecimal number CC into its binary equivalent. We get $CC_{HEX} = 1100\ 1100_2$. Therefore, the bits b7, b6, b3, and b2 of PORTC will act as output, and the remaining bits b5, b4, b1, and b0 as input.
(c) With DDRC set to $0000\ 1111_2$, the upper nibble (bits b7 through b4) will act as input bits, and the lower nibble (bits b3 through b0) will act as output bits.

Example 8.7

Write a fragment of code in order to configure Port C with the DDRC values as in Example 8.6.

SOLUTION

To configure Port C, we will use the binary format to set each bit of DDRC.

(a) For DDRC set to $1111\ 0000_2$, the following is the code in assembly language:

```
LDAA    #%11110000   ; Set up DDRC register
STAA    $1007        ; Program Port C
```

(b) Again, for DDRC set to $1100\ 1100_2$, the following is the code in assembly language:

```
LDAA    #%11001100   ; Set up DDRC register
STAA    $1007        ; Program Port C
```

(c) Again, for DDRC set to 0000 1111$_2$, the following is the code in assembly language:

```
LDAA    #%00001111  ; Set up DDRC register
STAA    $1007       ; Program Port C
```

Section 8.5 Review Quiz

How many bidirectional pins are there in Port C?

Common Misconception: Students often tend to avoid using binary types in assembly programming, thinking that it would be difficult to read. However, binary type is more readable in place of hex when specific bits are being emphasized.

Helpful Hint: Since the DDRC register is cleared after reset, it forces the Port C register to be configured as eight input bits.

8.6 Port D and Port E

There are no dedicated input or dedicated output bits in Port D. As shown in Table 8.1, Port D is an 8-bit I/O port with six programmable input or output bits, and the two unused bits. Figure 8.6 illustrates the eight general-purpose I/O bits of PORTD labeled PD7–PD0. The bits can be grouped according to their function as:

1. Input or outputs (PD5 through PD0)
2. Unused (PD7 and PD6)

As mentioned earlier, when the port bits are configured as output bits, they are capable of sampling a signal when the microcontroller executes

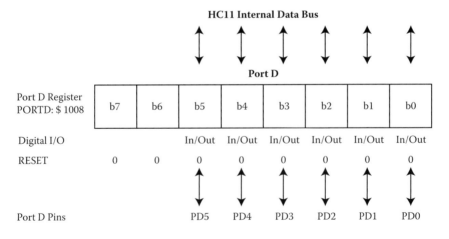

FIGURE 8.6
Details of Port D.

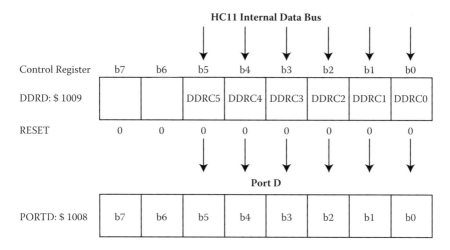

FIGURE 8.7
Details of DDRD.

an input instruction. Similarly, when programmed as input bits, they can hold the output data for an indefinite time—until the program changes it. These programmable bits are controlled by data direction control bits in the Data Direction register for Port D (DDRD). DDRD specifications are presented in Figure 8.7. Recall that a data direction control bit is a bit whose value decides if the corresponding bit of the port register will act as the input or output. If DDRDx bit = 1, the corresponding I/O pins acts as output. Otherwise, if DDRDx bit = 0, the corresponding I/O pin acts as input. For example, if all the DDRD bits are set to 0, then Port D will act as input port. After reset, bits in PORTD are cleared. Consequently, pins are forced to become input pins.

Example 8.8

How will Port D behave if the DDRD is set to the following?

(a) 240_{10}
(b) CC_{HEX}
(c) $0000\ 1111_2$

SOLUTION

Since the bits of the DDRD decide if the corresponding bits of the PORTD act either as input or output, we will pay attention to the 0 or 1 bit values of the DDRD register. We know that 0 represents an input, and 1 represents an output. Since there are two unused bits (i.e., b7 and b6) in PORTD, we will ignore the corresponding bits in the DDRD.

(a) First, we convert the decimal number 240 into its binary equivalent. We get $240_{10} = 1111\ 0000_2$. Therefore, the bits b5 through b4 of PORTD will act as output bits, and bits b3 through b0 (lower nibble) will act as input bits.

(b) Again, we convert the hexadecimal number CC into its binary equivalent. We get $CC_{HEX} = 1100\ 1100_2$. Therefore, the bits b3 and b2 of PORTD will act as output, and the remaining bits b5, b4, b1, and b0 as input.

(c) With DDRD set to $0000\ 1111_2$, the bits b5 and b4 will act as input bits, and the lower nibble (bits b3 through b0) will act as output bits.

Example 8.9

Write a fragment of code in order to configure Port D with the DDRD values as in Example 8.6.

SOLUTION

To configure Port D, we will use the binary format to set each bit of the DDRD.

(a) For the DDRD set to $1111\ 0000_2$, the following is the code in assembly language:

```
LDAA    #%11110000    ; Set up DDRD register
STAA    $1009         ; Program Port D
```

(b) Again, for the DDRD set to $1100\ 1100_2$, the following is the code in assembly language:

```
LDAA    #%11001100    ; Set up DDRD register
STAA    $1009         ; Program Port D
```

Helpful Hint: The Port D pins are alternatively used as the on-chip serial communication functions.

Helpful Hint: Out of all the ports in HC11, Port D is the only port that has two unused bits.

(c) Again, for the DDRD set to $0000\ 1111_2$, the following is the code in assembly language:

```
LDAA    #%00001111    ; Set up DDRD register
STAA    $1009         ; Program Port D
```

Port E is an 8-bit input port. Figure 8.8 illustrates the structure of Port E. The eight general-purpose input pins are labeled PE7–PE0. After reset, the bits in the Port E register (PORTE) are cleared. We will study more about Port E in Chapter 10.

Helpful Hint: PORT E can serve as an 8-bit input port or as the 8 channels for an 8-bit analog to digital converter (ADC).

Section 8.6 Review Quiz

How many pins are there in Port D and Port E?

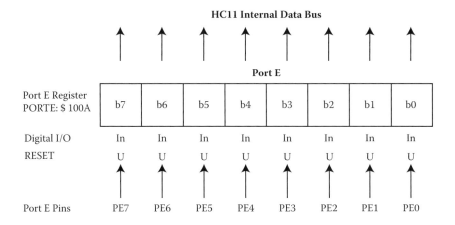

FIGURE 8.8
Details of Port E.

8.7 I/O Using Handshaking

Up to now we have seen that various HC11 ports act as junctions between the processor and the outside world for information transfer. Usually, this information transfer is in the form of a sequence of bytes transferred in parallel to achieve fast input and output. Typically, such parallel communication interfaces are used in printers and parallel buses for peripheral communication. The interfaces in such devices employ a controlled data transfer technique called Strobed I/O to coordinate the transfer of data. When an external device writes data for the microcontroller to read, the microcontroller has to know when the next data is sent. In the absence of this knowledge, the microcontroller would not know if the data sent is the same or there is no data to be sent by the external device. Similarly, when the data is output to a microcontroller port, for it to send the data to the external device, the microcontroller has to tell the external device that data is available. Again, in the absence of this knowledge, the external device would not know if the data at the output port of the microcontroller is the same or there is no data to be sent by the microcontroller. The microcontroller sends a strobe signal to the external device so that the flip-flops in the device can clock or latch in data. The data transfer where control signals are used between the computer and the peripheral device is called handshaking. These control signals are the code components of the handshaking protocol for information exchange between the microcontroller and the external device. The HC11 provides parallel I/O with handshaking. The handshaking subsystem consists of PORTB, Port C, and two strobe lines, that is, strobe A (STRA) and strobe B (STRB). Figure 8.9

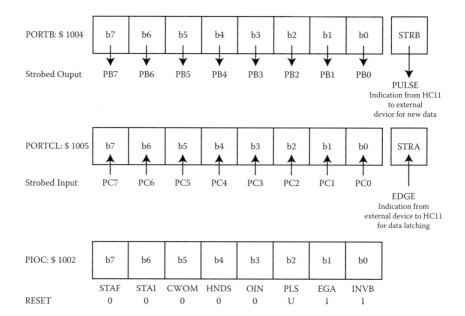

FIGURE 8.9
Elementat of the handshaking subsystem.

TABLE 8.2

Parallel I/O Handshaking Modes Setup

Mode Name	HNDS	OIN
Simple Strobed Mode	0	X
Full-Input Handshaking	1	0
Full-Output Handshaking	1	1

presents the elements of the handshaking subsystem. In Section 8.4, just for simplicity, we mentioned that Port C has only a data register called PORTC at $1003. In fact, Port C has another data register called port C latched data register (PORTCL) at $1005. The register PORTCL is employed for strobed and handshake I/O because it latches input data. Figure 8.9 shows the register PORTCL (input), PORTB (output), and parallel I/O control register (PIOC) located at $1002 memory location.

Table 8.2 presents the three modes of handshaking supported on the HC11. Only the Handshake Mode Select (HNDS) and Output or Input Handshake Select (OIN) bits in the PIOC register control the setup of these modes.

Common Practice: In the context of I/O, terms such as external device, external world, or outside world are used as synonyms for peripherals.

8.7.1 Simple Handshaking

Handshaking concerns the exchange of data between the microcontroller and an external device. Earlier, we discussed the significance of the handshaking protocol. *Protocol* is a term often used in computer networks. The information exchanged between devices on a network or other communications medium is governed by rules set out in a technical specification called a communication protocol. A simple handshaking protocol might only involve the receiver sending a message meaning "I received your last message, and I am ready for you to send me the next message." There can be more complex handshaking protocols. Simple handshaking in HC11 is called the simple strobed mode. The microcontroller (sender) uses a strobe to signal to the external device (receiver) that the data present on the parallel bus is new data for latching. As shown in Table 8.2, the simple strobed mode is configured when the HNDS bit of PIOC is cleared.

For data output, Port B is used to output data using the simple strobed mode. The data written by the program executed on HC11 on PORTB is transferred to the Port B pins. This data transfer is immediately followed by a PULSE (the strobe of two E-clocks length) generated on the Strobe B (STRB) pin. This indicates to the peripheral that new data is available. The Invert Strobe B (INVB) bit in the PIOC register controls the polarity of the STRB PULSE. INVB = 1 and INVB=0 represent the active state of the pulse as logic one and logic zero, respectively.

For data input, Port C is used as an input port during simple strobed mode. For this, the DDRC register is cleared. Port C Latched (PORTCL) is used to input data using the simple strobed mode. The data written by the external device on the Port C pins is transferred to the PORTCL register when an EDGE is detached on the Strobe A (STRA) pin. As soon as the active EDGE is detected on the STRA pin, the Strobe A Flag (STAF) bit of the PIOC becomes 1, indicating that an active edge has been detected on STRA and the data has been latched into the PORTCL register. When the PIOC register and PORTCL register are read one after the other, STRA becomes 0. In other words, the STRA is 0 when the data on the PORTCL is old. The Active Edge for Strobe A (EGA) bit in the PIOC register controls the polarity of the STRA EDGE. EGA = 1 and EGA = 0 represent that the STRA is active on the rising and falling edge, respectively. The Strobe A Interrupt Enable (STAI) bit will be covered in the next chapter, on interrupts.

Example 8.10

Write a fragment of code to reset the PIOC register and then output data $FF to PORTB.

SOLUTION

Referring to Figure 8.9, the RESET state for the PIOC is 00000011_2. We will first load this value to PIOC (at location $1002) and then deal with the PORTB register.

```
LDAA    #%00000011  ; RESET PIOC register
STAA    $1002       ; Setup PIOC
LDAA    #$FF        ; Load data to AccA
STAA    $1004       ; Output data to PORTB
```

Common Misconception: Students often think that the HC11 cannot simultaneously send and receive data. Actually, it can, because the simple strobed mode send and receive functions are handled by independent hardware.

8.7.2 Full-Input Handshaking

As shown in Table 8.2, the full-input handshaking mode is configured when the HNDS bit of the PIOC is equal to 1 and the OIN bit is equal to 0. In full-input handshaking mode, Port C is used as an input port for receiving data from an external device. For this, the DDRC register is cleared.

The full-input handshaking operation can be described in just a few sentences. The receiver writes "I am ready to receive data" on his wall. Let us call this the "ready" message. The sender looks at the "ready" message, sends the data, and writes "I have sent data" on his wall. Let us call this the "sent" message. As soon as the receiver looks at the "sent" message, he takes the data, and erases his "ready" message, that is, "I am ready to receive data" from his wall. With no message on the receiver's wall, the sender understands that the receiver is not ready to receive any more data. This is illustrated in Figure 8.10.

The messages "I am ready to receive data" and "I have sent data" are equivalent to active STRB and STRA signals, respectively. The peripheral waits for the STRB signal to be active. As soon as the STRB is active, the peripheral understands that the microcontroller is ready to receive data. The peripheral sends the data to the Port C pins and makes STRA active. With STRA active, the microcontroller understands that the sender has sent the data on the Port C pins, and it latches the data into the PORTCL register. The HC11 performs the erasing of the ready message by removing the STRB ready signal. With no STRB ready signal, the peripheral understands that the microcontroller is not ready to receive any data. When the PIOC register and the PORTCL register are read one after the other, STRA becomes 0. Additionally, a read of the PORTCL makes STRB active so that another full-input handshake operation can occur.

The Invert Strobe B (INVB) bit in the PIOC register controls the polarity of the STRB PULSE. INVB = 1 and INVB = 0 represent the active state of the pulse as logic one and logic zero, respectively. The Active Edge for the Strobe A (EGA) bit in the PIOC register controls the polarity of the STRA EDGE. EGA = 1 and EGA = 0 represent that the STRA is active on the rising and falling edge, respectively. As soon as the active EDGE is detected on the STRA pin, the Strobe A Flag (STAF) bit of PIOC becomes 1, indicating that an

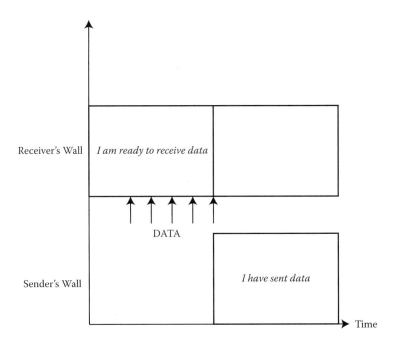

FIGURE 8.10
Simulating a full-input handshaking operation.

active edge has been detected on STRA and the data has been latched into the PORTCL register.

Example 8.11

Write a program to provide a message to the terminal window notifying the user as soon as the external device sends new data to the HC11 in the full-input handshaking mode.

SOLUTION

For notifying the user, we will use the BUFFALO utility subroutine OUTSTRG. Since the mode of operation is full-input, the HNDS and OIN must be set to 1 and 0, respectively. We assume that the STRA rising edge is selected. Therefore, the EGA = 1. Reset of the PIOC bits are assumed to be in their default state. We get PIOC = 00010111_2.

```
* ***************************************************
* Program Name: FIHand.asm
* Objective: Notify users when data is sent by peripheral
* Usage: Buffalo CALL 140 command
* Output: On Screen
* ***************************************************
```

```
MAIN        EQU   $0140           ; Equate MAIN to $0140
DATA        EQU   $0100           ; Equate DATA to $0100
OUTSTRG:    EQU   $FFC7           ; Jump table address

            ORG   DATA            ; Data origin
Messg       FCC   'Available!'    ; Message to display
            FCB   $04             ; EOT Character - Delimiter

            ORG   MAIN            ; Program origin
START       LDAA  #%00010111      ; Configure PIOC
            STAA  $1002           ; Set up PIOC
TEST        LDAA  $1002           ; look for STAF
            BPL   TEST            ; If not STAF, keep testing
            LDX   #Messg          ; Message pointer
            JSR   OUTSTRG         ; Send message
            SWF                   ; Quit program
```

8.7.3 Full-Output Handshaking

As shown in Table 8.2, the full-output handshaking mode is configured when the HNDS bit of the PIOC is equal to 1 and the OIN bit is equal to 1. In full-output handshaking mode, Port C is used as an output port for sending data to an external device. For this, the DDRC register must be set.

Similar to the full-input handshaking operation, the full-output handshaking operation can be described in just a few sentences. The sender writes "I am ready to send data" on his wall. The receiver looks at the "ready" message, receives the data, and writes "I have received data" on his wall. Let us call this the "receive" message. As soon as the sender looks at the "receive" message, he erases his "ready" message "I am ready to send data" from his wall. With no message on the sender's wall, the receiver understands that the send is not ready to send any more data. This is illustrated in Figure 8.11.

The messages "I am ready to send data" and "I have received data" are equivalent to active STRB and STRA signals, respectively. When the program writes to the PORTCL, the STRB signal automatically becomes active by the write to PORTCL. This is followed by the transfer of data to the Port C pins. The peripheral that was waiting for the STRB signal to be active looks at the active STRB now and understands that the microcontroller is ready to send data. The peripheral receives or latches the data from the Port C pins and makes the STRA active. With the STRA active, the microcontroller understands that the receiver has received the data from the Port C pins. The HC11 performs the erasing of the ready message by removing the STRB ready signal and sets the STAF bit. With no STRB ready signal, the peripheral understands that the microcontroller is not ready to send any data. When the PIOC register is read followed by a write to the PORTCL register, the STAF becomes 0. Additionally, a write to the PORTCL makes the STRB active so that another full-input handshake operation can occur. The Invert Strobe B (INVB) bit in the PIOC register controls the polarity of the STRB PULSE. INVB = 1 and INVB = 0 represent the active state of the pulse as logic one

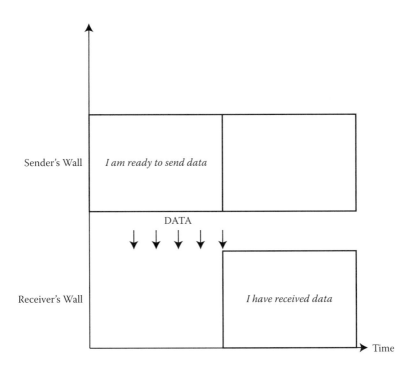

FIGURE 8.11
Simulating a full-output handshaking operation.

and logic zero, respectively. The Active Edge for Strobe A (EGA) bit in the PIOC register controls the polarity of the STRA EDGE. EGA = 1 and EGA = 0 that represent the STRA is active on the rising and falling edge, respectively.

Section 8.7 Review Quiz

What are the two types of I/O handshaking employed in the HC11?

8.8 A Project Using Port B

We have seen in Chapter 3 that seven-segment displays are used in a wide range of applications. Clocks, watches, digital instruments, and many household appliances already have such displays.

8.8.1 Objective

In this EVBU-based project, we will gain hands-on experience in displaying any of the ten decimal digits 0–9 on demand.

8.8.2 Requirements Analysis

Requirements analysis is the first step in the software development process. General statements are made as the first step of requirements analysis, and then the technical specifications are derived from them. An example of such a general statement is the one we just made above in "Objective." The problem is then decomposed into several smaller problems and represented by suitable notations that describe the essence of the problem.

Let us first focus on the hardware requirements. We know that we will need a seven-segment display and an EVBU for this experiment. When prompted, the user shall type a digit on the keyboard that will act as an input for the HC11. The HC11 shall output a signal or data corresponding to this digit to the seven-segment display. Since the HC11 has to output data for the seven-segment display unit, we will use Port B for that purpose. The Port B of the HC11 shall be connected to the seven-segment display via the 7406 (or 7404/7405) inverter IC and resistors. The resistors shall be used in order to limit the current to an appropriate level for each seven-segment display segment.

The eight pins of Port B shall be connected to eight segments (including the dp) of the seven-segment display. We plan to use a common anode type of seven-segment display. When working with a common anode seven-segment display, power shall be applied externally to the anode connection that is common to all the segments. This is shown as the V_{cc} connection to the seven-segment display in Figure 8.12. In order to light up the appropriate segment, a LOW (0) will be needed to a particular segment connection. Since we plan to provide HIGH (1) to the Port B pins for a digital representation of the input number, we will have to invert that output signal before it reaches the seven-segment display. Therefore, we will need eight inverters. Refer to Figure 8.12 to trace the connections from Port B to the seven-segment displays. The IC 7406 has only six NOT gates. To overcome this shortage of two gates, we will use another IC7406.

The major hardware circuit components can be listed as following:

1. EVBU
2. Two 7406s ICs
3. Seven-segment LED display
4. Eight 330 ohm resistors

With these hardware circuit components, we can build the EVBU connections to the listed circuit components, as shown in Figure 8.13.

With hardware requirements nailed down for this project, we now proceed to the software requirements. Let us first analyze the user interface for this experiment. The program that we will develop will supply a prompt to the user to enter a character. The program will read a character provided

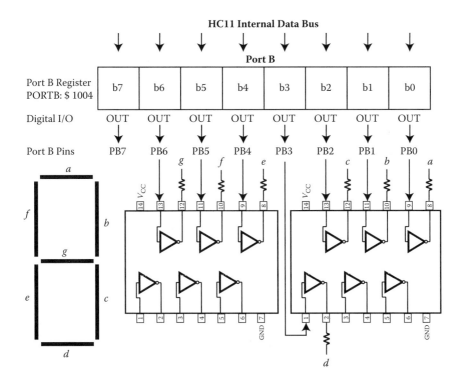

FIGURE 8.12
Schematic diagram for the seven-segment display project.

by the user. Like any other user interface, here also at any point of time, the program will allow the user a way to quit and return to the BUFFALO monitor. Since it is a customer requirement (see the objective of the project) that the character to be displayed must be a digit from 0 to 9, a product requirement arises for the program to determine if the input character is a digit or not. If the input character is a digit, the program will decode its corresponding ASCII code to determine an appropriate binary data value. This binary data will be sent as output data to the seven-segment display. Also, so that the user can use the program multiple times in a run, the program will send a carriage return to the monitor after each display and await the next character from the user.

8.8.3 Design Development

Design provides an engineering representation of the system to be built. Here, we will outline the organization of various elements such as memory locations, algorithms, control structure, state transition, BUFFALO utility subroutines, data, and program flow. We will also design the interfaces

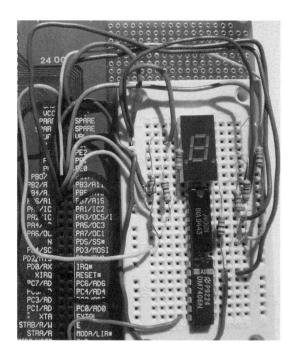

FIGURE 8.13
EVBU connections for the seven-segment display project.

for integrating components such as data input from the user and delivering to the seven-segment display. The design of each component to achieve all the hardware and software requirements is the key here. In microcontroller programming, optimization of memory usage is one of the most important objectives for any designer at the component level.

We discussed seven-segment displays in Chapter 3. There we discussed that to display a decimal digit, input bits of the seven-segment display have to be set to cause it to light up the appropriate segments. An example was also illustrated using Figure 3.9 and Table 3.2, where, in order to display the digit '9', the bit at 'e' must be LOW (0) and the remaining bits must be set to HIGH (1). We also obtained in Table 3.3 the logical states and seven-segment display status to display pattern for digits from 0 to 9. Finally, Table 3.4 was created to include the ASCII equivalent for each display digit.

As shown in the flowchart of Figure 8.14, the program first populates the seven-segment display Hex code array. This array is taken from Table 3.3. Recall that these values result in the display of the corresponding decimal number. Next, the program prompts the user to input a character on the keyboard. As soon as the user enters a character, the program evaluates to check if it is either 'Q'/'q' or not. If it is 'Q'/'q', then the program simply exits. If not, the program again evaluates to check if the character entered by the user is a nondigit. This is performed similar to Example 7.12. If the character

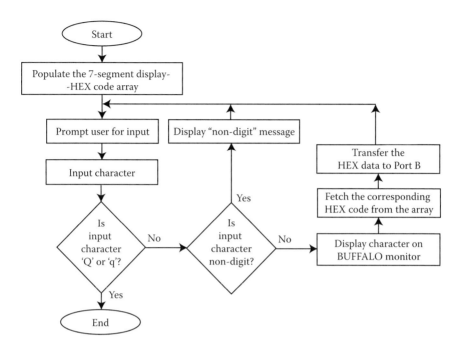

FIGURE 8.14
Flowchart for the seven-segment display project.

is a nondigit, then the user is shown a relevant message, and program control goes back to the user prompt a new input character. When the character entered by the user is a digit (i.e., 0–9), then the corresponding hex code is fetched from the stored array and transferred to Port B. The program control is again sent back to the user input prompt.

The following are the BUFFALO utility subroutines that are used in the program:

- OUTA—Output the ASCII character in accumulator A
- INCHAR—Input an ASCII character from the terminal, and place it in accumulator A
- OUTSTRG—Output the EOT-terminated ASCII string pointed to by X

The comments in the following program are provided for a better understanding of each line/segment of the code and its functionality.

```
*  ***********************************************************
*  Program Name: 7sd.asm
*  Objective: Output on demand on a 7-segment display digits 0-9
*  Usage: Buffalo Call 100 command
*  Output: 7-segment Display and Computer Monitor
*  ***********************************************************
```

```
MAIN        EQU  $0100   ; Assign address $0100 to label MAIN
DATA        EQU  $0150   ; Assign address $0150 to label DATA
OUTA        EQU  $FFB8   ; Assign address $FFB8 to label OUTA
OUTSTRG     EQU  $FFC7   ; Assign address $FFC7 to label OUTSTRG
INCHAR      EQU  $FFCD   ; Assign address $FFCD to label INCHAR
PORTB       EQU  $1004   ; Assign address $1004 to label PORTB


            ORG  DATA    ; Data starts here
* Populate an array with the 7-segment HEX code
            FCB  $3F, $06, $5B, $4F, $66, $6D, $7D, $07, $7F, $6F

* Message to enter a character
MSG1        FCC  'Enter a character (q or Q to quit): '
            FCB  $04

* Message for the display character
MSG2        FCC  'Now displaying: '
            FCB  $04

* Message for a non-digit alert
MSG3        FCC  'Entered a non-digit character'
            FCB  $04

* Message for program exit
MSG4        FCC  'Quiting program!'
            FCB  $04


            ORG  MAIN    ; Main program starts here
LOOP        LDX  #MSG1   ; Load base-address of MSG1
            JSR  OUTSTRG ; Output MSG1
            JSR  INCHAR  ; Input a character to AccA
            PSHA         ; Save it in Stack for future use

*************************************************************
* Check if the character entered by the user is 'q' or 'Q'
* if so, jump to label quit
*************************************************************
            CMPA #$71    ; Compare (AccA) with $71
            BNE  IFQ     ; Branch to IFQ if it not equal to $71
            JMP  END     ; If equal to $71, jump to END

IFQ         CMPA #$51    ; Compare (AccA) with $51
            BNE  IFDGT   ; Branch to IFDGT if not equal to $51
            JMP  END     ; If equal to $51, jump to END


*************************************************************
```

```
* Check if the character entered by the user is non-digit
* if so, then branch to NODGT
*************************************************************
IFDGT     CMPA #$30    ; Compare (AccA) with $30
          BLO  NODGT   ; Branch to NODGT if less than $30
          CMPA #$39    ; Compare (AccA) with $39
          BHI  NODGT   ; Branch to NODGT if greater than $39


*************************************************************
* Display the "displaying" message and the digit that will be
* displayed to the 7-segment display to the user on the monitor
*************************************************************
          LDX  #MSG2   ; Load base-address of MSG2
          JSR  OUTSTRG ; Output MSG2
          PULA         ; Retrieve the saved value of input
          JSR  OUTA    ; Display the digit on monitor


*************************************************************
* For the input digit, fetch the corresponding (7-segment)
HEX code
*************************************************************
          TAB          ; Make (AccB) equal to (AccA)
          LDX  #$0120  ; Load initialization address ($0120) to X
          ABX          ; Add (AccB) to (X) for pointing to HEX code
          LDAA 0,X     ; Load the corresponding HEX code to AccA


*************************************************************
* Output the 7-segment HEX code to the Port B
*************************************************************
          STAA PORTB   ; Save the HEX code to the Port B
          JMP  LOOP    ; Keep looping


*************************************************************
* NODGT: Display the non-digit error message to the user
*************************************************************
NODGT     LDX  #MSG3   ; Load base-address of MSG3
          JSR  OUTSTRG ; Output MSG3
          JMP  LOOP    ; Keep looping


*************************************************************
* END: Display quitting message to the user and quit program
*************************************************************
END       LDX  #MSG4   ; Load base-address of MSG4
          JSR  OUTSTRG ; Output MSG4
          SWI          ; End the program
```

```
>CALL 100

Enter a character (q or Q to quit): !
Entered a non-digit character
Enter a character (q or Q to quit): a
Entered a non-digit character
Enter a character (q or Q to quit): A
Entered a non-digit character
Enter a character (q or Q to quit): 0
Now displaying: 0
Enter a character (q or Q to quit): 6
Now displaying: 6
Enter a character (q or Q to quit): @
Entered a non-digit character
Enter a character (q or Q to quit): 5
Now displaying: 5
Enter a character (q or Q to quit): J
Entered a non-digit character
Enter a character (q or Q to quit): 9
Now displaying: 9
Enter a character (q or Q to quit): q
Quiting program...
P-0146 Y-FFFF X-01C0 A-0D B-39 C-D4 S-003B
```

FIGURE 8.15
Sample execution of the seven-segment display project.

One way to perform the tests on our program is through entering various input values and inspecting the display of the seven-segment display. Figure 8.15 shows a sample execution of the program. Note how the program responds when a nondigit character is entered. Also, notice how the user quits the program.

Group Discussion: What would happen i any two connectors to the seven-segmen display are swapped?

Self-Learning: Change the configuration by employing a common cathode type seven-segment display in your project.

Section 8.8 Review Quiz

Why are IC 7406 used in this project?

8.9 Summary

1. There are three ways of transferring data between the microcontroller and the outside world: programmed I/O, interrupt-driven I/O, and direct memory access (DMA).

2. There are five I/O ports on the HC11. Table 8.1: Summary of HC11 ports.

3. Thirty-eight general-purpose I/O pins are supported in HC11.

4. The dedicated output bits are capable of sampling a signal when the microcontroller executes an input instruction.

5. The dedicated input bits hold the output data for an indefinite time—until the program changes it. On the other hand, the programmable bits are controlled by two data direction control bits.

6. A data direction register controls the direction of its corresponding port's bits.

7. Simple handshaking in HC11 is called simple strobed mode.

8. The full-input handshaking mode is configured when the HNDS bit of the PIOC is equal to 1 and the OIN bit is equal to 0.

9. The full-output handshaking mode is configured when the HNDS bit of the PIOC is equal to 1 and the OIN bit is equal to 1.

10. PORTB and Port C support two types of handshaking for interface control: simple strobed mode and full handshaking.

11. The PIOC control register contains control and status flag bits relevant to the handshaking operations.

12. PORTA, PORTD, and PORTE support alternative functions.

Glossary

Bidirectional: Data flow in both directions.

Data Direction Control Bit: A bit that determines the direction of the corresponding bit in the data register.

Data Direction Register: A register whose bits determine the direction of the corresponding data register.

External World: See *Peripherals*.

Full-Input Handshaking Mode: Handshaking when the HNDS bit of the PIOC is equal to 1 and the OIN bit is equal to 0.

Full-Output Handshaking Mode: Handshaking when the HNDS bit of the PIOC is equal to 1 and OIN bit is equal to 1.

Handshaking: A method of data transfer that uses control signals between the computer and the peripheral device.

Input/Output (I/O): Communication between two entities or systems.

Outside World: See *Peripherals*.

Peripherals: A device attached to a host microcontroller (or computer), but not part of it, and is more or less dependent on the host.

Ports: An interface between the processor and the outside world. Ports support a variety of input/output functions.

Protocol: A technical specification consisting of a set of rules that govern the information exchanged between devices.

Simple Strobed Mode: Simple handshaking in HC11.

Unidirectional: Data flow in only one direction.

Answers to Section Review Quiz

8.1 Programmed input/output, interrupt-driven input/output, and direct memory access (DMA)

8.2 2

8.3 8

8.4 6

8.5 6 and 8, respectively

8.6 Simple and full

8.7 For NOT gates

True/False Quiz

1. After reset, PORTA becomes 00000000_2, that is, cleared.
2. Port B is composed of an 8-bit data register (PORTB) and eight Port B pins.
3. Port C is composed of an 8-bit data register (PORTC), an 8-bit data direction register (DDRC), and eight Port C pins.
4. The DDRC register is an output register and can be used to configure the bits in the Port C as inputs or outputs.
5. Since the DDRC register is cleared after reset, it forces the Port C register to be configured as eight input bits.
6. If DDRDx bit = 1, the corresponding I/O pins in Port D act as the output.
7. If DDRDx bit = 0, the corresponding I/O pin in Port D act as the input.
8. After reset, the bits in the Port E register (PORTE) are cleared.
9. The port C latched data register (PORTCL) is located at address $5001.
10. For data input, Port G is used as an output port during the simple strobed mode.
11. For data input, Port B is used to input data using the simple strobed mode.

12. The full-input handshaking mode is configured when the HNDS bit of PIOC is equal to 0 and the OIN bit is equal to 0.

13. In the full-input handshaking mode, Port D is used as an input port for receiving data from an external device.

14. The full-output handshaking mode is configured when the HNDS bit of the PIOC is equal to 0 and the OIN bit is equal to 0.

15. In full-output handshaking mode, Port E is used as an output port for sending data to an external device.

Questions

QUESTION 8.1

Comment on the PORTA bit configuration if both the data direction control bits (PA7 and PA3) are set to 0.

QUESTION 8.2

Comment on the PORTC bit configuration if the DDRC register is equal to $F0.

QUESTION 8.3

Comment on the configuration of Port C bits after the DDRC is reset.

QUESTION 8.4

Comment on the configuration of Port D bits after the DDRD is reset.

QUESTION 8.5

Which bits of the PIOC register play a crucial role in configuring full-input and full-output handshaking modes in HC11?

QUESTION 8.6

Why do we refer to the term *protocol* when talking about handshaking operations?

QUESTION 8.7

List the HC11 ports that have pins with bidirectional capability for data transfer.

QUESTION 8.8

List the HC11 ports that do not depend on data direction control bits for determining the direction of information flow.

QUESTION 8.9

During one experiment on an EVBU, Port B became defective. The experiment requires an output of eight bits. Which port can substitute for Port B? What kind of configuration setup will be required from the program designer for such a substitute?

QUESTION 8.10

Out of all the ports in the HC11, which is the only port that has two unused bits?

Problems

PROBLEM 8.1

Write a fragment of code that will load data from the memory locations $00EE and output the result on the Port B pins. Do not use the assembler directive EQU in your code.

PROBLEM 8.2

Use the assembler directive EQU for Port B as well as $00EE, and rewrite the fragment of code in Problem 8.1.

PROBLEM 8.3

Write a fragment of code in order to configure Port C with the DDRC value set to $F0.

PROBLEM 8.4

Describe the operation performed by the following fragment of code:

```
PORTC   EQU     $1003
DDRC    EQU     $1007

        LDAA    #%00000001
        STAA    DDRC
        CLRA
        STAA    PORTC
```

PROBLEM 8.5

How will Port D behave if the DDRD register is set to $88? Write a fragment of code in order to configure Port D with the DDRD register equal to $88.

PROBLEM 8.6

The DDRC register is set up as shown in Figure 8.16. List the input and output pins of the corresponding PORTC.

Control Register	b7	b6	b5	b4	b3	b2	b1	b0
DDRC: $ 1007	0	0	0	1	1	1	1	0

PORTC: $ 1003	?	?	?	?	?	?	?	?

FIGURE 8.16
The DDRC register for Problem 8.6.

PROBLEM 8.7

The individual bits of PORTC are shown to perform either input or output in Figure 8.17. How would the DDRC look like for such a configuration?

PROBLEM 8.8

The DDRD register is set up as shown in Figure 8.18. List the input and output pins of the corresponding PORTD.

Control Register	b7	b6	b5	b4	b3	b2	b1	b0
DDRC: $ 1007	?	?	?	?	?	?	?	?

PORTC: $ 1003	OUTPUT	OUTPUT	OUTPUT	INPUT	OUTPUT	INPUT	INPUT	OUTPUT

FIGURE 8.17
PORTC for Problem 8.7.

Control Register	b7	b6	b5	b4	b3	b2	b1	b0
DDRD: $ 1009			1	1	0	0	1	1

	b7	b6	b5	b4	b3	b2	b1	b0
PORTD: $ 1008			?	?	?	?	?	?

FIGURE 8.18
The DDRD register for Problem 8.8.

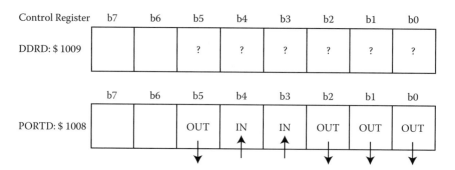

FIGURE 8.19
PORTD for Problem 8.9.

PROBLEM 8.9

The individual bits of PORTD are shown to perform either input or output in Figure 8.19. How would the DDRD look like for such a configuration?

PROBLEM 8.10

By adding comments, describe the operation performed by the following program:

```
MAIN    EQU    $0140
PORTB   EQU    $1004
PORTC   EQU    $1003
DDRC    EQU    $1007
        ORG    MAIN
        CLR    DDRC
        LDAB   PORTC
        STAB   PORTB
        SWI
```

PROBLEM 8.11

Draw a flowchart to show the program flow for Problem 8.10.

PROBLEM 8.12

Execute the Problem 8.10 program on EVBU. How would you demonstrate the results of such a program?

PROBLEM 8.13

By adding comments, describe the operation performed by the following program:

```
MAIN    EQU    $0140
PORTC   EQU    $1003
DDRC    EQU    $1007
PORTD   EQU    $1008
```

```
DDRD     EQU   $1009
         ORG   MAIN
         CLR   DDRC
         LDAA  #$0F
         STAA  DDRD
         LDAB  PORTC
         STAB  PORTD
         SWI
```

PROBLEM 8.14

Draw a flowchart to show the program flow for Problem 8.13.

PROBLEM 8.15

Execute the Problem 8.13 program on EVBU. How would you demonstrate the results of such a program?

PROBLEM 8.16

By adding comments, describe the operation performed by the following program:

```
MAIN     EQU   $0140
PORTC    EQU   $1003
DDRC     EQU   $1007
PORTD    EQU   $1008
DDRD     EQU   $1009
         ORG   MAIN
         CLR   DDRC
         LDAA  #$0F
         STAA  DDRD
         LDAB  PORTC
         STAB  PORTD
         SWI
```

PROBLEM 8.17

Draw a flowchart to show the program flow for Problem 8.16.

PROBLEM 8.18

Execute the Problem 8.16 program on EVBU. How would you demonstrate the results of such a program?

9

Interrupts

"Other people's interruptions of your work are relatively insignificant compared with the countless times you interrupt yourself."

—Brendan Francis (poet, novelist)

OUTLINE

OBJECTIVES

Upon completion of this chapter, you should be able to

1. Understand the basic terms using to describe interrupts.
2. Understand the classification of various interrupts.
3. Understand the operational requirements of interrupts.
4. Describe the process of servicing an interrupt.
5. Enable and mask interrupts.
6. Cite the advantages and disadvantages of maskable and nonmaskable interrupts.
7. Understand local and global control bits.
8. Use the vector table and vector jump table.
9. Distinguish between input and output events.
10. Identify the three basic modules of output-compare system.
11. Configure and use basic timing registers like TCNT, TCTL1, TMSK1/2, TFLG1/2, etc.

12. Use the output-compare register TOCx.
13. Understand the output-compare function of HC11.
14. Use HC11 interrupts in real-time problem solving.

Key Terms: Baud Rate, Embedded System, Event, Global Control Bit, Hard Real-time System, Interrupt, Interrupt Request (IRQ), Interrupt Service Routine (ISR), Interrupt Vector, Local Control Bit, Maskable Interrupt, Mission Critical, Nonmaskable Interrupt, Non-real-time System, Output-Compare, Output Compare Registers, Polling, Reactive System, Reset, Real-time System, Real-Time Constraint, Soft Real-time System, TCNT, TCTL1, TCTL2, TMSK1, TMSK2, Transition, Vector Address.

9.1 Introduction

Embedded systems are generally reactive systems. A reactive system is a system whose role is to maintain an ongoing interaction with its environment rather than produce some final value upon termination. An air traffic control system and programs controlling mechanical devices such as a train, a plane, or ongoing processes such as a nuclear reactor are some of the typical examples of reactive systems. In many applications they must be viewed as real-time systems. Real-time systems are called real-time because they are subject to operational deadlines from event to system response. By contrast, a non-real-time system is one for which there is no deadline, even if fast response or high performance is desired or preferred. Thus, the success of a real-time system greatly depends upon the time in which its functions are performed. A real-time system may be one where its application can be considered to be mission critical. Mission critical refers to any factor that is essential to the core function of a system, and whose failure or disruption will result in the failure of the system. For example, the antilock brakes on a car are an example of a real-time system. Here, the deadline or real-time constraint is the short time in which the brakes must be activated to prevent the wheel from locking. Real-time systems can be considered to have failed if they do not respond before their deadline, where their deadline is relative to an event.

Real-time systems are divided into hard real-time and soft real-time. In a hard real-time system, the completion of an operation after its deadline is considered useless and may cause a critical failure of the complete system. Hard real-time systems are used when it is imperative that an event is reacted to within a strict deadline. Some examples of hard real-time embedded systems include medical systems, such as heart pacemakers, and industrial process controllers. Thus, in a hard real-time system, failure to react

by a given deadline can be catastrophic. On the
other hand, a soft real-time system will tolerate
some lateness, and may respond with decreased
service quality. An example of decrease service quality may be the video
output if some frames are omitted. Soft real-time systems are typically
used where there is some issue of concurrent access and the need to keep a
number of connected systems up to date with changing circumstances. For
example the system that maintains and updates the train plans belonging
to railway transportation system. The train plans must be kept reasonably
current but can operate to a latency of seconds. Live
audio–video systems are also usually soft real-time;
violation of constraints results in degraded service
quality, but the system can continue to operate.
Thus, soft real-time systems perform tasks which
should be completed within specified periods, but the consequences of not
meeting the deadlines are not severe.

Team Discussion: Discuss some examples of real-time applications related to embedded systems.

Self-Learning: Do some Internet research to find out the relationship between Quality of Service, real-time, and non-real-time systems.

We have seen in Chapter 8 that there are three ways of transferring data
between the microcontroller and the outside world:

1. Programmed Input/Output
2. Interrupt-driven Input/Output
3. Direct memory access (DMA)

We focused on programmed input/output in Chapter 8. Now we will explore
the interrupt-driven input/output. Recall from Chapter 4 that a computer
program is a complete sequence of instructions that direct a computer to
perform a specific task or solve a problem. In many applications, it is a basic
requirement to execute sets of instructions in response to some external
stimulus. Often, the origin of external stimulus is peripherals. These stimuli
are usually asynchronous to the program being executed by the processor.
Asynchronous events occur independently of the main program flow. For
example, the opening of a refrigerator door can be an example of an external stimulus to the main program monitoring the overall status of a refrigerator. The door can be opened at any point of time. The external stimulus
interrupts the processor and forces it to save its state of execution and begin
execution of another set of instructions (ISR). Upon completion of this set
of instructions (ISR), the processor returns back to what it was doing prior
to the external stimulus. Therefore, an interrupt is an asynchronous signal
indicating the need for attention or a synchronous event in software indicating the need for a change in execution. Interrupts can be of two types: hardware interrupts and software interrupts. While a hardware interrupt causes
the processor to save its state of execution and begin execution of an interrupt handler, a software interrupt is usually implemented as instructions in
the instruction set, causing a context switch to an interrupt handler similar

to hardware interrupt. A peripheral requests the microcontroller to transfer data by activating a signal on the microcontroller's interrupt line during interrupt I/O. In response, the microcontroller executes a set of instructions called the ISR to carry out the function desired by the external device, again by way of one or more I/O ports.

As discussed earlier, microcontrollers keep track of various external and internal events by manipulating flags. There are two methods for the microcontroller to interact with the events (or tasks) through the flags. The first approach is by *polling*. The polling approach results in a wait loop. For example, in a program, a character is sent over the serial port. The program also checks a "flag B" in a loop. If the flag B is set, it would mean that the serial buffer is empty. With flag B set, the microcontroller comes out of the loop and transfers a byte to the serial buffer register. Obviously, there are drawbacks in this method.

The processor does nothing other than checking the status of the flag B until it is set, at which point the transfer of data is performed. In other words the computer waits until the device is ready. Even for high baud rates, there is rate variation between the processor and the serial port. The second approach is to use the interrupt method. This method implements a "hidden" check of flags. We say it is hidden because it is performed by hardware, concurrently with the program execution. An important point to note here is that it is done concurrently, thus enabling multitasking. When the hardware finds an assert flag, an interrupt request is generated. The processor completes the instruction in progress, stacks the current processor registers like the program counter (PC) so that it can return later and starts executing a specific set of instructions called *ISR*. When the ISR is over, the processor comes back to the main program by relocating the PC with the saved address. Figure 9.1 illustrates this flow of control in a program in response to an interrupt request. The interrupt method facilitates the processor with timing parameters that are very difficult to obtain with a polling approach. The main goal of this chapter is to explore HC11 interrupts. The project at the end will help readers in understanding various design aspects involved in using interrupts on EVBU.

9.2 Basics of an Interrupt

Let us start with an example from our day-to-day life. Let us assume that Bob is reading a book in his house. His sister Alice, who left home for work, kept some cookies to bake in an oven (located in the kitchen) before leaving from home. Alice texts Bob on his cell phone to tell him to go and switch off the oven in the kitchen. This text from Alice to Bob is like an interrupt. As soon as Bob listens to the ringtone of the arrival of the text message, he

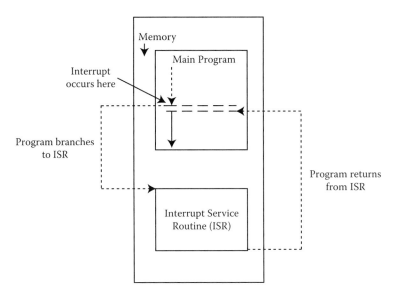

FIGURE 9.1
Flow of control in response to an interrupt request.

stops reading the book by finishing the sentence he was reading. This is like the processor finishing the instruction it is executing. The interrupt (Alice's text) has forced him to stop reading the book briefly and take some necessary actions. He places a bookmark on the page that he was reading, closes the book, and puts it aside. We all know why he placed the bookmark on the page. The bookmark keeps one's place in a book and enables the reader to return to it with ease. This is like the processor when it saves the address of the next instruction in the stack so that it can resume the program execution later. Bob then picks up the cell phone and opens the inbox. First, he looks at who the sender of the text message is. He recognizes that it has come from Alice, his sister. He opens the text message and reads the message which says "Switch off the cooking oven as soon as possible." In the context of a microcontroller, the ability to recognize the source and objective of the interrupt is built into the microcontroller hardware. Remember here that the source is Alice, and the objective is to switch off the oven. After reading the message, Bob immediately goes to the kitchen and switches the oven off. This is like a microcontroller taking some actions appropriate to the interrupt source. Bob then sends a confirmation of this action by sending another text to Alice. Finally, he comes back to his desk, opens the book and turns to the page where he placed the bookmark and resumes reading his book. In the context of a microcontroller, it is like the processor jumping back to the next instruction when the interrupt occurred and resuming program execution.

An act of interrupting is referred to as an interrupt request (IRQ). In other words, the external stimulus is called an interrupt request. In our Alice and Bob example, the ringtone of the arrival of the text message is the IRQ. Thus, it alerts the user for immediate attention. The process of responding to an interrupt request is called *servicing the interrupt*. As mentioned earlier, the interrupt (i.e., Alice's text message) has forced Bob to stop reading the book briefly and take some necessary actions. These actions were placing a bookmark on the page that he was reading, closing the book, and putting it aside, etc. These actions constitute servicing the interrupt. The actions taken by Bob after reading the message were that he immediately went to the kitchen and turned off the oven. If the message would have asked him to turn off the garage lights, he would have gone to the garage and would have turned off the lights there. So the actions or tasks taken by Bob are according to the source and objective of the interrupt. In the context of a microcontroller, upon receiving an interrupt, the task is accomplished by the execution of a subroutine called an *interrupt service routine* (ISR). It is also referred to as an interrupt handler. An ISR is a set of instructions that is triggered by the reception of an interrupt.

Team Discussion: Discuss another analogy for interrupts and relate the actions with IRQ and ISR.

Section 9.2 Review Quiz

IRQ stands for (a) Interrupt Request, (b) Immediate Request, (c) Internal Resource Query, (d) Interrupt Reset Query. (Choose one.)

9.3 Servicing an Interrupt

Figure 9.2 illustrates the sequence of steps to service an interrupt and the interrupt states involved in servicing an interrupt. When an interrupt request occurs, the processor first completes execution of the current instruction before servicing the interrupt. The status of the interrupt changes from "pending" to "being serviced" as soon as the processor is finished executing the current instruction. For example, the LDAB immediate instruction requires two machine cycles to execute. Now think of an interrupt request being made during the fetch or execution of LDAB instruction. The microcontroller hardware will wait until the execution of LDAB instruction is complete before servicing the interrupt.

When the execution of the current instruction is complete, the status of the interrupt changes from "pending" to "being serviced," and the hardware interrupt mechanism comes into play. The first step in servicing the interrupt involves preparing proper grounds for the return of the interrupt. This is done by saving the program context so that when the interrupt servicing is complete, the program can resume from where it left off before the interrupt was serviced. The processor registers constitute the program

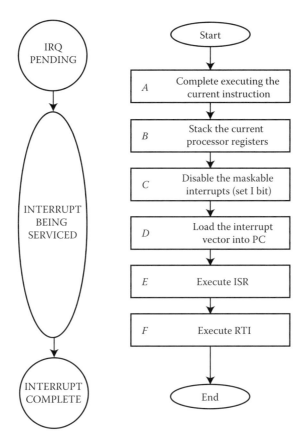

FIGURE 9.2
Steps and states involved in servicing an interrupt.

context. Recall from Chapter 6 that when a subroutine is called from the main program, the contents of the PC are stored or pushed on the stack. This way the processor knows where to come back to after it has finished the subroutine. A similar process occurs with the processor registers to save the program context. The only difference here is that unlike subroutine where only the return address is saved, here the entire context is saved by stacking all of the processor registers. The only register that is not saved is the stack pointer. Figure 9.3 illustrates the order in which the registers are stacked. The first address at which the first register is to be stored is identified by the stack pointer (SP). In Figure 9.3, the direction of stacking implies that PCL is the first register to be stored and CCR is the last register to be stored in the stack. The unstacking process is done when RTI (Real Time Interrupt) is executed, which obviously has the reverse direction due to the FILO (first in/last out) property of the stack.

Self-Learning: Do some Internet research to write a paragraph on the way items can be stored in some data structures. One such example you know by now is FILO.

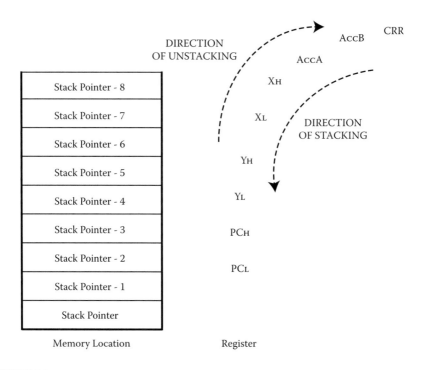

FIGURE 9.3
Saving and retrieving the program context.

Example 9.1

Assume that the stack pointer (SP) is set to $1009. Determine the memory location of the processor registers when the program context is stacked in preparation of executing interrupt service routine.

SOLUTION

From Figure 9.3, we can easily determine the memory locations on which the processor registers will be saved. The first register to be stacked is PC_L. This register is stored at the memory location pointed by the SP, which is $1009. The SP will get decremented so that it always points to the next available stack location.

Common Misconception: Students often think that in the process of stacking of the program context, the upper byte should be followed by the low byte of the same register. In reality the reverse happens, that is, the low byte of each 16-bit register is followed by the upper byte in the stacking process.

Therefore, SP now becomes $1008. Next register to be stacked will be PC_H. This register is stored at memory location pointed by the SP, which is $1008. Again the SP will get decremented. If we continue in this way, we will end up with the mapping of the stacked processor registers and their corresponding memory location as shown in Table 9.1.

TABLE 9.1

Saved Program Context For Example 9.1

Memory Location	Processor Register
$1001	CCR
$1002	AccB
$1003	AccA
$1004	X_H
$1005	X_L
$1006	Y_H
$1007	Y_L
$1008	PC_H
$1009	PC_L

Stacking of the current processor registers is followed by disabling maskable interrupts. The I control bit of CCR is set so that no interrupts can occur during an ISR. The interrupts are again enabled after the program context is restored at the end of the process. We will discuss more about this I control bit in the sections to follow.

The next step in servicing an interrupt is to execute the ISR. The address of the first instruction of the ISR is called the *interrupt vector*. Programmers often call this interrupt vector the vector because it simply acts as a pointer. It is up to the programmer to locate the ISR in the memory. This means that according to the processor memory map, the interrupt vector can have any appropriate value. However, the interrupt vector must be stored at a specific memory location. This specific memory location where the interrupt vector is stored is called the *vector address*. The vector addresses are fixed memory locations and are designed into the hardware. These vector addresses cannot be changed by the programmer. The interrupt vector is an address that occupies two bytes in memory. The vector address identifies the high byte, whereas the vector address plus one identifies the low byte of the interrupt vector. Figure 9.4 illustrates the vector address, interrupt vector, and ISR for a Timer Input Capture 1 (TIC1) interrupt. The vector address for TIC1 is $FFEE. The high byte of the interrupt vector is stored at this address. The address of the low byte of the interrupt vector is obtained by just adding 1 to $FFEE. Therefore, the low byte of the interrupt vector is stored at $FFEF. Now the processor will see what is stored at these two memory locations. As seen in Figure 9.4, the contents of $FFEE and $FFEF are $00 and $CC, respectively. This means that $00CC is the interrupt vector for the ISR for the TIC1. Table 9.2 lists the frequently used interrupt sources with their vector addresses.

Helpful Hint: There is some flexibility in choosing the interrupt vector. However, the vector addresses cannot be changed by the programmer.

The final step in servicing an ISR is the processing of the RTI instruction. This is a mandatory step for servicing any interrupt. The RTI instruction

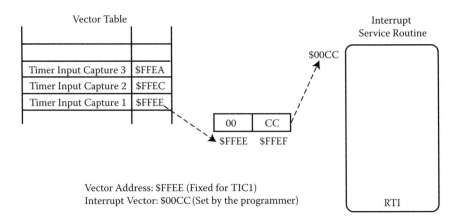

FIGURE 9.4
Linking of vector address, interrupt vector and ISR.

causes the nine bytes of processor register data to be pulled from the stack. They are pulled in the reverse order in which they were stacked, as shown in Figure 9.3. This way, the context of the processor is restored to the point it was when the interrupt occurred. Returning from an ISR is similar to a normal subroutine in the way that the program continues the execution with the next instruction. An ISR is similar to a normal subroutine in many ways. Table 9.3 presents a comparison between an ISR and a subroutine. An important point to note about the ISR is that it is executed only when an interrupt request is recognized, and it returns to the main program with the RTI instruction. The last instruction in an ISR is always the RTI instruction. The program context, that is, the nine bytes of processor register data, is restored from the stack with the execution of RTI instruction. We have seen the saving and retrieving of the program context in Figure 9.3. As with any stack operation, this saving and retrieving also follows FILO.

Example 9.2

Using Table 9.2, determine the vector address and the interrupt vector in Figure 9.5. The interrupt source is IRQ.

SOLUTION

From Table 9.2, the vector address for interrupt source IRQ is $FFF2. The high byte of the interrupt vector is stored at the vector address, that is, $FFF2. The low byte of the interrupt vector is stored at the vector address + 1, that is, $FFF3. From Figure 9.5, we can observe that the contents of memory location $FFF2 and $FFF3 are $00 and $A0, respectively. This $00A0 is the interrupt vector, which is the address of the first instruction of the ISR. Figure 9.6 illustrates this relationship.

Helpful Hint: For beginners, using pencil and paper to draw the pictorial relationship between vector address, interrupt vector, etc. makes understanding them quite easy.

TABLE 9.2

Frequently Used HC11 Interrupts and Their Vector Addresses

Interrupt Source	Vector Address	Type	Global Mask	Local Enables	Local Control Register	Priority
SCI Serial System	FFD6	Maskable	I bit in CCR	RIE, TIE, TCIE, & ILIE	SCCR2	15 in maskable
SPI Serial Transfer Complete	FFD8	Maskable	I bit in CCR	SPIE	SPCR	14 in maskable
Pulse Accumulator Input Edge	FFDA	Maskable	I bit in CCR	PAII	TMSK2	13 in maskable
Pulse Accumulator Overflow	FFDC	Maskable	I bit in CCR	PAOVI	TMSK2	12 in maskable
Timer Overflow	FFDE	Maskable	I bit in CCR	TOI	TMSK2	11 in maskable
Timer Input Capture 4/ Timer Output Compare 5	FFE0	Maskable	I bit in CCR	I4/O5I	TMSK1	10 in maskable
Timer Output Compare 4	FFE2	Maskable	I bit in CCR	OC4I	TMSK1	9 in maskable
Timer Output Compare 3	FFE4	Maskable	I bit in CCR	OC3I	TMSK1	8 in maskable
Timer Output Compare 2	FFE6	Maskable	I bit in CCR	OC2I	TMSK1	7 in maskable
Timer Output Compare 1	FFE8	Maskable	I bit in CCR	OC1I	TMSK1	6 in maskable
Timer Input Capture 3	FFEA	Maskable	I bit in CCR	IC3I	TMSK1	5 in maskable
Timer Input Capture 2	FFEC	Maskable	I bit in CCR	IC2I	TMSK1	4 in maskable
Timer Input Capture 1	FFEE	Maskable	I bit in CCR	IC1I	TMSK1	3 in maskable
Real Time Interrupt	FFF0	Maskable	I bit in CCR	RTII	TMSK2	2 in maskable
IRQ (External Pin)/Strobe A	FFF2	Maskable	I bit in CCR	None/ STAI	NA/ PIOC	1 in maskable
XIRQ (External Pin)	FFF4	Nonmaskable	X bit in CCR	None	NA	4 in nonmaskable
Software Interrupt (SWI)	FFF6	Nonmaskable	None	None	NA	6 in nonmaskable

Continued

TABLE 9.2 (continued)

Frequently Used HC11 Interrupts and Their Vector Addresses

Interrupt Source	Vector Address	Type	Global Mask	Local Enables	Local Control Register	Priority
Illegal Opcode Trap	FFF8	Nonmaskable	None	None	NA	5 in nonmaskable
COP Failure	FFFA	Nonmaskable	None	NOCOP	CONFIG	3 in nonmaskable
Clock Monitor Fail	FFFC	Nonmaskable	None	CME	OPTION	2 in nonmaskable
RESET	FFFE	Nonmaskable	None	None	NA	1 in nonmaskable

TABLE 9.3

Interrupt Service Routine versus Subroutine

Characteristic	Interrupt Service Routine	Normal Subroutine
Reason for execution	Interrupt request	JSR or BSR instruction
End Instruction	RTI	RTS
Address of first instruction	Loaded by hardware to the PC	JSR or BSR results in PC update
Execution Order	Sequential	Sequential

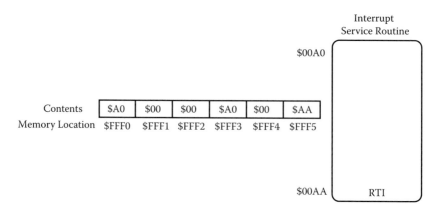

FIGURE 9.5

Basic setup for Example 9.2.

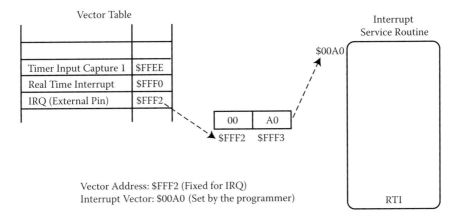

FIGURE 9.6
Vector address, interrupt vector, and ISR for IRQ entry in the vector table.

Example 9.3

Using Table 9.2, determine the interrupt source in Figure 9.7.

SOLUTION

The key here is to find the interrupt vector. From the right most block of ISR, we can see that the address of the first instruction is given as $00C6. This is the interrupt vector. Also in Figure 9.7, the high and low bytes of the interrupt vector are stored at memory location $FFE4 and $FFE5, respectively. Therefore, $FFE4 is the vector address here. If we trace these memory locations in Table 9.2, we can see that vec-

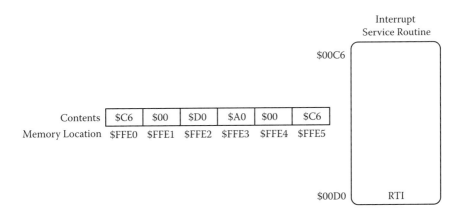

FIGURE 9.7
Basic setup for Example 9.3.

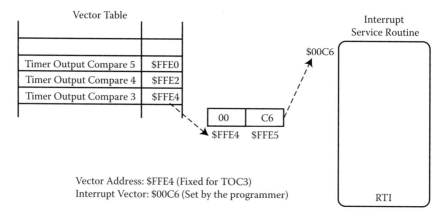

Vector Address: $FFE4 (Fixed for TOC3)
Interrupt Vector: $00C6 (Set by the programmer)

FIGURE 9.8
Referring to the ISR using the interrupt vector.

tor address $FFE4 corresponds to the interrupt source Timer Output Compare 3 (TOC3). Hence, the interrupt source is TOC3. Figure 9.8 illustrates this relationship.

Example 9.4

Saved data and various processor registers are set to the following values. If an interrupt is requested while the following STAB instruction is executed, what will be the locations and the values of the various processor registers stored in the stacking operation?

Data:

```
00A0    88    FF    8E    8F    FF    00    EE    FF
00B0    11    00    EE    FF    CA    00    8E    8F
AccA    $8E
AccB    $F0
X       $FF00
Y       $FF00
SP      $00EE
CCR:    $D1
```

Memory Location	Machine Code	Source Code
0000	D7 A0	STAB $A0

SOLUTION

First we find the updated values of the processor registers. Since the operand field is a single byte ($A0), and there is no # sign, this store instruction is in DIR addressing mode. In DIR addressing mode, the effective address is obtained by putting 00 in front of the operand field of A0. Therefore, the effective address is $00A0. After the execution of STAB instruction, the PC becomes $0002. The store instruction STAB makes the contents of the memory location $00A0 equal to the contents of

TABLE 9.4

Saved Program Context For Example 9.4

Memory Location	Processor Register	Register Contents
$00E7	CCR	$D9
$00E8	AccB	$F0
$00E9	AccA	$8E
$00EA	X_H	00
$00EB	X_L	FF
$00EC	Y_H	00
$00ED	Y_L	FF
$00EE	PC_H	00
$1009	PC_L	02

AccB ($F0). Note that the original value of CCR value is $D1 = 1101 0001_2. Hence, the sequence of SXHINZVC is 11010001. Since the operand is $F0 = 1111 0000_2, it makes N = 1 (MSB of operand is one), Z = 0 (operand is not zero), and V = 0 (this is always clear for any store instruction). Therefore, the value of CCR changes from 1101 0001_2 to 1101 1001_2. Hence, the new value of CCR is 1101 1001_2 = $D9.

Next, from Figure 9.3, we can determine the memory locations on which the processor registers will be saved. The first register to be stacked is PC_L. This register is stored at the memory location pointed by the SP, which is $00EE. The SP will be decremented so that it always points to the next available stack location. Consequently, SP now becomes $00ED. Next register to be stacked will be PC_H. This register is stored at memory location pointed by the SP, which is $00ED. Again the SP will be decremented. If we continue in this way, we will end up with the mapping of the stacked processor registers and their corresponding memory location as shown in Table 9.4. This table also shows the updated contents of the processor registers.

Common Misconception: Students often think that interrupt vector and vector address are the same. Examples in Section 9.2 are given to make their distinction clear to the students.

Section 9.3 Review Quiz

What instruction is executed as a last step in servicing any interrupt?

9.4 Interrupt Control

The ISR is executed when an interrupt service request arrives. With the execution of RTI instruction, the execution of the ISR is completed. The process of enabling or disabling interrupts is often called *masking*. For a programmer, it is important to understand interrupt classification based on masking. Interrupts are classified into two categories: maskable interrupts and nonmaskable interrupts. Maskable interrupt is an interrupt that may be ignored (or disabled) by setting a control bit. There are some standard interrupt masking techniques in the system to achieve the disabling of maskable interrupts. We will study these techniques along with the control bits in detail shortly. A nonmaskable interrupt is an interrupt that usually lacks an associated bit-mask, so that it can never be ignored (or disabled). Nonmaskable interrupts are often used for timers, especially watchdog timers. They are used to signal attention for nonrecoverable hardware errors also.

The HC11 uses control bits in various registers to allow various kinds of controls for interrupts. In this section, we will look into the HC11 instructions that are used in designing interrupts. Additionally, we will study the various registers that allow various kinds of controls for interrupts, collectively or individually.

9.4.1 Interrupt Related Instructions

In Table 9.3 we compared an ISR with a normal subroutine. One of the main differences between them is the reason for their execution. An ISR is executed when an interrupt request arrives. On the other hand, a normal subroutine is executed after the program encounters a JSR or a BSR instruction. The HC11 provides four instructions used for direct control of maskable interrupts. These instructions are summarized in Table 9.5. To interpret this table, follow the method we used for most of the instruction set tables in Chapter 5. Recall from the previous section that the last instruction of every ISR must be a return from interrupt instruction (RTI). As shown in the Table 9.5, the CLI and SEI instructions control the maskable interrupt mask control bit (I bit) in the CCR. SEI sets the control bit, disabling maskable interrupts. CLI clears the control bit, enabling maskable interrupts. We will be using the CLI instruction in the project covered at the end of this chapter. Both CLI and SEI instructions are inherent mode instructions. As shown in the Table 9.5, they have no effect on any other CCR except the I bit.

Common Misconception: Students often make the mistake of using RTS as the last instruction in ISR. The last instruction for any ISR must be RTI.

Since the process of stacking the registers is time consuming and requires many machine cycles, it leads to unwanted delays for the actual execution of the ISR. For programmers designing time critical applications, these delays might become overhead for their solutions. To overcome this delay, HC11 has the WAI instruction. The WAI instruction supports the hardware in getting ready for an interrupt. This is how it works: When the program encounters a WAI instruction, the registers are stacked following Figure 9.3. Next, the processor enters an idle state, waiting for the actual interrupt to occur. As soon as the interrupt occurs, the execution of the ISR will start immediately because all the required registers were already stacked. This way, there is

TABLE 9.5

Set of Maskable Interrupt Control Instructions

Instructions	IMM	DIR	INDX	INDY	EXT	INH	Function	H	I	N	Z	V	C
Wai	—	—	—	—	—	X	Refer to Section 9.3	—	—	—	—	—	—
Rti	—	—	—	—	—	X	Refer to Section 9.3	×	×	×	×	×	×
Sei	—	—	—	—	—	X	$1 \to I$	—	1	—	—	—	—
Cli	—	—	—	—	—	X	$0 \to I$	—	0	—	—	—	—

no time gap between the interrupt request and the actual execution of ISR. The WAI instruction uses the inherent addressing mode and has no effect on the CCR.

9.4.2 Local and Global Control

The maskable interrupts can only be enabled or disabled using the control bits. To understand the control bits, we first look into the classification of the controls. In HC11, the interrupt controls are classified into two categories: local and global. With the global control, the entire class of interrupts is controlled with a single bit. The I bit in the CCR is used to mask all maskable interrupts. If I bit in the CCR is set to 1, all maskable interrupts are disabled, irrespective of the state of the local interrupt control bits. This I bit of the CCR is often referred to as a *global mask*. On the other hand, local control bits provide individual control of each interrupt within the class. Local control can only be done when the global mask is cleared, that is, I is set to zero. With I bit in CCR set to zero, each individual interrupt source can be enabled or disabled using local control bits. The local control bits are often called

Helpful Hint: Maskable interrupts are enabled with LOW global control bit (I) (enable is active low). Local interrupts are enabled with HIGH local control bit (enable is active high).

local enables. An interrupt source is said to be enabled when both the local and global control bits are enabled. A summary of the local and global interrupt control bits is presented in Table 9.2.

The concept of enabling or disabling an interrupt can be explained through our example of Alice and Bob. Recall that Alice texts Bob on his cell phone to tell him to go and switch off the oven in the kitchen. This text from Alice to Bob is like an interrupt. As soon as Bob listens to the ringtone of the arrival of the text message, he stops reading the book by finishing the sentence he was reading. What if Bob wants to ignore all the text messages and puts his cell phone in silent mode? With the cell phone in silent mode, even if Alice keeps sending the text messages to Bob, Bob will not get disturbed and would naturally ignore all the interrupt requests that arrive to him. Note that this does not mean that Alice is deprived of her ability to interrupt. However, her interrupts will not get Bob's attention. Setting the phone in silent mode is an example of disabling interrupt.

Section 9.4 Review Quiz

    (a) The normal delay that occurs during the process of stacking the registers is completely avoided by using the instruction WAI. (True/False)

    (b) Nonmaskable interrupt is an interrupt that usually lacks an associated bit-mask, so that it can never be ignored. (True/False)

9.5 Maskable Interrupts

Interrupts are categorized as maskable interrupts or nonmaskable interrupts. In the previous section, we saw that the maskable interrupt is a hardware interrupt that may be ignored by masking technique using control bits. For the entire class of interrupt, the task of global interrupt mask is handled by the I bit in the CCR. When I bit in the CCR is HIGH, all the maskable interrupts are disabled. The global mask takes precedence over local interrupt enable masks. The control is transferred to the local bit if and only if the I bit in CCR is LOW.

9.5.1 Serial Communication System Interrupt Sources

Referring to Table 9.2, the serial communication system consists of two interrupt sources: SCI serial system and SPI serial transfer complete. Both of these have I bit in CCR as a global mask. The local control register for SCI serial system is SCCR2. This SCCR2 register contains four local enable bits: RIE, TIE, TCIE, and ILIE. When a SCI serial system interrupt request is made, these four local enable bits are examined. Within the serial communication interface, four events can be mapped to the SCI system interrupts. Each of these bits must be set to enable the corresponding interrupt request to be made. The local control register for the interrupt source SPI serial transfer complete is SPCR. SPCR register contains a local enable bit: SPIE. The SPIE bit must be set to enable the SPI interrupt.

Team Discussion: What does serial communication mean in general? How is it related to HC11?

9.5.2 Timer System Interrupt Sources

Referring to Table 9.2, the timer system interrupt sources consists of 12 interrupt sources: Pulse accumulator input edge, pulse accumulator overflow, timer overflow, timer input capture 4/Timer Output Compare 5, Timer Output Compare 4, Timer Output Compare 3, Timer Output Compare 2, Timer Output Compare 1, timer input capture 3, timer input capture 2, timer input capture 1, and real-time interrupt. Let us examine each of them in order.

The first one is the pulse accumulator input edge. PAII bit in the TMSK2 register enables the pulse accumulator input edge interrupt. This interrupt occurs whenever a valid pulse accumulator input edge is detected.

Next is the pulse accumulator overflow. The PAOVI bit in the TMSK2 register enables the pulse accumulator overflow interrupt. When the PACNT register rolls over from $FF to $00, the pulse accumulator overflow is triggered. For the timer overflow interrupt, the TOI bit in the TMSK2 register is the enabling bit. When TCNT register rolls over from $FFFF to $0000, timer overflow occurs.

Next in line is the timer input capture 4/Timer Output Compare 5. I4/O5I enables this interrupt. Finally, the RTII in the TMSK2 register enables the real-time interrupt. For the Timer Output Compare (1 to 4), the OC1I, OC2I, OC3I, and OC4I enable interrupts for the output-compare function OC1-OC4, respectively. These bits are contained in the TMSK1 register.

Likewise, for the timer Input Capture (1 to 3), the IC1I, IC2I, and IC3I enable interrupts for the input-capture function IC1-IC3, respectively. These bits are also contained in the TMSK1 register. We will take a closer look at input capture in Chapter 11.

Helpful Hint: I4/O5I enables the interrupt for the IC4 or the OC5 function according to the configuration.

9.5.3 External Interrupts Using IRQ

Earlier in this chapter, we discussed that the phrase "interrupt request" (or IRQ) is used to refer to either the act of interrupting the bus lines used to signal an interrupt, or the interrupt input lines on a microcontroller. Until now we have also seen that maskable interrupt sources in the time and communication subsystems are tied to some internal hardware system in HC11. But the IRQ interrupt source is different from those in that respect. It is not tied to any internal hardware system in the HC11. Rather, the IRQ interrupt source is actually an external hardware pin dedicated to the interrupt system. If an interrupt is required to be triggered from an outside source, the IRQ pin can be connected to that source to achieve this outside request. As seen in Table 9.2, the IRQ interrupt is only controlled by the I bit in the CCR. There is no local control ability associated with IRQ interrupt.

To achieve more functionality from IRQ interrupt, IRQ Edge bit (IRQE) in the OPTION register is configured for edge and level sensitivity. Figure 9.9 shows the structure of the OPTION register. When IRQE is HIGH (set or 1), IRQ is configured for edge-sensitive operation. When IRQE is LOW (cleared or 0), IRQ is configured for level-sensitive operation. The default for IRQE is LOW as shown in Figure 9.9. We will revisit the OPTION register in Chapter 10 in context with analog-to-digital conversion.

Helpful Hint: IRQ interrupt has no local enables. The only control is the global mask, that is, the I bit in the CCR.

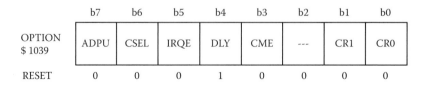

	b7	b6	b5	b4	b3	b2	b1	b0
OPTION $ 1039	ADPU	CSEL	IRQE	DLY	CME	---	CR1	CR0
RESET	0	0	0	1	0	0	0	0

FIGURE 9.9
IRQ edge control bit in the OPTION register.

9.5.4 Maskable Interrupt Priority

There are 15 maskable interrupts. The likelihood that two interrupts may occur at the same time is quite high. If a pulse accumulator overflow occurs at the same time as the timer input capture 1, the processor will have to decide what to process first and what to processes next or to ignore. This is like a situation when Bob gets a text message from Alice as well as hears someone knocking on the door at the same time. There are two tasks for Bob now. Attend text message from Alice and attend the door. What would Bob do first? The problem of concurrence of interrupts is resolved by allotting the interrupts with priority ranking. When two interrupts occur at the same time, the higher priority interrupt will take precedence over the lower priority one. Table 9.2 shows the priority for interrupts.

Example 9.5

Refer to the priority ranking in Table 9.2 and find out which of the following interrupt sources will be served first.

(a) IRQ (External Pin)
(b) SCI Serial System
(c) Timer Output Compare 1

SOLUTION

Referring to Table 9.2, IRQ (External Pin) will be served first as it has the highest priority.

Example 9.6

Refer to the priority ranking in Table 9.2 and find out which of the following interrupt sources will be served last.

(a) Real-Time Interrupt (RTI)
(b) SPI Serial Transfer Complete
(c) Timer Output Compare 2

Helpful Hint: When designing programs with multiple interrupts, keep in mind the priority ranking.

SOLUTION

Referring to Table 9.2, SPI Serial Transfer Complete will be served last as it has the lowest priority.

The highest priority register (HPRIO), shown in Figure 9.10, is used to elevate the interrupt priority of one maskable interrupt to the highest level. When an interrupt priority is elevated to the highest level, the priorities of the rest of the interrupts do not change. Table 9.6 shown the default priority as well as the PSELx bits to be set in the HPRIO register in order to configure

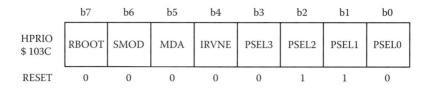

	b7	b6	b5	b4	b3	b2	b1	b0
HPRIO $ 103C	RBOOT	SMOD	MDA	IRVNE	PSEL3	PSEL2	PSEL1	PSEL0
RESET	0	0	0	0	0	1	1	0

FIGURE 9.10
The HPRIO register with PSELx bits.

TABLE 9.6

PSELx Bits For Highest Interrupt Priority

Interrupt Source	Priority	PSEL[3:0]
IRQ (External Pin)/Strobe A	1	0110
Real Time Interrupt	2	0111
Timer Input Capture 1	3	1000
Timer Input Capture 2	4	1001
Timer Input Capture 3	5	1010
Timer Output Compare 1	6	1011
Timer Output Compare 2	7	1100
Timer Output Compare 3	8	1101
Timer Output Compare 4	9	1110
Timer Input Capture 4/Timer Output Compare 5	10	1111
Timer Overflow	11	0000
Pulse Accumulator Overflow	12	0001
Pulse Accumulator Input Edge	13	0010
SPI Serial Transfer Complete	14	0011
SCI Serial System	15	0100

the highest priority for an interrupt. All the nonmaskable interrupts carry higher priority than the maskable interrupts.

Section 9.5 Review Quiz

List a few maskable interrupts.

9.6 Output Compare

For any embedded system to achieve its functional objective, time is a critical factor. Applications require several functions to be performed at various points in time. An example of such an application function is the digital clock found in computers. This clock is basically a counter that increments at every one-second interval. In this way, it produces what we call as seconds, minutes, and hours. The incrementing of the counter is the event that occurs in a timely fashion. An event may be defined as a change in behavior such as signal transition. The change in the state of a signal from high to low, or low to high is called *signal transition*. Events are categorized into internal and external events. If an event is created from within HC11 and sent to some external device, it is called an *output event*. On the other hand, if an external source sends a signal that is sensed by the HC11, it is called an *input event*.

If an edge transition occurs in the output signal, the system can detect it as an output event. For example, an airplane control system is required to switch off the seat belt sign 10 minutes after the start of its takeoff. An embedded system employed for such a task would load a control register with the time the event is to take place. In this case, 10 minutes after the start of its takeoff. The system will compare the system clock to the event time. When the system time becomes equal to the specified event time, the output event will occur. The output event will be the state change in terms of voltage at the output signal port. Before the system time is not equal to the designated event time, the voltage would be low. However, as soon as the match will occur, the voltage at the output signal port would change its state from low to high. This change of state, or transition, would be considered as an output event. By utilizing this output event, the application will switch off the seat belt sign. In this section, we will discuss some features of the interrupt that are used to manage such output events.

Group Discussion: Discuss a few examples of embedded applications based on output events.

9.6.1 Internal Timing Devices

The HC11 has several built-in timing devices that measure time periods and constitute the HC11 timing system. From a higher level, there are five timing chains that combine to form the HC11 timing system. These chains originate from the internal bus clock. The internal bus clock runs at the same frequency as the E clock. For a precise and easy-to-implement measurement processes, the timing devices include dedicated counters, scalars, flags, and interrupts. One of the basic counters in the timing system is the Timer Counter register (TCNT). It is driven by the main clock divider chain of the HC11 timing system. TCNT is also called a "free-running" counter because the user cannot make any change to its contents. By free-running, what we mean is that it cannot be modified by the user but it will be incremented by every tick of the E clock. As shown in Figure 9.11, it is located at the memory addresses $100E and $100F. When TCNT is reset, all its bits become 0.

Common Misconception: Students often get confused about the size of TCNT and consider it as an 8-bit register. TCNT is a 16-bit free-running counter.

Helpful Hint: All Port A timing functions are synchronized to the phase of the internal bus clock.

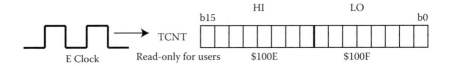

FIGURE 9.11
The TCNT register.

	b7	b6	b5	b4	b3	b2	b1	b0
TFLG2 Register Address $ 1025	TOF	RTIF	PAOVF	PAIF	-	-	PR1	PR0
RESET	0	0	0	0	0	0	0	0

FIGURE 9.12
Structure of TFLG2.

When all the bits of TCNT are 0, the counter is at $0000 or 0_{10}. When all its bits are 1, the counter is at $FFFF or $65,535_{10}$. Thus, TCNT can go from 0 to 65,535. When it reaches its maximum limit of 65,535, it rolls over and starts again from 0. Whenever a roll over occurs with TCNT, the Timer Overflow Flag (TOF) is set in the Timer Interrupt Flag Register 2 (TFLG2), indicating that the TCNT has rolled over (timer overflow). With TOF set, the value of TCNT is interpreted as the number of counts since the latest TCNT overflow. As a matter of fact, the TCNT register always shows the number of counts or number of E clocks since the latest overflow. Figure 9.12 shows the structure of TFLG2.

In the event of a timer overflow, if the Timer Overflow Interrupt Enable (TOI) bit of the Timer Interrupt Mask Register 2 (TMSK2) is set, an interrupt will be requested. Figure 9.13 shows the structure of TMSK2. If a user wants a slower E clock rate, a programmable prescalar can be set to values as shown in Table 9.7. The 68HC11 comes with a programmable prescalar that allows the user to extend the time of each count by dividing the E clock before it is applied to the input of the TCNT register. The available factors are 1, 4, 8, and 16, and can be selected using the bits 0 and bit 1 in the TMSK ($1024) control register. The TMSK2 has bits b_0 and b_1 as Timer Prescaler Select bits PR1

	b7	b6	b5	b4	b3	b2	b1	b0
TMSK2 Register Address $ 1024	TOI	RTII	PAOVI	PAIF	-	-	-	-
RESET	0	0	0	0	0	0	0	0

FIGURE 9.13
Structure of TMSK2.

TABLE 9.7

TCNT Register Timing For 2 MHz E Clock

PR1	PR0	Prescale Factor	Bus Frequency (E clock) 2 MHz Resolution (µsec)/Overflow (msec)
0	0	1	0.5/32.77
0	1	4	2/131.1
1	0	8	4/262.1
1	1	16	8/524.3

and PR0, respectively. By setting these bits to values shown in the Table 9.7, one can slow the clock rate, which consequently will slow the TCNT register count rate and overflow rate. The smallest time interval that can be measured in a timing system is referred to as resolution in Table 9.7. In the 68HC11, the default value for the TCNT register increment is once every 0.5 µs. This 0.5 µs is the default resolution of the microcontroller system.

Example 9.7

In the 68HC11, the standard clock frequency is 2 MHz. How much time (in ms) will the TCNT register require to roll over?

SOLUTION

Since the standard clock frequency is 2 MHz, the time it takes for the TCNT register to increment will be 1/(2 MHz) = 0.5 µs. Because the size of the TCNT register is 16 bits, it takes 2^{16} (or 65,536) counts to go from $0000 to $FFFF. At the standard clock frequency (2 MHz), the TCNT register requires 32.77 ms to roll over (65,536 × 0.5 µs). This 32.77 ms is the figure in the last column of Table 9.7 for the prescale factor equal to 1.

Example 9.8

If the prescale factor is set to 4, how much time (in ms) will the TCNT register require to roll over?

SOLUTION

The default roll-over time of the TCNT register is 0.5 µs. With prescale factor set to 4, the standard roll-over time is multiplied by 4. Therefore, the new roll-over time with prescale factor will be 4 × 0.5 µs = 2 µs. Using this roll-over time, we can compute that the TCNT register requires 131.1 ms to roll over (65,536 × 2 µs). This 131.1 ms is the figure in the last column of Table 9.7 for the prescale factor equal to 4.

The flag bits in the 68HC11 control registers are often cleared by writing a 1 into the respective bit position of the register. Writing a 0 into the bit position has no effect on the flag. Programmers often build a control word and then

write that control word to the corresponding flag register. This is performed
by first loading an accumulator with a mask that has bits set corresponding
to the flags that are required to be cleared. The next step is then to write this
mask value to the timer input register. The write operation is just like an
XOR operation. The contents of the flag register are XORed with the mask.
Those bit positions in the mask that have 1's are cleared in the flag.

Example 9.9

Write a subroutine named ABC to clear the Timer Overflow Flag (TOF) in Timer
Interrupt Register 2 (TFLG2).

SOLUTION

The subroutine named ABC to clear the TOF in TFLG2 is as follows:

```
ABC:    LDAA    #%10000000    ; Set MSB = 1
        STAA    $1025         ; Write it to TFLG2
        RTS                   ; Return
```

The first statement creates the binary control word %10000000 and loads the
AccA with this control word. This way the MSB is set to 1, and rest of the bits
are 0. Next statement stores this control word to the address $1025, which is the
address for TFLG2.

Example 9.10

The HC11 uses the inverse mask to clear the bit when using BCLR instruction.
Write a subroutine named XYZ that utilizes the BCLR instruction to clear the Timer
Overflow Flag (TOF) in Timer Interrupt Register 2 (TFLG2).

SOLUTION

A flag can be cleared using a BCLR instruction. The important point to remember
is that since the inverse mask is used to clear the bit when using BCLR instruction, a 0 in the bit position will be required for
clearing. Therefore, we will build a mask of $7F (or
%01111111) so that the bit is identified by 0's in the
mask and it will clear the MSB of TOF.

Team Discussion: How are the AND and
XOR operations performed by the processor in Example 9.10?

```
XYZ:    LDX     #$1000        ; Load register X with $1000
        BCLR    $25, X $7F    ; Use mask $7F
        RTS                   ; Return
```

The two Timer Interrupt Flag registers (TFLG1 and TFLG2) contain 12 status flag bits. Each of these status flag bits correspond to an interrupt enable
bit located in two Timer Interrupt Mask registers (TMSK1 and TMSK2). In
other words, there is bit-to-bit mapping between the flag registers and mask

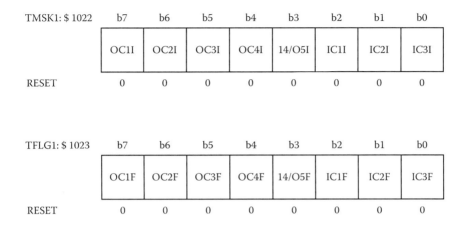

FIGURE 9.14
Timer system flag and interrupt enable bits.

registers. Figure 9.14 illustrates these registers with their internal structure. An example of the bit-to-bit mapping is the b_5 of TFLG1 and TMSK1. The b_5 of TFLG1 is OC3F (status flag), which has the relationship with the b_5 of TMSK1, which is OC3I (interrupt enable flag). We will discuss these bits and their applications in detail in the subsequent sections of this chapter. One important point to remember about the flag bits is that when an interrupt is enabled, the corresponding status flag bit indicates to the interrupt control logic the status of the event, that is, if the event has occurred or not. An interrupt is requested if this flag bit is enabled. Therefore, during the ISR, the programmer must clear the flag bit. If this is not done, the hardware will keep requesting other interrupts, and the program will not behave in the required fashion.

Helpful Hint: During the ISR, the programmer must clear the flag bit corresponding to the interrupt. Not doing so will cause the program to behave incorrectly.

9.6.2 Output Compare Interrupts

In the beginning of Section 9.6, we discussed briefly about an embedded system used in an airplane that caused an event to occur on the output pins at a specified time. The events are pin actions or signal transitions. Output compare is dedicated to cause such transitions on the OCx output pins. The HC11 output-compare functions are listed in Table 9.2. These are abbreviated as OC1, OC2, OC3, OC4, and OC5. The TCNT acts as a reference of time. The output-compare functions are realized by three basic modules. These modules are shown in Figure 9.15 and are listed as follows:

- An output-compare register
- A dedicated 16-bit comparator
- Interrupt generation logic

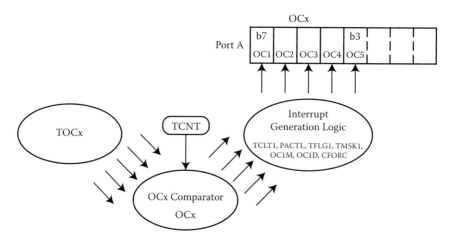

FIGURE 9.15
Basic modules of output compare.

We will discuss these modules in detail in the subsequent subsections. During each time count, the comparator compares the value of the TCNT register with the 16-bit compare register (TOCx and TI4/O5). If both the values come out to be the same, the corresponding Output-Compare Flag (OCxF) is set. Consequently, the corresponding pin action or transition is triggered. This completes the output event. Now if the interrupt is enabled, this will also trigger an interrupt request.

Common hardware is allotted to OC5 and IC4 functions. This sets OC5 apart from the rest of the OCx functions. We will discuss OC5 function as and when needed in our future programming problems and projects. Let us first discuss the common functions of OC2 through OC4. We will briefly discuss OC1 at the end of this section so that the differences between OC1 and others may be highlighted.

9.6.2.1 Output-Compare Registers

The registers used in output-compare function are listed in Table 9.8. Out of these registers, we will only focus on the TOx (i.e., TOC1, TOC2, TOC3, and TOC4) and TI4/O5. The basic structure of these registers is shown in

Helpful Hint: TOx registers set with same 16-bit value can cause an output event or the pin action to happen simultaneously.

Best Practice: Programmers always prefer to use a double-byte write instruction such as store D (STD) when modifying data in the TOCx registers.

Figure 9.16. These registers are independent of each other and can operate simultaneously. All of them are 16-bit read/write registers. Their values represent the relative time at which the pin action or the output event has to take place. Therefore, they are primarily used to store a 16-bit value that is compared to the TCNT registers during every

TABLE 9.8

Output-Compare Registers

Name	Address	Brief Description
CFORC	$100B	Timer compare force register
OC1M	$100C	Output-compare 1 mask register
OC1D	$100D	Output-compare 1 data register
TOCx	$1016 to $101D	Timer output-compare registers
T14/05	$101E to $101F	Shared timer output-compare register
TCTL1	$1020	Timer control register 1
TMSK1	$1022	Timer interrupt mask register 1
TFLG1	$1023	Timer interrupt flag register 1
PACTL	$1026	Pulse accumulator control register

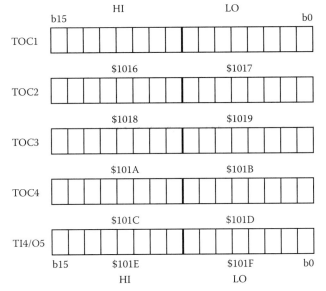

FIGURE 9.16
Structure of TOC registers.

tick of the E clock. If the result of the comparison is true, the corresponding pin action or transition is triggered. If the interrupt is enabled, this will also trigger an interrupt request.

9.6.2.2 Dedicated Comparators

Dedicated comparators are provided to the output-compare functions such that when active, they compare the value in the output-compare register to the value in the TCNT register every time the TCNT counter ticks. If the

TCTL1 Register Address $ 1020	b7	b6	b5	b4	b3	b2	b1	b0
	OM2	OL2	OM3	OL3	OM4	OL4	OM5	OL5
RESET	0	0	0	0	0	0	0	0

FIGURE 9.17
Layout of the TCTL1 register.

comparison is successful, that is, if both the values match, the output event is triggered. The effect on the output pins (i.e., output event with OC2, OC3, OC4, and OC5 events) is determined by the Timer Control Register 1 (TCTL1). The layout for the TCTL1 register is given in Figure 9.17.

As shown in the figure, the bits in TCTL1 register have the logical names: OMx and OLx. Again, here x takes values between 2 and 5. Table 9.9 itemizes the effect that the bits in TCTL1 have on the output pins. There are four different output events that can be configured using the TCTL1. They are: no effect, toggle OCx pin, clear OCx pin, and set OCx pin. At reset, all the bits in the TCTL1 register are cleared. This is the default state. This means that with OMx and OLx all set to zero, OCx is disabled. This is shown in Table 9.9 where OMx and OLx are equal to zero. Let us look at the settings a programmer will have to make in order to activate OC3 function to set the OC3 pin on each successful compare. As per Table 9.9, each OM3 (b_5) and OL3 (b_4) must be set to 1. Similarly, for OC4 function to toggle the OC4 pin on each successful compare, OM4 (b_3) and OL4 (b_3) must be set to 0 and 1, respectively. The OC5 function has an additional setting that is set outside TCTL1. Figure 9.18 shows the structure of PACTL register. The bit I4/O5 located at b_2

TABLE 9.9

Basic Rules For TCTL1 Configuration

OMx	OLx	Pin Action For OC2 through OC5 Event (when compare result is true)
0	0	No effect, OCx disabled
0	1	Toggle OCx pin
1	0	Clear (low) OCx pin
1	1	Set (high) OCx pin

PACTL Register Address $ 1026	b7	b6	b5	b4	b3	b2	b1	b0
	DDRA7	PAEN	PAMOD	PEDGE	DDRA3	14/O5	RTR1	RTR0
RESET	0	0	0	0	0	0	0	0

FIGURE 9.18
The structure of PACTL register.

in PACTL register determines which function out of OC5 and IC4 are active. When I4/O5 bit is equal to zero, the OC5 function is activated and IC4 function is disabled.

Example 9.11

Write a code fragment to demonstrate the configuration of TCTL1 such that a clear signal is sent at the OC3 output pins upon successful comparison between TOC3 and TCNT registers.

SOLUTION

From Table 9.9, we can get the values of OM3 and OL3 when a clear signal is needed at the output pins. We get the settings OM3 = 1 and OL3 = 0. Making the rest of the bits in TCTL1 equal to zero, we build the control word as 0010 0000$_2$. This control word should be saved to the address \$1020 which is the address of TCTL1. To save the control word, we use load-and-store method as shown here:

```
LDAA    #%00100000    ; OC3 clear on successful compare
STAA    $1020         ; Activate OC3
```

Example 9.12

Figure 9.19 shows some bits that were set by a programmer to activate OC4 such that clear signal is sent at the OC4 output pins upon successful comparison between TOC3 and TCNT registers. Find out the mistake made by the programmer.

SOLUTION

From Table 9.9, we can simply take the values of OM3 and OL3 when clear signal is needed at the output pins. Since we are dealing with OC4, we will set OM4 (b_3) and OL4 (b_2) to 1 and 0, respectively. Rest of the TCTL1 bits will be zero. This generates a control word 0000 1000$_2$. However, the control word that was set by the programmer is 1000 0000$_2$. With b_7 and b_6 set to 1 and 0, respectively, it appears that the programmer configured it for OC2 instead of OC4.

Helpful Hint: Setting all the bits in TCTL1 to zero disables all the OC2 to OC5 functions.

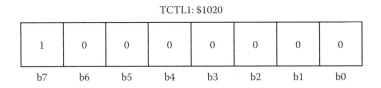

TCTL1: \$1020

1	0	0	0	0	0	0	0
b7	b6	b5	b4	b3	b2	b1	b0

FIGURE 9.19
TCTL1 for Example 9.12.

Recall from Figure 9.14 where we discussed the timer system flags and interrupt enable bits. The TFLG1 register contains the output-compare status flags (OCxF, I4/O5F). Every time a valid comparison is performed, these flags are set automatically. The status flags are cleared by writing a 1 to the corresponding bit position. Likewise, the TMSK1 register contains the output-compare local interrupt enables (OCxI, I4/O5I). Output-compare interrupts are locally controlled by these local interrupt enable bits. Whenever the status flag in TFLG1 is set, its corresponding interrupt enable bit (if set), will request that interrupt. Let us take a look at an example where the OC3F flag in TFLG1 is cleared. Also, let us look into how this OC3F can request an OC3 interrupt.

Example 9.13

Write a code fragment to clear the OC3F bit in TFLG1 register.

SOLUTION

We just read that the status flags are cleared by writing a 1 to the corresponding bit position. Given that the OC3F is to be cleared, we will write a 1 to the OC3F bit, which is the b_5. The rest of the bits in TFLG1 will remain at 0. We will use the load-and-store method. First, we will load the control word ($0010\ 0000_2$) to AccA and then store the contents of AccA to the address $1023. The memory location of TFLG1 is $1023.

```
LDAA    #%00100000    ;Set up OC3 flag control word
STAA    $1023         ;Clear OC3F in TFLG1
```

Example 9.14

Write a code fragment to enable the interrupt system for OC3 function. Use the code from Example 9.13 if needed.

SOLUTION

The two bits that need to be set to 1 are the OC3I (b_5) in TMSK1 and OC3F (b_5) in TFLG1. The rest of the bits in both these registers should be cleared to 0. We will get two identical control word $0010\ 0000_2$. We just have to save these values to the TMSK1 ($1022) and TFLG1 ($1023) registers. Finally, the CLI instruction will enable the system interrupt. The code fragment is as follows:

Best Practice: Programmers often use equate directives as much as possible. It improves the readability of the code.

```
LDAA    #%00100000    ; Set up OC3 interrupt control word
STAA    $1022         ; Enable OC3I mask in TMSK1
LDAA    #%00100000    ; Set up OC3 flag control word
STAA    $1023         ; Enable OC3F in TFLG1
CLI                   ; Enable system interrupt
```

This is also called interrupt initialization. We will be using this code in our project at the end of this chapter.

Port A

PORTA: $ 1000	b7	b6	b5	b4	b3	b2	b1	b0
	PA7	PA6	PA5	PA4	PA3	PA2	PA1	PA0
Digital I/O	In/Out	Out	Out	Out	In/Out	In	In	In
RESET	0	0	0	0	0	0	0	0

Data direction control bits DDRA7 and DDRA3

PACTL: $ 1026	b7	b6	b5	b4	b3	b2	b1	b0
	DDRA7				DDRA3			
RESET	0	0	0	0	0	0	0	0

FIGURE 9.20
A complete picture of Port A.

The OC events OC2, OC3, and OC4 are related to pins PA4 to PA6, respectively, on PORTA. Recall from Chapter 8 that these pins only have the direction state of "output." Figure 9.20 provides a complete picture of Port A. The other output-compare event, OC5, is related to pin PA3 on PORTA. This pin is bidirectional in nature. Therefore, to use TCTL1 to affect PA3, the direction state needs to be configured. The OC1 event uses the output pin somewhat differently than OC2-OC5. Functions are quite easily defined for the output-compare events OC2-OC5. In a nutshell, when the event OCx occurs (i.e., a valid comparison occurs), an output event at pin PAx takes place, depending upon how the bits OMx and OLx are set. This mechanism is widely used since an embedded application often has to generate output voltages in response to output events. The output voltage generation is quite fast due to the fact that the pin state changes are handled in hardware instead of in software.

Helpful Hint: The input only pins of Port A (PA0/IC3, PA1/IC2, PA2/IC1) also serve as edge-sensitive timer input capture pins.

Helpful Hint: Output-only pins of Port A (PA3/OC5, PA4/OC4, PA5/OC3, PA6/OC2) also serve as main timer output-compare pins.

Helpful Hint: Whenever an o/p compare function is enabled, that pin cannot be used for general purpose o/p.

Section 9.6 Review Quiz

Fill in the blanks:

(a) The Timer Counter register (TCNT) is also called _____-running counter. (closed/free/open/looped)
(b) All Port A timing functions are a function of the _____ clock frequency. (A/B/C/D/E)
(c) _____ is located at locations $100E and $100F. (TCNT/TFLG2/TMSK2)

(d) ____ and ____ are located at locations $1024 and $1025, respectively. (TCNT/TFLG2/TMSK2/TFLG1/TMSK1)

(e) Bits OMx and OLx are found in the ____ register. (TCNT/TFLG1/TMSK1/TCTL1)

9.7 Nonmaskable Interrupts

Figure 9.2 lists six nonmaskable interrupts. Nonmaskable interrupts are usually the interrupts that cannot be disabled using the masking technique. Three out of six of these interrupts (SWI, Illegal opcode trap, and Reset) are totally free of any control by the user. The rest (XIRQ, COP failure, and clock monitor fail) can be controlled to a small degree by the users. The SWI, illegal opcode trap, and reset have "none" in the global mask as well as local mask fields in Table 9.2.

9.7.1 External Interrupts Using XIRQ

The nonmaskable external interrupt using XIRQ is similar to maskable IRQ interrupt in two ways. First, the XIRQ interrupt source is an external hardware pin ($\overline{\text{XIRQ}}$) dedicated to the interrupt system. The dedicated pin can be connected to an outside source in order to receive an interrupt from outside. Second, the default priority of this interrupt is the highest among the interrupts, similar to the highest default priority of the IRQ interrupt in the maskable category. Remember that all the nonmaskable interrupts carry higher priority than the maskable interrupts. Following are the important properties of the XIRQ:

1. XIRQ is disabled immediately after reset.
2. The user can enable the XIRQ by using a TAP instruction.
3. Once the XIRQ is enabled, it cannot be disabled by the user.
4. XIRQ has the highest priority among all the interrupts.

Upon the request of an XIRQ, the bit X and I in the CCR are set by the hardware. This disables further interrupts from taking place. When the last instruction (i.e., RTI instruction) is encountered, the original values of the X and I bits are restored.

Helpful Hint: Once XIRQ is enabled, it cannot be disabled by the user. Only system reset can disable it again.

9.7.2 Other Nonmaskable Interrupts

Here we briefly outline some of the other nonmaskable interrupts. Earlier in this chapter we discussed SWI instruction. The illegal opcode fetch is an interrupt requested when an illegal opcode is detected. The computer operating properly (COP) failure is a nonmaskable interrupt designed to detect software processing errors. The clock monitor failure is responsible for detecting problems in the proper running of the system clock. The Power-On Reset (POR) is designed to initialize the internal microcontroller circuits. The nonmaskable interrupt priority cannot be changed. The priority rankings are shown in Table 9.2. Overall, the nonmaskable priority is higher than the maskable interrupts. We will discuss in detail the nonmaskable interrupts and when we will use them in our programming projects.

Section 9.7 Review Quiz

List a few nonmaskable interrupts.

9.8 Interrupts on the EVBU

In Section 9.2, we discussed the vector table. On a standard HC11 system, the vectors found in the vector table are the addresses of the ISR. However, the contents of the ROM-based vector table are fixed during the manufacturing process. This means that they cannot be changed later. Since the programmer cannot alter the vector table contents, it creates a problem for the EVBU programmers in pointing to the right ISR in the vector table. When writing programs to run on EVBU, the location of the ISR are not decided until the programmer actually creates the program that employs interrupts. However, this problem is solved by using vector jump tables. Here is how it works: The unalterable vectors in the ROM-based vector table are address locations in the RAM. Each vector in the vector table refers to a JMP instruction in RAM. The set of these JMP instructions in the RAM is referred to as a *vector jump table*. The JMP instruction is initialized with a default ISR memory location located in BUFFALO. Therefore, the first instruction to be executed for an ISR is this JMP instruction. The programmer can overwrite the default effective address of the JMP instruction with the actual start address of the ISR. The default effective address of the JMP instruction is replaced by the ISR's start address, which is determined after creating the program. This way, the programmer gets some room for locating the ISR in the RAM, and points to this ISR in the vector jump table for it to be executed with the correct IRQ. Figure 9.21 illustrates this phenomenon. Important point to remember here is that each of the interrupt sources has a corresponding JMP instruction RAM except the RESET.

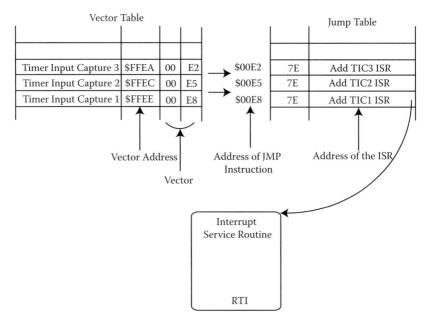

FIGURE 9.21
Association between interrupt vector table and interrupt vector jump table.

Example 9.15

Write a code fragment to initialize the jump table in EVBU. The interrupt used is OC3. The ISR starting address is $01EA.

SOLUTION

We will first use the equate directive to label the ISR starting address ($01EA) and address of jump instruction for OC3 ($D9) to OC3_SVC and OC3_VEC, respectively. We obtained $D9 from the basic vector table that came with the EVBU. If the interrupt were TOC2, this would have been $DC.

```
OC3_SVC    EQU $01EA    ; stating address of ISR
OC3_VEC    EQU $D9      ; Address of jump instruction (vector)
```

We know that $7E is the JMP instruction. Therefore, we will use accumulator A (AccA) through load-and-store instructions to make the contents of the memory location $00D9 equal to $7E. Next, we will use register D through LDD and STD to store the starting address of ISR ($01EA) at locations $00DA and $00DB. This is achieved by just adding 1 to the $00D9 in the STD instruction. The code for jump table initialization for OC3 is as follows:

```
LDAA    #$7E           ;SET UP OC3 JUMP TABLE
STAA    OC3_VEC        ;FOR JMP
LDD     #OC3_SVC       ;TO
STD     OC3_VEC+1      ;OC3 INTERRUPT SERVICE
```

This code fragment may be the first part of our main program in the project covered at the end of this chapter.

Section 9.8 Review Quiz

Each of the interrupt sources has a corresponding JMP instruction RAM except the RESET. (True/False)

9.9 A Project with Interrupts

In Chapter 8 we had sufficient hands-on exposure in creating an HC11 program that interfaced with a seven-segment display. Now we will combine that knowledge and code with the concept of interrupts. A standardized pedestrian signal uses three different displays to communicate the appropriate crossing activity to pedestrians. These displays represent "Walk," a clearance interval, and "Don't Walk." "Walk" is intended to indicate the period when a pedestrian may begin to cross the street. A white crossing man symbol is displayed during the "Walk" interval. The clearance interval, which displays a flashing orange hand, is meant to provide sufficient time for a pedestrian to cross the roadway at an average walking pace. An additional digital countdown display often provides additional information to users by showing how much time is left in the pedestrian clearance interval.

9.9.1 Objective

In this EVBU-based project, we will gain hands-on experience in displaying any of the first nine natural numbers 1–9 on demand as a digital countdown display. Each seven-segment symbol is to be displayed for exactly 0.7 s and off for 0.3 s. For example if an "8 "is entered via the keyboard, the output should appear as the sequence "7","6","5","4","3","2","1","0".

9.9.2 Requirements Analysis

The EVBU is to be configured in single chip mode. All hardware requirements are exactly the same as the previous project covered in Section 8.8. The seven-segment display shall be wired to port B of Motorola 68HC11. PB0 shall be assigned to segment a, PB1 to segment b, and so on. The setup shall use the recommended 74ls06 interface. To display a decimal digit using even segments, output bits from PORTB shall be fed into the seven-segment display to cause it to light up the appropriate segments for displaying the proper digit. The connection shall be tested by memory modifying Port B ($1004) and assigning a bit pattern to test each bit-to-segment connection. Refer to the Figure 8.12 to trace the connections from Port B to the seven-segment displays. The major hardware circuit components can be listed as the following:

1. EVBU
2. Two 7406s ICs
3. Seven-segment LED display
4. Eight 330 ohms resistors

Note that they are exactly the same as the previous project (Section 8.7). The circuit board connections for this project would look exactly like Figure 8.13.

With hardware requirements exactly the same as the previous project, we now proceed to the software requirements. Let us first analyze the user interface for this experiment. The program that we will develop shall supply a prompt to the user to enter a character. The program shall read a character provided by the user. Like any other user interface, here also at any point of time, the program shall allow the user a way to quit and return to the BUFFALO monitor. Since it is a customer requirement (see the objective of the project) that the character to be displayed must be a digit from 1 to 9, a product requirement arises for the program to determine if the input character is a digit or not. If the input character is a digit, the program shall decode that character's corresponding ASCII code to determine an appropriate binary data value. This binary data shall be sent as output data to the seven-segment display. Also, in order for the user to use with the program multiple times in a run, the program shall send a carriage return to the monitor after each display and shall await the next character from the user. All these look similar to the software requirements of the previous project. These steps are shown in Figure 9.22. However, there is a product integration of the ISR and the previous project. To display the 0.7 s and 0.3 s pattern of the countdown, we will employ Timer Output Compare 3 interrupt source.

9.9.3 Design

Basic design components to display seven-segment symbols are taken from the previous project. Here, we will mainly focus on the interrupt to be used and its integration with the previous project. The key to this project is to have a memory location that can be accessed by both the ISR, as well as the main program. We call this memory location CNTR. It will act as a message that is accessible by both main and ISR. The main program will set the value of the CNTR as per the desired delay. Remember that the delays that are needed here are 0.3 and 0.7 seconds. The delays are set in the Blink subroutine as shown in Figure 9.23. The ISR will keep decrementing the CNTR at regular intervals in parallel with the main program because of the OC3 interrupt.

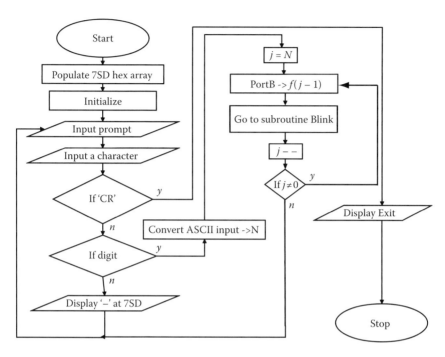

FIGURE 9.22
Main flowchart for the digital countdown display project.

The regular intervals are set in the TOC3 register as shown in the Figure 9.24. It also shows the CNTR as a message accessible to the two important portions of the code. Port B is used to interface with the seven-segment display. As shown in the left hand side of Figure 9.22, the main program executes a loop such that the digital value to be displayed on the seven-segment display is decremented. The program fetches the appropriate hex code in order to display the digit on the seven-segment display.

Following are the BUFFALO utility subroutines that are used in the program:

- OUTA—Output the ASCII character in accumulator A
- INCHAR—Input an ASCII character from the terminal and place in accumulator A
- OUTSTRG—Output the EOT-terminated ASCII string pointed to by X

9.9.4 Description of OC3 ISR

The ISR for OC3 starts at an address that is adjusted by the programmer. Here, we use OC3_SVC label to denote that memory location.

```
ORG     OC3_SVC     ; origin of ISR
```

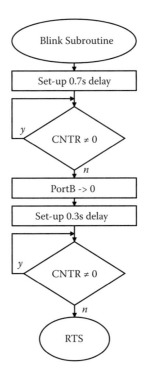

FIGURE 9.23
Flowchart for the blink subroutine.

Refer to Example 9.13 for the code fragment to clear the OC3F bit in TFLG1 register.

```
LDAA    #OC3F       ; set OC3 flag
STAA    TFLG1       ; clear OC3F
```

Now, we set up the time when the next OC3 interrupt should occur. This is done by updating the TOC3 register by the addition of the content of the TOC3 register and adding a time-base of 10 ms. This would mean that interrupt should occur after 10 ms.

```
LDD     TOC3        ; set up for next interrupt
ADDD    TIME_VAL    ; add time-base (10 ms)
STD     TOC3        ; add store
```

The counter that acts as an information transfer medium between the main program and this ISR is decremented next. The last instruction for any ISR is RT.

```
DEC     CNTR        ; decrement time-base counter
RTI                 ; exit
```

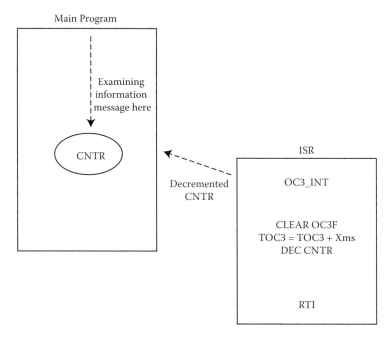

FIGURE 9.24
Information flow between main program and ISR.

The comments in the following program are provided for better understanding of each line/segment of the code and its functionality.

```
*  ***********************************************************
*  Program Name: 7sdTimer.asm
*  Objective: Countdown 7-segment display for digits 1-9
*  Usage: Buffalo Call 13A command
*  Output: 7-segment Display and Computer Monitor
*  ***********************************************************
MAIN    EQU     $013A       ; equating main
DATA    EQU     $0100       ; equating data
DATAs EQU       $0120       ; equating DATAs for 7-segment arrays
ONE_CNT EQU 70              ; 10 MS * 70 = 0.7 SEC
ZERO_CNT EQU 30             ; 10 MS * 30 = 0.3 SEC

OC3_SVC EQU     $01EA   ; ISR starting address, set by programmer
OC3_VEC EQU     $D9     ; for OC3 interrupt source
TCNT    EQU     $100E   ; address of TCNT register
TOC3    EQU     $101A   ; address of TOC register
TMSK1   EQU     $1022   ; address of TMSK1 register
TFLG1   EQU     $1023   ; address of TFLG1 register
OC3F    EQU     %00100000   ; control word - OC3F
OC3I    EQU     %00100000   ; control word - OC3I
```

```
OFFSET    EQU   10000 ; 0.5us *10000 = 5 ms
PORTB     EQU   $1004 ; address of PORTB
************************************************
* Buffalo Utility Subroutines
************************************************
INCHAR    EQU   $FFCD  ; BUFFALO subroutine inchar reference
OUTA      EQU   $FFB8  ; BUFFALO subroutine outa reference
OUTSTRG   EQU   $FFC7  ; BUFFALO subroutine outstrg reference

          ORG   DATA   ; origin of DATA
Indata RMB   1        ; input value through MM, address $0010
CNTR   RMB   1        ; time-base counter for message exchange
iCNTR  RMB   1        ; an i counter
jCNTR  RMB   1        ; a j counter
************************************************
* Set a message for the user to enter a character
************************************************
MSG1      FCC        'Input'
          FCB        $04
************************************************
* Set a message for the user to display that
* the entered character was not a digit
************************************************
MSG3      FCC        'Out of range'
          FCB        $04
************************************************
* Set a message for the user to display that
* the program is exiting
************************************************
MSG4      FCC        'Quit'
          FCB        $04
TIME_VAL  FDB        20000 ; 0.5 microSec *20000 = 10 MS
          ORG        DATAs ; origin of data for 7-segment arrays
************************************************
* Populate an array with the 7-segment hex code
************************************************
DISP FCB     $3F, $06, $5B, $4F, $66, $6D, $7D, $07, $7F, $6F


************************************************
* Populate an array with the 7-segment decimal code
************************************************
NUM  FCB     $00, $01, $02, $03, $04, $05, $06, $07, $08, $09

          ORG     MAIN   ; origin of main program


************************************************************
* Jump Table Initialization
************************************************************
          LDAA    #$7E           ; load JMP
          STAA    OC3_VEC        ; store JMP
```

```
        LDD    #OC3_SVC        ; load OC3 ISR starting address
        STD    OC3_VEC+1       ; store OC3 ISR starting address
*****************************************************************
* TOC3 Initialization
*****************************************************************
        LDD    TCNT            ; get current value of TCNT
        ADDD   #OFFSET         ; add 5 ms offset
        STD    TOC3            ; set up TOC3
*****************************************************************
* OC3 Interrupt Initialization
*****************************************************************
        LDAA   #OC3I           ; get OC3I control word
        STAA   TMSK1           ; enable OC3 in mask register
        LDAA   #OC3F           ; get OC3F control word
        STAA   TFLG1           ; Enable OC3F
        CLI                    ; Enable system interrupt
*****************************************************************


LOOP:       LDX  #MSG1      ; display msg1
            JSR  OUTSTRG    ; call subroutine OUTSTRG
            JSR  INCHAR     ; input character to AccA
            PSHA            ; save value in A for future use
            TAB             ; transfer (AccA) to (AccB)
            LDX  #$0100     ; load the base-address i.e. $0100
to X
            ABX             ; add B to X so that it becomes
address
            LDAA 0,X        ; load the respective value to AccA
            STAA Indata     ; Indata = input data in decimal
system
            PULA
*************************************************
* Check if the character entered is CR
* if so, go to label quit
*************************************************
if_Q:       CMPA #$D        ; compare A with #$D
            BNE  if_dgt     ; branch if not equal to CR
            BRA  quit       ; if equal to #$51 then go to quit
*************************************************
* Check if the character entered belongs 1-9
*Refer to project covered in Chapter 8 for details
*************************************************
if_dgt:     CMPA #$31       ; compare A with #$31
            BLO  no_dgt     ; branch to no_dgt if less than #$30
            CMPA #$39       ; compare A with #$39
            BHI  no_dgt     ; branch to no_dgt if greater than
#$39
*************************************************
            PULA            ; restore saved value of A
*************************************************
```

```
               LDAA Indata      ; jcounter = Indata, initialization
               STAA jCNTR
LoopN          TAB              ; transfer accumulator A to B
               LDX  #$0120      ; load the ini address i.e. $0121
to X
               ABX              ; add B to X, =s address of DISP
               LDAA 0,X         ; load the value from DISP to AccA
               STAA PORTB       ; transfer the hex to port B
               JSR  Blink       ; call blink srt
               LDAA jCNTR       ; increment jcounter
               DECA
               STAA jCNTR
               CMPA #$0         ; check to see if it is the end
               BNE  LoopN       ; go to LoopN
               BRA  LOOP        ; keep looping
*************************************************
* If the entered character is a non-digit, then
* display non-digit error message to the user
*************************************************
no_dgt:   LDX  #MSG3      ; display message MSG3
          JSR  OUTSTRG    ; call subroutine OUTSTRG
          LDAA #$40
          STAA PORTB
          BRA  LOOP       ; keep looping


*************************************************
* If the entered character is CR
* then display exiting message
*************************************************
quit:     LDX  #MSG4      ; display message MSG4
          JSR  OUTSTRG    ; call subroutine OUTSTRG
          SWI

Blink     LDAA #ONE_CNT   ; set up 0.7s delay
          STAA CNTR
LOOP1     LDAA CNTR       ; check counter
          BNE  LOOP1      ; wait if not zero
          LDAA #$0
          STAA PORTB
          LDAA #ZERO_CNT  ; set up 0.3s delay
          STAA CNTR
LOOP0     LDAA CNTR       ; check counter
          BNE  LOOP0      ; wait if not zero
          RTS
**********************************************************************
* The OC3 ISR interrupts every 10 msec.
* It also decrements a counter
* And provides a time-base event at every 10 MS
**********************************************************************
```

FIGURE 9.25
Sample execution of the digital countdown display project.

```
ORG     OC3_SVC    ; origin of ISR
LDAA    #OC3F      ; set OC3 flag
STAA    TFLG1      ; clear OC3F
LDD     TOC3       ; set up for next interrupt
ADDD    TIME_VAL   ; add time-base (10 msec)
STD     TOC3       ; add store
DEC     CNTR       ; decrement time-base counter
RTI                ; exit
```

One way to perform the tests on our program is through entering various input values and inspecting the display of the seven-segment display. Figure 9.25 shows a sample execution of the program. Note how the program responds when a nondigit character is entered. Also, notice how the user quits the program.

Group Discussion: What would happen if the instructions BHI and BLO are swapped by mistake in the program?

Section 9.9 Review Quiz

Which port was used in this project? What is the last instruction used in the ISR?

9.10 Summary

1. Most of the embedded systems are reactive systems.
2. Port A may be used as an 8-bit I/O port or it may be associated with timer and pulse accumulator functions.
3. I/O functions can be interlaced with timer and pulse accumulator functions.
4. Maskable interrupts allow the interrupt request to be ignored if the interrupt disable flag is set.
5. Nonmaskable interrupts are those types that force the processor to respond to the interrupt request.
6. The routine specified by the interrupt is called an *interrupt service routine (ISR)*.

7. An interrupt request (IRQ) is the occurrence of an interrupt signal.

8. All resets and interrupts are vectors.

9. A vector indicates the start address of reset or interrupt routines.

10. A vector address is a 2-byte memory location that stores a vector.

11. HC11 uses a single area in memory to store the vectors called the *vector table*.

12. TCNT is a 16-bit built-in register that is driven by the main clock and is read only for the user.

13. Whenever a roll over occurs with TCNT, the Timer Overflow Flag (TOF) is set in the Timer Interrupt Flag Register 2 (TFLG2) indicating that the TCNT has rolled over (timer overflow).

14. With TOF set, the value of TCNT is interpreted as the number of counts since the latest TCNT overflow.

15. In the event of a timer overflow, if the Timer Overflow Interrupt Enable (TOI) bit of the Timer Interrupt Mask Register 2 (TMSK2) is set, an interrupt will be requested.

16. The 68HC11 comes with a programmable prescalar that allows the user to extend the time of each count by dividing the E clock before it is applied to the input of the TCNT register.

17. The available factors in prescalar are: 1, 4, 8, and 16, and can be selected using the bits 0 and bit 1 in the TMSK ($1024) control register.

18. Output-compare register, dedicated 16-bit comparator, and interrupt generation logic are the three basic building blocks of output-compare system in HC11.

19. The four output events that can be set by configuring TCTL1 are: no effect, toggle OCx pin, clear OCx pin, and set OCx pin.

20. The OC1 event uses the output pin somewhat differently than OC2-OC5.

21. For OC2-OC5, when the event OCx occurs (i.e., valid comparison occurs), output event at pin PAx takes place, depending upon how the bits OMx and OLx are set.

22. Reset is a special kind of interrupt.

Glossary

Baud Rate: The measurement of the number of times per second a signal in a communications channel changes.

Embedded Systems: A computer system designed to perform one or a few dedicated functions, often with real-time computing constraints.

Event: Change in behavior such as signal transition.

Global Control Bit: The entire class of interrupts is controlled with a single bit.

Hard Real-Time System: Tasks must be completed within specified periods.

Interrupt: An asynchronous signal indicating the need for attention, or a synchronous event in software indicating the need for a change in execution.

Interrupt Request (IRQ): The external stimulus.

Interrupt Service Routine (ISR): A function called when a particular interrupt occurs. The instructions executed during an interrupt.

Interrupt Vector: The address of the first instruction of the ISR.

Local Control Bit: Provide individual control of each interrupt within the class.

Maskable Interrupt: Type of interrupt procedure that allows the interrupt request to be ignored if the interrupt disable flag is set.

Mission Critical: Any factor whose failure or disruption is essential to the core function of an organization, and will result in the failure of business operations.

Nonmaskable Interrupt: Type of interrupt procedure that forces the processor to respond to the interrupt request.

Non-real-time System: Systems not subject to operational deadlines from event to system response.

Output Compare: Mechanism that causes output events like pin actions and signal transition to occur on the OCx output pins at specified times.

Output-Compare Registers: Five 16-bit built-in output-compare counters to control output-compare functions. They are TOC1, TOC2, TOC3, TOC4, and TI4/O5.

Polling: Actively sampling the status of an external device by a client program as a synchronous activity.

Reactive System: Maintain an ongoing interaction with their environment rather than produce some final value upon termination.

Real-time Constraint: Operational deadline.

Real-time System: System subject to operational deadlines from event to system response.

Reset: Sets initial conditions within the system and begin executing instructions from a predetermined address.

Soft Real-time System: Tasks may not be completed within specified periods.

TCNT: A 16-bit built-in register that is driven by the main clock and is read only for the user.

TCTL1 (Timer Control Register 1): An 8-bit built-in register used to control the action of each of the output-compare-functions of the HC11.

TCTL2 (Timer Control Register 2): An 8-bit built-in register used to program the edge(s) that the input-capture-functions of HC11 will react to.

TFLG1 (Timer Flag Register 1): An 8-bit built-in register that contains five Output Compare Flags (OC1F to OC5F) and three Input Capture Flags (IC1 to IC3F).

TMSK1 (Timer Interrupt Mask Register 1): An 8-bit built-in register that contains among other control bits, the three Input Capture Interrupt Enable control bits (IC1I to IC3I) primarily used by the Input Capture function of the HC11.

TMSK2 (Timer Interrupt Mask Register 2): An 8-bit built-in register located at $1024. It contains the bit TOI, RTII, PAOVI, PAII, PR1, and PR0.

Transition: The change in state of a signal from high to low, or low to high.

Vector Address: A specific address in memory where the interrupt vector is stored.

Answers to Section Review Quiz

9.2 (a)

9.3 RTI

9.4 (a) True, (b) True

9.5 See Table 9.2

9.6 (a) Free, (b) E, (c) TCNT, (d) TMSK2, TFLG2, (e) TCTL1

9.7 See Table 9.3

9.8 True

9.9 Port B, RTI

True/False Quiz

1. Hard real-time systems are used when it is imperative that an event is reacted to within a strict deadline.
2. The OC events OC2 through OC4 are linked to pins PA4 through PA6, respectively, on PORTA.
3. Generally, asynchronous events do not occur independently of the main program flow.
4. Microcontrollers cannot keep track of various external and internal events by manipulating flags.
5. The bits OMx and OLx in TCTL1 have no role in the generation of any output event in HC11.

6. The bit I4/O5 located at b_2 in PACTL resister determines which function out of OC5 and IC4 are active.

7. In the context of microcontroller, upon receiving an interrupt, the task is accomplished by the execution of a subroutine called interrupt service routine (ISR).

8. Stacking of the current processor registers is never followed by disabling maskable interrupts.

9. Using each time count, the comparator compares the value of the TCNT register with the 16-bit compare register (TOCx and TI4/O5).

10. Interrupts are classified into two categories: maskable interrupts and nonmaskable interrupts.

11. All the output-compare registers like TOx and TI4/O5 are 16-bit read/write registers.

12. The WAI instruction supports the hardware in getting ready for an interrupt.

13. In HC11, the interrupt controls are classified into two categories: local and global.

14. There are four different output events that can be configured using the TCLT1.

15. The PAOVI bit in the TMSK2 register disables the pulse accumulator overflow interrupt.

Questions

QUESTION 9.1

Define software and hardware interrupts. State a few typical uses of interrupts.

QUESTION 9.2

In the computing world, what is multitasking?

QUESTION 9.3

Differentiate hard and soft real-time systems.

QUESTION 9.4

(a) What is an ISR? (b) What is a nonmaskable interrupt? (c) List a few nonmaskable interrupts.

QUESTION 9.5

What is the meaning of an edge triggered interrupt?

QUESTION 9.6

List all the HC11 maskable interrupts from highest priority to lowest priority. Use the default priority in your ranking.

QUESTION 9.7

What are some of the main differences between an ISR and a normal subroutine?

QUESTION 9.8

How would you modify the countdown display project if you were required to display digits in increasing order instead of decreasing order?

QUESTION 9.9

How does the program in the countdown display project check if the character entered is carriage return?

QUESTION 9.10

Which port in HC11 often directly relates to timer and pulse accumulator functions?

QUESTION 9.11

What is the role of TOF flag in TFLG2 register?

QUESTION 9.12

With TOF in TFLG2 set, what should be concluded from the value of TCNT register at any given time?

QUESTION 9.13

What is the role of TOI flag in TMSK2 register?

QUESTION 9.14

How can a programmer reduce the E clock rate in a microcontroller? What options are available to the user to achieve it?

QUESTION 9.15

Name the three basic modules of output-compare system. Name the four different output events that can be configured using the TCTL1.

Problems

PROBLEM 9.1

Let the value of stack pointer (SP) be equal to $E09F. Determine the memory location of the processor registers when the program context is stacked in preparation of executing interrupt service routine.

PROBLEM 9.2

Using Table 9.2, determine the vector address and the interrupt vector in Figure 9.26 The interrupt source is IRQ.

PROBLEM 9.3

Using Table 9.2, determine the interrupt source in Figure 9.27.

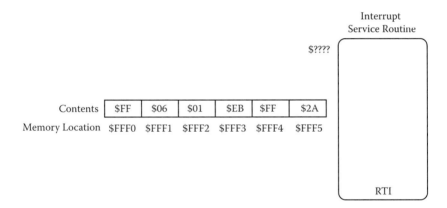

FIGURE 9.26
Basic setup for Problem 9.2.

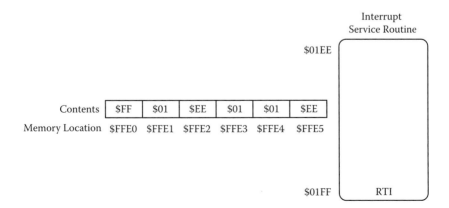

FIGURE 9.27
Basic setup for Problem 9.3.

PROBLEM 9.4

If all the timer input capture interrupt sources occur at the same time, which one will the processor serve first? Will your answer change if all the timer input capture and timer output-compare interrupt sources occur at the same time?

PROBLEM 9.5

Refer to the priority ranking in Table 9.2 and find out which of the following interrupt sources will be served first.

(a) Timer Overflow
(b) SCI Serial System
(c) Timer Output Compare 1

PROBLEM 9.6

Refer to the priority ranking in Table 9.2 and find out which of the following interrupt sources will be served last.

(a) Timer Overflow
(b) SPI Serial Transfer Complete
(c) Real Time Interrupt

PROBLEM 9.7

If the prescale factor is set to 8, how much time (in msec) will the TCNT register require to roll over?

PROBLEM 9.8

The following is a subroutine written by a programmer to clear the Timer Overflow Flag (TOF) in Timer Interrupt Register 2 (TFLG2). Find out the problem in this subroutine.

```
CAT:    LDAA    #%01111111    ; Set MSB = 0 for clearing
        STAA    $1025         ; Write it to TFLG2
        RTS                   ; Return
```

PROBLEM 9.9

The following is a subroutine written by a programmer to clear the Timer Overflow Flag (TOF) in Timer Interrupt Register 2 (TFLG2). The programmer has used the BCLR instruction. Find out the problem in this subroutine.

```
HAT:    LDX     #$1025        ; $1025 is the address of TFLG2
        BCLR    $25, X $7F    ; Use mask $7F
        RTS                   ; Return
```

PROBLEM 9.10

How can you configure TCTL1 register such that the OC3 output pins toggles for an output event?

PROBLEM 9.11

The control word for TCTL1 register is 0100 0000$_2$. What will be the automatic pin action if such a control word is saved to TCTL1 register?

PROBLEM 9.12

Write a code fragment to clear the OC2F bit in TFLG1 register.

PROBLEM 9.13

Write a code fragment to enable the interrupt system for OC2 function. Use the code from Problem 9.12 if needed.

PROBLEM 9.14

Write a code fragment to initialize the jump table in EVBU. The interrupt used is OC2. The ISR stating address is $01EE.

PROBLEM 9.15

Figure 9.28 shows an altered flowchart for the digital countdown display project. What change in the program output will this update bring?

PROBLEM 9.16

Figure 9.29 shows an altered flowchart for the Blink subroutine. What change in the program output will this update bring?

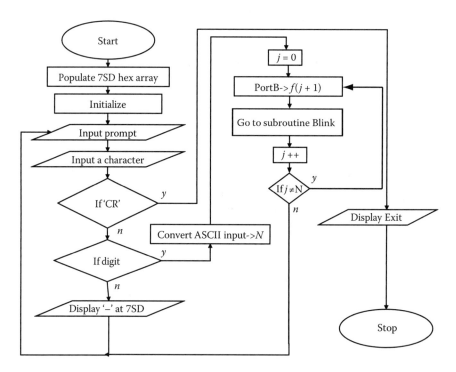

FIGURE 9.28
Altered flowchart for the digital countdown display project.

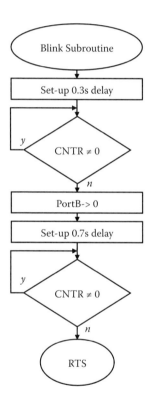

FIGURE 9.29
Altered flowchart for the blink subroutine tables.

10

Analog Capture

"The most important thing in communication is to hear what isn't being said."

—Peter F. Drucker (writer, professor)

OUTLINE

OBJECTIVES

Upon completion of this chapter you should be able to

1. Appreciate the differences between analog and digital quantities.
2. Draw a block diagram for a general data acquisition system.
3. Understand the process of analog-to-digital conversion.
4. Calculate range, step voltage, and resolution.
5. Calculate the digital quantity translated from any analog input.
6. Verify the computed digital output code for an analog-to-digital converter.
7. Express the HC11 hardware used for analog-to-digital conversion.
8. List all the PORTE pin designations and cite the function assignment.
9. Configure the HC11 for single and multichannel conversions.
10. Interface HC11 microcontroller for external hardware for signal transfer.
11. Implement an HC11 solution using analog data acquisition.

Key Terms: Analog, ADC, Analog-To-Digital (A/D), Binary, Channel, DAC, Digital, Digital-To-Analog, Hexadecimal, Maximum Range, Range, Resolution, Sensor, Step, Step Boundaries, Transducer, Voltage Range

10.1 Introduction

Finger counting, or dactylonomy, is the art of discrete counting along one's fingers. The Latin word for finger is *digitus*. The English word *digital* is derived from the same source as the word *digit* or *digitus*. Recall from Chapter 1 the discussion where we compared analog and digital systems. A system that deals with continuously varying physical quantities such as voltage, temperature, pressure, or velocity is called an *analog system*. Most quantities in nature occur in analog, yielding an infinite number of different levels. Most importantly, most quantities in nature that we are interested in measuring are in analog form. However, in Chapter 3 we saw that an HC11 processor is a digital system that can process data only in digital format. A system that deals with discrete digits or quantities is called a *digital system*. Digital electronics deals exclusively with 1's and 0's, or ONs and OFFs. Unlike a continuous-time signal (analog), a discrete-time signal (digital) is not a function of a continuous argument. Therefore, it is easy to conclude that a discrete signal has a countable domain (e.g., the natural numbers). Most of the computing and digital electronics systems are digital. To bridge the gap between the analog forms of the quantities that we are most interested in measuring and their measuring technologies that work only with digital quantities, special mechanisms are needed that can translate the analog signal to some form of digital signal. These mechanisms are required to convert real-world information to binary numeric form so that the processor can use the converted data for processing.

Often a signal is a measured response to changes in physical phenomena, such as sound, temperature, light, position, or pressure. These analog quantities may not be in electrical form. A transducer or sensor is a device that is used to convert the physical quantity into electrical quantity. An analog-to-digital (A/D) conversion is a process where a continuous quantity is converted to a discrete digital number. The reverse operation is performed in digital-to-analog conversion. Microphones, Geiger meters, potentiometers, pressure sensors, thermometers, and antennae are some common examples of transducers. A microphone, for example, converts sound waves that vibrate its diaphragm into an analogous electrical signal.

A/D conversion tools and operation are the main ingredients of this chapter. The main goal of this chapter is introduce the concept of A/D conversion with HC11 in order for the readers to acquire enough knowledge to implement a project that interfaces HC11 microcontroller to a sensor. A project covered at the end

Helpful Hint: All microcontroller-based devices are essentially of digital nature.

of this chapter will serve as a problem-solving opportunity for the students such that the implementation of a solution would require the use of a combination of the knowledge gained from this chapter as well as all the previous chapters.

10.2 Analog-to-Digital Conversion

In all the chapters that we have covered until now in this text, we were dealing with the digital world. We talked in terms of bits, bytes, and digital gates. However, in order to benefit from the advantages that digital signals offer us, we need to recognize that most real-world information is analog in nature. It is necessary, therefore, for a digital system like a microcontroller to be able to read values that are analog. The process of converting an analog signal to digital, along with all the attendant signal manipulation, is usually called *data acquisition*. With the acquired data converted into digital form, the digital system gains the tremendous power of being able to make a variety of measuring devices. An analog-to-digital converter (ADC) is used to convert a signal from analog-to-digital form. The circuits available to do this translation are relatively complex. Their design is a mature art form; however, ready-to-use ICs or modules in the microcontrollers are available to serve as ADC. A designer designing embedded applications must understand the characteristics of the ADC in order to provide a correct program design. Figure 10.1 illustrates a simplified flow of data from analog-to-digital form.

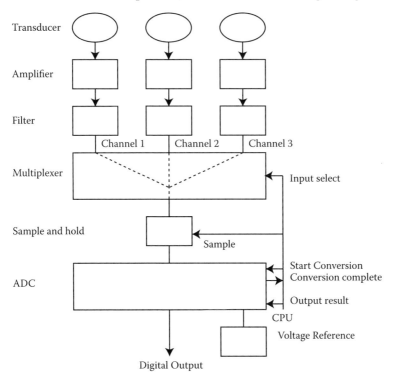

FIGURE 10.1
Elements of a basic data acquisition system.

Note that ADC comes at the end of this block diagram. The first stage is comprised of transducers that generate signals. In this figure, three transducers are shown. These form three channels. A channel is a path for analog data to be converted to digital codes. Transducers usually are electric or electronic devices that transform energy from one manifestation into another. Next, this signal is amplified. In this stage, a DC voltage is added as an offset in order for the signal to match ADC input range. We will study the details of ADC range soon. Next block is the filter that removes unwanted signal components. The next stage is a multiplexer. It is a device that selects one of several analog or digital input signals (three in the figure) and forwards the selected input into a single line. The single line becomes an input for the next block, i.e., sample and hold. This block samples its input signal and holds that voltage as a steady value at its output. Finally, the ADC converts its analog input to digital output. The CPU controls the input select signals for the multiplexer.

Helpful Hint: There are two types of sensors (or transducers). First type outputs a frequency proportional to the measurement. The second type outputs a voltage proportional to the measurement. The latter ones require ADC for data processing.

Helpful Hint: A multiplexer of 2^n inputs has n select lines, which are used to select the input line to send to the output.

Helpful Hint: The primary purpose of the multiplexer is for the several signals to share one device or resource instead of having one device per input signal.

The input for ADC is analog. This means that ADC accepts an input voltage that is infinitely variable. When this input is converted to a digital output, the input is converted to a fixed number of output values. Figure 10.2 illustrates ADC conversion characteristic for conversion of analog voltage to its digital values. As seen in the plot, the input voltage is represented on the

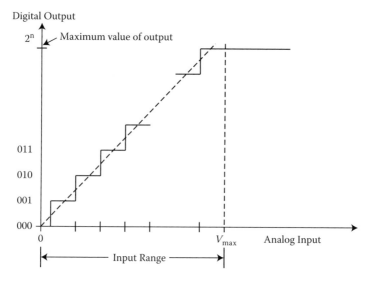

FIGURE 10.2
The Ideal ADC input/output characteristic.

horizontal axis and digital output on the vertical. We begin our observation from the origin (i.e., at the point where analog input and digital output both are zero). To start with, we gradually increase the input voltage from zero, and observe that the output is also initially zero. If we keep moving, at a certain value of input the output changes to 001. It remains at this same value as the input keeps increasing further, until at another input value the output switches to 010. Again, if we keep increasing the input voltage, the output stays at this same value, until at another input value the output switches to 011. If we keep increasing the input voltage like this, the output at some point reaches its maximum value and saturates. At the maximum value of the output, the input has traversed its full range. As shown in Figure 10.2, the input range starts from zero and moves up to the value V_{max}.

10.2.1 Range

In general, the analog inputs are limited to values that fall between a high and a low reference voltage. The voltage range is defined as the difference between the high and low reference voltages. With HC11, we use V_{RH} and V_{RL} to represent voltage-reference high and voltage-reference low, respectively. Thus, the range can be written as:

$$\text{Range} = V_{RH} - V_{RL} \qquad (10.1)$$

The important point to remember is that V_{max} is located such that the horizontal axis is divided into exactly 2^n equal sections, each centered on an output transition. Maximum and minimum reference voltages are the most common specifications found in the manufacturer data sheets of ADC. The maximum range can be easily calculated by subtracting the minimum reference voltage from the maximum reference voltage. Let us take an example of an ADC that accepts an input in the range from –18 volts to +18 volts. Here, +18 volts is the maximum voltage level and –18 volts is the minimum voltage level. We can find out that the maximum range of this system to be equal to 36 volts, (+18 V) – (–18V). Note that while designing applications with ADC, V_{RH} can never exceed maximum reference voltage, and V_{RL} can never go below minimum reference voltage. However, a designer can use the same system and design an application such that $V_{RH} = 10$ volts and $V_{RL} = 0$ volts. The range for such design is now 10 volts, (+10 V) – (0 V). The maximum range remains the same (i.e., 36 volts). The designer now has the opportunity to just observe a small portion (10 volts) of the maximum range (36 volts). This lowering of the range provides the designer better resolution. Resolution with reference to ADC will be discussed shortly.

Helpful Hint: The voltage range is defined as a difference between the high and low reference voltage.

Helpful Hint: Range is commonly associated with input voltage to ADC.

Common Misconception: Students often confuse range and maximum range. While maximum range is the difference between the maximum and minimum reference voltages, the range is the difference between the high and low reference voltages.

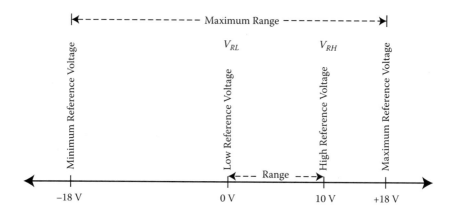

FIGURE 10.3
Graphical representation of some ADC parameters.

Example 10.1

We just discussed an ADC with the following settings:

  Maximum reference voltage = +18 V
  Minimum reference voltage = –18 V
  Maximum voltage range = 36 V
  High reference voltage = +10 V
  Low reference voltage = 0 V
  Range (or voltage range) = 10 V

The relationship between the parameters, along with their values, can be represent graphically on a numbered line.

SOLUTION

Figure 10.3 illustrates the relationship between maximum reference voltage, minimum reference voltage, maximum voltage range, high reference voltage, low reference voltage, and range (or voltage range) with the given data.

Example 10.2

Determine the minimum reference voltage of an ADC with maximum voltage range of 30 volts and maximum reference voltage of +15 volts.

SOLUTION

The maximum voltage range is equal to 30 volts, and maximum reference voltage is equal to +15 volts. We use the formula:

  Maximum voltage range = maximum reference voltage –
  minimum reference voltage
  30 = (+15) – (minimum reference voltage in volts)
  Minimum reference voltage = –15 volts.

Team Discussion: Discuss why th
ramp/counter method similar t
Figure 10.2 where output move
stepwise up to its maximum valu
is generally not considered an eff
cient method.

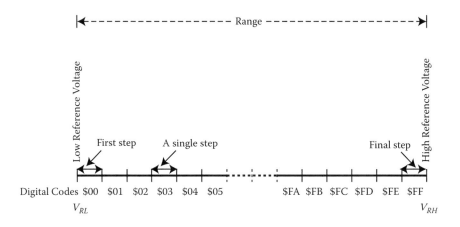

FIGURE 10.4
Illustration of steps, digital codes, and range.

10.2.2 Steps

It is quite clear from Figure 10.2 that the output moves stepwise up to its maximum value. A step is a small portion of the range that has a name or a digital code associated with it. Steps are a sequence of hexadecimal numbers that start from the lowest value and increase to the highest value in the range. In Figure 10.2, each of 000, 001, 010, 011, … are steps. If we redraw Figure 10.2 into a number line, we get Figure 10.4. The digital codes corresponding to the steps are shown in the Figure 10.4. The digital codes are mapped to the corresponding analog input values.

Table 10.1 shows the relation between an n-bit ADC and the number of steps. For an n-bit ADC, the number of steps will be 2^n. Here n is the number of bits in the resulting digital code. For example, for an 8-bit ADC, the number of steps will be 2^8, or 256_{10}. Note that for some ADC the calculation for the number of steps is $2^n - 1$ instead of 2^n. Since the HC11 is designed to use the zero code for the first step, it eliminates the "minus one" from $2^n - 1$.

TABLE 10.1

n-Bit ADC With Their Maximum Output Values

Bits (n)	Computation (2^n)	Maximum Value In Decimal
1	2^1	2
2	2^2	4
3	2^3	8
4	2^4	16
5	2^5	32
6	2^6	64
7	2^7	128
8	2^8	256

Example 10.3

Determine the number of steps that will be formed in an HC11 ADC with number of bits equal to seven. What will be the digital code for the first step and the final step?

SOLUTION

With the number of bits equal to seven, we have $n = 7$. Therefore, the number of steps $= 2^7$, or 128_{10}. The first step will be $00. The last step will be $0111\ 1111_2$, or $7F.

10.2.3 Step Voltage and Digital Code

We know by now that the unit of range is voltage. Many times it is referred to as voltage range for clarity. If we divide the range by the total number of steps in that range, we will get the size of a step in terms of voltage, commonly referred to as *step voltage*. This step voltage (V_{STEP}) is an important parameter used by the converter and the processor. To calculate step voltage, we will be using the following formula:

$$V_{STEP} = (\text{Range})/(\text{Number of steps}) = (V_{RH} - V_{RL})/(2^n) \qquad (10.2)$$

The next example shows the use of this formula to determine step voltage for an ADC.

Example 10.4

If $n = 8$, $V_{RH} = 10$ V, and $V_{RL} = 0$ V, determine (a) range, (b) number of steps, and (c) V_{STEP}.

SOLUTION

(a) Range $= V_{RH} - V_{RL} = 10 - 0 = 10$ V.
(b) Number of steps $= 2^n = 2^8 = 256$.
(c) Using the formula for V_{STEP}, we get

$$V_{STEP} = (10\ V - 0\ V)/(2^8)$$

$$= 10V/256 = 39.06\ mV.$$

In the example above, each step is equal to a range of 39.06 mV. The whole range of 10 V is divided into 256 steps. The magnitude of each of these steps is 39.06 mV. In other words, each of these steps represents a unique 39.06 mV subset of range. Every step has a starting point and an ending point. If we move from lowest voltage to highest voltage in the range of a step, we actually move from the starting point to the ending point of that step. These upper and lower limits to each step are called *step boundaries*. Figure 10.4 is

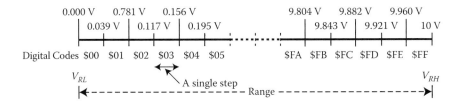

FIGURE 10.5
Illustration of step boundaries.

enhanced to Figure 10.5 in order to show step boundaries. The digital code $00 is assigned to the first step, which covers the section of the range from V_{RL} up to 39.06 mV. The digital code $01 is assigned to the second step, which occupies the portion of the range from 39.06 mV up to 78.12 mV. This process is continued for all the 256 steps to ensure that the entire range of input voltage (analog) is assigned a digital code. The lower boundary of the final step (digital code $FF) can be computed by multiplying 255 (i.e., the second last step) and 39.06 mV (i.e., the step voltage). The result of this multiplication comes out to be 9.96 V. Therefore, the lower boundary of the final step is equal to 9.96V. Also, the upper boundary of the final step will be V_{RH}. The advantage of categorizing in term of steps within the range is to obtain a convenient way of mapping input analog values to their output digital code. By mapping, we mean that any input voltage in the range will have a corresponding digital code at the output of the ADC. Let us take the example of input voltages 9.95 V, 9.97 V, 9.98 V, and 9.99 V. Since the voltages 9.97 V, 9.98 V, and 9.99 V fall within the range of 9.96 V to 10 V, which are the lower boundary and upper boundary of the final step, respectively, they will belong to the final step, which has the digital code $FF. Likewise, input voltage 9.95 V will have the digital code $FE.

Helpful Hint: The zero code is used in HC11 for the first step.

Helpful Hint: The upper boundary of a step is equal to the lower boundary of the next step (if the next step exists).

Often, when designing with ADC, a designer has a known input voltage for the ADC, but its corresponding step is unknown. To find the step to which an input voltage is mapped, a simple method is to divide the input voltage by the step voltage and ignore decimal points. Here, input voltage means the input voltage with reference to V_{RL}. In the formula to follow, we cast the answer from $((V_{IN} - V_{RL})/V_{STEP}$ to an integer in order to truncate the decimal portion of the computation.

$$STEP = (Integer) ((V_{IN} - V_{RL})/V_{STEP})$$

The truncation of the decimal portion is done because steps do not occur in fractions. The next example shows the use of this formula to determine the step for a given input voltage.

Example 10.5

Let $V_{RL} = 0$ V, $V_{IN} = 6.66$ V, and $V_{STEP} = 39.06$ mV. To what step does this input voltage correspond? Convert your answer to hexadecimal in order to find the digital code for that step. Confirm your answer by calculating the lower and upper step boundaries.

SOLUTION

We simply use the formula (Equation 10.2) to find the step.

$$(V_{IN} - V_{RL})/V_{STEP} = (6.66 \text{ V} - 0 \text{ V})/(39.06 \text{ mV}) = 170.506 \text{ steps}$$

Next, we truncate the decimal portion of this answer. We get 170 steps. The reason to truncate is because steps are integers and cannot exist in fractions. When 170 is converted to hex, we get the digital code of $AA. Next, we want to confirm our answer by determining the lower and upper step boundaries. We will use the fact that the upper boundary of a step is equal to the lower boundary of the next step.

$$\text{Lower step boundary for step } 170 = 170 \times V_{STEP} = 6.640 \text{ V}$$

The next step that comes after 170 is step 171. We will use this step to find the upper step boundary of the step 170.

$$\text{Upper step boundary for step } 170 = 171 \times V_{STEP} = 6.679 \text{ V}$$

What is left to confirm is if V_{IN} falls within the step range (i.e., is bound by the lower step boundary and upper step boundary). Since 6.6 V is between 6.64 V and 6.67 V, the step number 170 (or $AA) is the right answer.

10.2.4 Resolution

Most of us know that the display resolution of a digital television is the number of distinct pixels in each dimension that can be displayed. It is commonly believed that the higher the resolution, the better the quality of the image because more pixels can fit in each dimension. Optical resolution describes the ability of an imaging system to resolve detail in the object that is being imaged. But the resolution that we are concerned about in the discussion of ADC is not optical in nature. It is the resolution that is the relationship of each step to the range. Recall the computation of step voltage (V_{STEP}). Since we compute the step voltage by dividing the range (in volts) to the total number of steps in that range, we can easily say that step voltage is the resolution of the system in terms of volts. We repeat the V_{STEP} formula that we discussed earlier.

$$V_{STEP} = (\text{Range})/(\text{Number of steps}) = (V_{RH} - V_{RL})/(2^n)$$

What if we just want a resolution not in terms of volts but in terms of percentage? Such a resolution will be independent of the voltage range and will provide designers information in a scale from 0 to 100. To formulate such a resolution, we just have to replace the voltage range by 100%.

$$\text{Resolution (\%)} = 100\%/\text{Number of steps} = 100\%/2^n \qquad (10.3)$$

The percentage resolution is inversely proportional to the number of steps. This means that systems with more steps within the range will have a lower value of percentage resolution. A lower value of percentage resolution is considered better as it increases the number of possible codes (or more number of steps) and reduces the conversion errors.

Example 10.6

Determine the percentage resolution of an n-bit ADC with (a) $n = 4$ and (b) $n = 8$. Relate n-bit converter with step and range.

SOLUTION

We will use Equation 10.3 to find the resolution.

(a) Resolution (%) = $100/2^4 = 100/16 = 6.25\%$

In this 4-bit ADC, each step of a 16-step system represents 1/16th of the range.

(b) Resolution (%) = $100/2^8 = 100/256 = 0.39\%$

In this 8-bit ADC, each step of a 256-step system represents 1/256th of the range.

Section 10.2 Review Quiz

Fill in the blanks:

(a) A picture scanner while scanning takes the _____ information provided by the light and converts it into _____. (digital/analog/thermal/chemical/incremental)
(b) When a user uses a VoIP solution on his or her computer, a _____ is used to convert his or her voice, which is analog, into digital signal. (A/D, D/A)
(c) An electronic multiplexer is generally considered as a _____-input, _____-output switch. (multiple/single/dual/zero)

10.3 A/D Tools

Recall from Figure 10.1 the flow of data from analog-to-digital form. Figure 10.6 illustrates the HC11 analog-to-digital (A/D) subsystem. Port E is an 8-bit port. The HC11 A/D subsystem inputs an analog (or digital) signal through Port E to a multiplexer. Each bit of Port E acts as a channel. Therefore, HC11 A/D subsystem is an 8-channel multiplexed input converter. It utilizes a sample and hold and an 8-bit successive approximation converter. We will not go in the details of successive approximation conversion here. Multiplexer selects one of several analog or digital input signals (up to eight in case of Port E) and forwards the selected input into a single line. The single line feeds signal as an input to the ADC. Finally, the ADC converts its analog input to digital output. The conversion control unit that consists of registers ADCTL and OPTION selects signals for the multiplexer. We will discuss conversion control in detail soon. The conversion results are stored in result storage registers. The purpose of this section is to look in detail at each of the blocks of Figure 10.6.

Helpful Hint: A successive approximation conversion converts an analog input into a digital representation by using a binary search through all possible quantization levels before finally converging upon a digital output for each conversion.

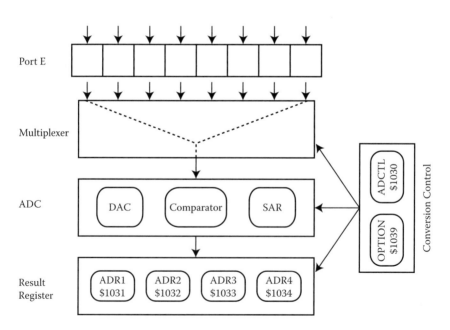

FIGURE 10.6
HC11 analog-to-digital (A/D) subsystem.

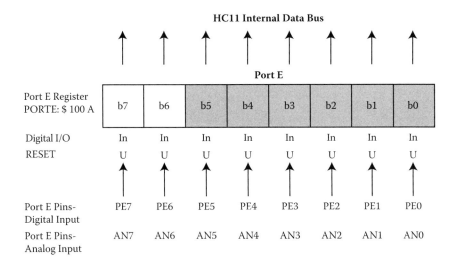

FIGURE 10.7
Details of Port E.

10.3.1 Port E

Recall from Chapter 8 where we briefly outlined Port E's ability to serve as an 8-bit input port or as the 8 channels for an 8-bit ADC. Figure 10.7 illustrates the structure of Port E. The eight general-purpose input pins are labeled PE7-PE0. After reset, the bits in the Port E register (PORTE) are cleared. Observe in the figure that when the input pins of Port E are used for digital input and analog input they are denoted by AN7 through AN0, and PE7 through PE0, respectively. Some or all the bits of Port E can be configured to receive digital signal. The rest can be configured to receive analog signal. Both these types of signals can be received simultaneously. Each input pin or input path is called an *analog channel*. The control bits CD, CC, CB, and CA in A/D Control Register (ADCTL) decide which channel is to be selected by the multiplexer in order to forward the signal from Port E to the analog converter. The multiplexer is required as the converter can convert only one signal at a time. We will study about the ADCTL register in the subsequent subsections.

10.3.2 ADC

This is the block between the multiplexer and result storage registers as shown in Figure 10.6. It converts the analog channel selected by the input multiplexer and transfers the converted digital output to one of the four result registers. ADC contains capacitive array digital-to-analog converter (DAC), a comparator, and a successive approximation register (SAR). DAC array plays two important roles here. The DAC array provides the sample and hold mechanism. It also provides a comparison voltage to the comparator during the successive approximation conversion process. The successive approximation technique starts from the most significant bit (MSB) and processes each bit to come up with an 8-bit result. As soon as the calculation on a bit is completed, it is stored in the SAR. Finally, when the entire 8-bit result is ready, the contents of the SAR are transferred to one of the four registers.

10.3.3 Conversion Control

The two registers that contain eight control bits and one status flag for controlling the A/D function are Option Register (OPTION) and the A/D Control Register (ADCTL). Figure 10.8 shows the relevant structure of OPTION. This register resides at the address $1039, and contains two control bits ADPU and CSEL at b7 and b6, respectively, that are relevant to the A/D function. The ADPU bit turns off the A/D hardware when equal to zero. The A/D hardware is turned on when the ADPU bit is equal to one. Therefore, ADPU bit acts like a master enable. A programmer has to wait for at least 100 microseconds after power-up (ADPU set) to allow the charge pump and comparator circuits to stabilize before the converter system can be used. This is 200 cycles at a 2MHz E-clock.

The CSEL bit selects the reference clock to be used by the A/D and the EEPROM hardware. If equal to zero, the system E-clock is made available for the A/D. On the other hand, if equal to one, a special internal A/D clock that runs at around 2MHz is made available.

Helpful Hint: If the E-clock is 750KHz or higher, CSEL should be 0. Otherwise, CSEL should be 1.

Figure 10.9 illustrates the ADCTL register. It contains a status flag CCF at b7, a unused bit at b6, and six control bits (b5 to b0). The ADCTL register is responsible for:

    I. Setting the time to start conversion.

    II. Setting the type of conversion.

    III. Deciding the channels to scan.

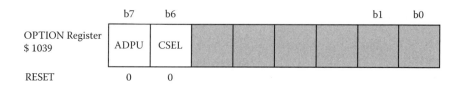

FIGURE 10.8
The OPTION register A/D control bits.

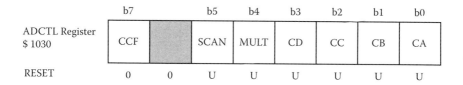

FIGURE 10.9
The ADCTL register control bits and status flag.

Let us start from the most significant bit of the ADCTL register. The CCF bit is a read-only flag. It is set only by the conversion hardware. CCF bit is cleared after reset. By writing to the ADCTL register, a conversion sequence is invoked, and the conversion processes continues without any software interference. The CCF bit is set as soon as the conversion is complete. CCF bit has no effect from a read performed on the ADCTL register. However, the CCF bit is cleared by any write performed to the ADCTL register.

Next is the SCAN bit, located at b_5. When SCAN bit control is cleared, a single conversion sequence will happen. With SCAN bit cleared, if a programmer wants to start a subsequent conversion sequence, he or she will have to write to the ADCTL register. When SCAN bit is set, a sequence conversion occurs continuously. With SCAN bit set, without the need of any write to the ADCTL register, when a sequence conversion ends, another sequence conversion starts automatically. Thus, the SCAN bit controls continuous or single scan modes.

With MULT bit set, four channels are converted, one each time. The CC control bit decides which group of four channels will be converted. With MULT bit cleared, a single channel is converted four consecutive times. The CC:CB:CA control bits decide which channel is selected. Table 10.2 describes the channel selection based on these control bits. HC11 has 16 channels numbered 1 to 16. Channels 1 to 8 correspond to the PORTE pins. CD is equal to zero during the normal operation. CB and CA are ignored when MULT is

TABLE 10.2

Channel Selection and Assignments

Channel Number	CD	CC	CB	CA	Channel Signal	ADRx with MULT=1
1	0	0	$0_{MULT=0}, X_{MULT=1}$	$0_{MULT=0}, X_{MULT=1}$	AN0	ADR1
2	0	0	$0_{MULT=0}, X_{MULT=1}$	$1_{MULT=0}, X_{MULT=1}$	AN1	ADR2
3	0	0	$1_{MULT=0}, X_{MULT=1}$	$0_{MULT=0}, X_{MULT=1}$	AN2	ADR3
4	0	0	$1_{MULT=0}, X_{MULT=1}$	$1_{MULT=0}, X_{MULT=1}$	AN3	ADR4
5	0	1	$0_{MULT=0}, X_{MULT=1}$	$0_{MULT=0}, X_{MULT=1}$	AN4	ADR1
6	0	1	$0_{MULT=0}, X_{MULT=1}$	$1_{MULT=0}, X_{MULT=1}$	AN5	ADR2
7	0	1	$1_{MULT=0}, X_{MULT=1}$	$0_{MULT=0}, X_{MULT=1}$	AN6	ADR3
8	0	1	$1_{MULT=0}, X_{MULT=1}$	$1_{MULT=0}, X_{MULT=1}$	AN7	ADR4

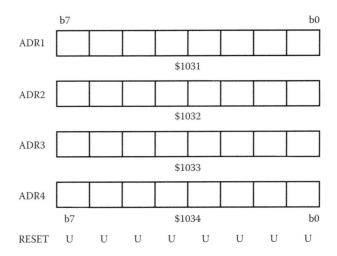

FIGURE 10.10
Result storage registers.

equal to one. CC plays the group selector's role. When CC is set, the upper group (channels 5 to 8) is selected. When CC is cleared, the lower group (channels 1 to 4) is selected. On the other hand, when MULT is equal to zero, CC, CB, and CA together play the group selector's role. We will see how to use this table in some examples in the next section.

10.3.4 Result Registers

During a conversion, the HC11 performs four conversions. The results of the four conversions are saved in a set of four result registers. These results are 8 bits each. Each of the result registers are named ADR1 to ADR4 and are located at $1031 to $1034, respectively. Figure 10.10 illustrates these ADRx registers. All of these are read-only registers.

Section 10.3 Review Quiz

Why is a multiplexer needed in the HC11 analog-to-digital (A/D) subsystem?

10.4 A/D Operation

The flowchart in Figure 10.11 provides a procedure for the A/D converter configuration in order to receive and process analog data from Port E. We will use this flow to solve few problems in the next few examples.

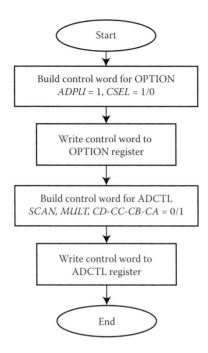

FIGURE 10.11
Procedure for configuring A/D subsystem.

Example 10.7

Build a control word for ADCTL register to start a single scan of Channel 8.

SOLUTION

As per the procedure from Figure 10.11, we start by building a control word for the OPTION register. On the EVBU, BUFFALO activates the A/D subsystem by setting the ADPU bit in the OPTION register. Therefore, we do not have to take any action on this part. So come straight to the building of a control word for the ADCTL register. Since CCF is read-only, and b6 unused, we won't worry about them. The SCAN bit should be equal to zero in order to perform a single scan. Also, to scan a single channel, MULT should be equal to zero. From Table 10.2, for MULT = 0 and channel = 8, we refer to the last row. Here, CD = 0, CC = 1, CB = 1, and CA = 1. Combining all these, we can build a control word as shown in Figure 10.12.

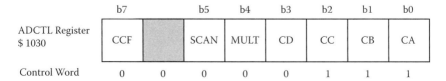

	b7		b5	b4	b3	b2	b1	b0
ADCTL Register $ 1030	CCF		SCAN	MULT	CD	CC	CB	CA
Control Word	0	0	0	0	0	1	1	1

FIGURE 10.12
Example of a control word for ADCTL register.

Example 10.8

Write a code fragment to wait for the conversion to complete after loading the control word built in the previous example.

SOLUTION

As soon as the control word is written to the ADCTL register, three things happen:

    I. A conversion sequence is initiated.
    II. The conversion sequence samples and converts the signal channel 8 (AN7) four successive times.
    III. The result is written into the result storage registers ADR1 through ADR4 respectively.

The code fragment shown here first loads the control word built in the previous example and stores it to the address $1030. Then it keeps looping by testing if CCF has become one or not. If CCF is read as zero, the ADCTL is read again in the loop. As soon as the CCF becomes one, the program execution control comes out of the loop.

```
        LDAA   #%00000111    ;build a control word
        STAA   $1030         ;start conversion sequence
LOOP    LDAA   $1030         ;read ADCTL
        BPL    LOOP          ;if CCF is zero, read ADCTL again
```

BPL stands for Branch if Positive. The BPL instruction checks if $N = 0$ before performing the branch.

Example 10.9

Following the steps shown in Example 10.7, build a control word for ADCTL register to start a single scan of channels 5–8.

SOLUTION

Bits b7 to b5, will remain the same as in Example 10.7. Since now the scan is related to multiple channels, the value of MULT should be one. From Table 10.2, with MULT = 1 and channel group equal to 5–8, we can observe that CC has to be set. CD is obviously zero, and CB and CA have no effect. Here, CD = 0, CC = 1, CB = 0, and CA = 0. Combining all these, we can build a control word as shown in Figure 10.13.

Section 10.4 Review Quiz

Why do we use the instruction BPL in waiting for the conversion to complete?

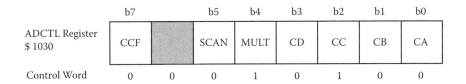

FIGURE 10.13
Example of a control word for ADCTL register.

10.5 A Project with Analog Capture

Temperature sensors are widely used in home, office, and industrial applications. The measurement of temperature is one of the basic requirements for environmental control, as well as certain chemical, electrical, and mechanical controls.

Objective: Interface a temperature sensor with EVBU's channel AN7, and write an HC11 program in order to build a thermometer.

10.5.1 Requirements Analysis

Many different types of temperature sensors are commercially available. The type of temperature sensor suitable for a particular application depends on several factors. The LM34 series are precision integrated-circuit temperature sensors whose output voltage is linearly proportional to the Fahrenheit temperature (10 mV/°F). They are suitable for remote applications or direct PCB mounting. The LM34 has a wide operating voltage range of 5 to 30 volts DC and a temperature range of −50° to +300°F. To display the temperature in Fahrenheit on the screen, the program uses BUFFALO subroutines. But before that, the program will sample and convert the input temperature to the decimal temperature. The A/D converter shall be setup by the program to perform the correct and complete conversion. The program shall also ensure that it waits for the conversion to complete. The program shall wait for one second in order to perform again the set of sample, conversion, and display for continuous repetition.

10.5.2 Hardware Design

As shown in Figure 10.14, LM34 requires three wires for power, ground, and signal. We connect the power and ground wires to the respective positions on the analog input terminal block. The LM34 output is connected to the desired analog input port. We have selected AN7 channel.

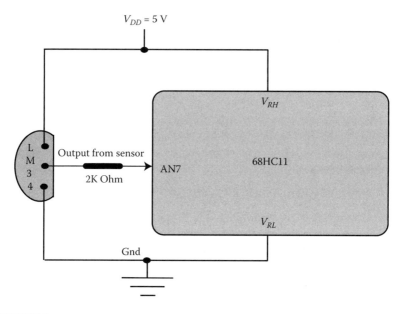

FIGURE 10.14
The circuit diagram for a microcontroller-based thermometer.

10.5.3 Software Design

The software design is not so straightforward as the hardware design. Figure 10.15 shows the program flow. The program starts with setting the A/D converter to convert that data on AN7 using a signal channel, single conversion. We have seen how to make such settings in previous examples. The program waits for the conversion to complete. Next, the program samples and converts the input temperature to a hex value. The hex value is not what we are interested in displaying. Therefore, we will have to convert it into its decimal equivalent. A common dividing and scaling scheme is used. The results that are in decimal are saved as a 4-bit BCD digits in the rightmost digits of a two-byte temporary storage for temperature values. We will call this a TTEMP variable. The two BUFFALO utility subroutines used are:

- OUT2BS: Converts a 2-byte binary number to four ASCII characters and displays them.
- OUTCRL: Outputs carriage return or line feed.

A BUFFALO subroutine DLY10MS located at $E2E5 is used for 10 ms delay. This can be repeated to obtain a 1-second delay in a loop. With a bigger loop we can repeat the program for continuous measurement and display of temperature.

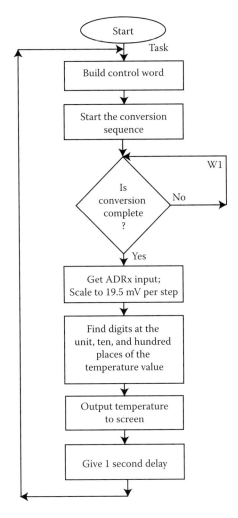

FIGURE 10.15
Flowchart for a microcontroller-based thermometer program.

The program is shown here:

```
*  ***************************************************
*  Program Name: Therm.asm
*  Objective: Measure and display temperature
*  Usage: Buffalo Call 100 command
*  Output: On Screen decimal temperature in Fahrenheit
*  Interface: LM34 temperature sensor and Port E
*  ***************************************************
```

```
DATA      EQU    $0000     ; Equate DATA
MAIN      EQU    $0100     ; Equate MAIN
ADCTL     EQU    $1030     ; Equate ADCTL
ADR1      EQU    $1031     ; Equate ADR1 result register
OUT2BS    EQU    $FFC1     ; Utility subroutines to o/p 2 bytes
OUTCRL    EQU    $FFC4     ; O/p carriage return or line feed
DLY10MS   EQU    $E2E5     ; Internal 10 ms delay
          ORG    DATA      ; Data origin
TTEMP RMB  2              ; Digital temperature temporary storage
          ORG    MAIN      ; Main origin
TASK  LDAA #$00000111     ; Control word for AN7 single scan
      STAA ADCTL          ; Start conversion sequence
W1    LDAA ADCTL          ; Test for CCF = 1
      BPL  W1             ; If CCF = 0 read ADCTL again
      LDAA ADR1           ; Get the digital temp from result reg
      LDAB #195           ; Scale to 19.5 mV step size
      MUL                 ; Perform multiplication
      LDX  #100           ; Remove hundreds scale
      IDIV                ; Integer division
      XGDX                ; Exchange
      LDX  #100           ; Compute hundreds digit
      IDIV                ; Hundreds digit -> X, remainder -> D
      XGDX                ; Hundreds -> AccB, remainder -> X
      STAB TTEMP          ; Save hundreds, use later for display
      XGDX                ; Exchange
      LDX  #10            ; Calculate tens and ones digits
      IDIV                ; Tens digit -> X, ones digit -> D
      STAB TTEMP+1        ; Store ones digit (temp)
      XGDX                ; Tens digit -> B
      LSLB                ; Move tens digit
      LSLB                ; Move tens digit
      LSLB                ; Move tens digit
      LSLB                ; Move tens digit
      ADDB TTEMP+1        ; Ones digit + tens digit
      STAB TTEMP+1        ; Save tens, ones
      LDX  #TTEMP         ; Load X with temp's location
      JSR  OUT2BS         ; Display the temp
      JSR  OUTCRL         ; O/p carriage return/line feed
      LDAB #100           ; 1 sec wait
WAIT1 JSR  DLY10MS        ; Utility 10ms delay
      DECB                ; Decrement AccB
      BNE  WAIT1          ; Test 100 X 10 ms delay completion
      BRA  TASK           ; Start again
```

The testing of this program is left as an exercise for the students. It is highly recommended to play with this program in order to get a better understanding of A/D.

Section 10.5 Review Quiz

What temperature sensor was used in this project?

10.6 Summary

1. Most real-world information is analog in nature. Embedded systems use sensors such as thermometers, barometers, voltmeters, etc.

2. The process of converting an analog signal to digital, along with all the attendant signal manipulation, is usually called *data acquisition*.

3. An analog-to-digital converter (ADC) is used to convert a signal from analog-to-digital form.

4. A channel is a path for analog data to be converted to digital codes.

5. Multiplexer is a device that selects one of several analog or digital input signals and forwards the selected input into a single line.

6. In general, the analog inputs are limited to values that fall between a high and a low reference voltage. The voltage range is defined as the difference between the high and low reference voltages.

7. A step is a small portion of the range that has a name or a digital code associated with it.

8. The number of steps for n-bit ADC is 2n.

9. Step voltage is computed by dividing the range by the total number of steps in that range.

10. The upper and lower limits to each step are called *step boundaries.*

11. To find the step to which an input voltage is mapped to, divide the input voltage by the step voltage and ignore decimal points.

12. Resolution is the relationship of each step to the range.

13. The HC11 has a built-in ADC subsystem.

14. The conversion control unit that consists of registers ADCTL and OPTION select signals for the multiplexer.

15. The control bits CD, CC, CB, and CA in A/D Control Register (ADCTL) decide which channel is to be selected by the multiplexer in order to forward the signal from Port E to the analog converter.

16. The ADC in HC11 converts the analog channel selected by the input multiplexer and transfers the converted digital output to one of the four result registers.

17. ADPU bit in OPTION register acts like a master enable for A/D hardware.

18. The CSEL bit in OPTION register selects the reference clock to be used by the A/D and the EEPROM hardware.
19. The ADCTL register is responsible for setting the time to start conversion, setting the type of conversion, and deciding the channels to scan.
20. The CCF bit in ADCTL register is set only by the conversion hardware and is a read-only flag.
21. The SCAN bit in ADCTL register controls continuous or single scan modes.
22. With MULT bit set, four channels are converted, one each time. With MULT bit cleared, a single channel is converted four consecutive times.
23. Bits CC to CA in ADCTL register play role in channel or channel group selection.
24. The conversion results are stored in result storage registers.

Glossary

ADC: An analog-to-digital converter is a device that converts a continuous quantity to a discrete digital number.

Analog: A system that deals with continuously varying physical quantities such as voltage, temperature, pressure, or velocity. Most quantities in nature occur in analog, yielding an infinite number of different levels.

Analog-To-Digital: See ADC.

Binary: The base 2 numbering system. Binary numbers are made up of 1's and 0's, each position being equal to a different power of 2.

Channel: With reference to ADC, it is a path for analog data to be converted to digital codes.

DAC: A digital-to-analog converter is a device that converts a digital code to an analog signal.

Digital: A system that deals with discrete digits or quantities. Digital electronics deals exclusively with 1's and 0's, or ONs and OFFs. Digital codes (such as ASCII) are then used to convert the 1's and 0's to a meaningful number, letter, or symbol for some output display.

Digital-To-Analog: See DAC.

Hexadecimal: The base 16 numbering system. The 16 hexadecimal digits are 0 to 9 and A to F. Each hexadecimal position represents a different power of 16.

Maximum Range: With reference to ADC, the maximum voltage range is defined as a difference between the maximum and minimum reference voltage.

Range: With reference to ADC, the voltage range is defined as a difference between the high and low reference voltage.

Resolution: With reference to ADC, resolution is the relationship of each step to the range.

Sensor: A device that measures a physical quantity and converts it into a signal that can be read by an observer or by an instrument.

Step Boundaries: Upper and lower limits to each step are called step boundaries.

Step: A step is a small portion of the range that has a name or a digital code associated with it.

Transducer: A transducer is a device that converts one type of energy to another.

Voltage Range: See Range.

Answers to Section Review Quiz

10.2 (a) analog, digital (b) A/D, (c) multiple, single.

10.3 The converter can convert only one signal at a time.

10.4 We use loops to wait for the CCF to set. BPL provides the test condition for the loop.

10.5 LM34.

True/False Quiz

1. Port E is an 8-bit input port.

2. In HC11 analog-to-digital subsystem, digital input, and analog input are selected on a bit-by-bit basis.

3. The number of steps for 8-bit ADC is 2^8, or 256_{10}.

4. Step voltage is a simple division of the total number of steps by the range.

5. The zero code is used in HC11 for the first step.

6. The upper boundary of a step can never be equal to the lower boundary of the next step.

7. The percentage resolution is inversely proportional to the number of steps.

8. The conversion control unit consists of result storage registers.

9. Usually, a programmer has to wait for at least 100 microseconds after power-up (ADPU set) to allow the charge pump and comparator circuits to stabilize before the converter system can be used.

10. The CSEL bit in OPTION register acts like a master enable for A/D hardware.

11. During program execution, CCF bit in ADCTL register changes from read-only to read-write flag.

12. The SCAN bit in ADCTL register is set only by the conversion hardware.

13. HC11 has 16 channels numbered 1 to 16. Channels 1 to 8 correspond to the PORTE pins.

14. When CC bit in ADCTL register is set, the upper group (channels 5 to 8) is selected. When it is cleared, the lower group (channels 1 to 4) is selected.

15. Port E can serve as an 8-bit input port or as the 8 channels for an 8-bit ADC.

Questions

QUESTION 10.1

In the calculation to determine a step, why is the decimal portion truncated?

QUESTION 10.2

What is the role played by the DAC array in HC11 ADC subsystem?

QUESTION 10.3

What would happen if the SCAN bit in ADCTL register is set?

QUESTION 10.4

How is the step voltage computed?

QUESTION 10.5

What are some characteristics of the result storage registers?

Problems

PROBLEM 10.1

Find the maximum voltage range and voltage range if an ADC has the following voltage settings:

Maximum reference voltage = +9 V
Minimum reference voltage = –9 V
High reference voltage = +5 V
Low reference voltage = 0 V

PROBLEM 10.2

Determine the minimum reference voltage of an ADC with maximum voltage range of 20 volts and maximum reference voltage of +10 volts.

PROBLEM 10.3

Determine the number of steps that will be formed in an HC11 ADC with number of bits equal to 4. What will be the digital code for the first step and the final step?

PROBLEM 10.4

If n = 4, VRH = 20 V, and VRL = 0 V, determine (a) range, (b) number of steps, and (c) VSTEP.

PROBLEM 10.5

Let VRL = 0 V, VIN = 4.0 V, and VSTEP = 19.53 mV. To what step does this input voltage correspond to? Convert your answer to hexadecimal in order to find the digital code for that step. Confirm your answer by calculating the lower and upper step boundaries.

PROBLEM 10.6

Determine the percentage resolution of an n-bit ADC with n = 16. Relate n-bit converter with step and range.

PROBLEM 10.7

Build a control word for ADCTL register to start a single scan of Channel 7.

PROBLEM 10.8

Write a code fragment to wait for the conversion to complete after loading the control word built in the previous problem.

PROBLEM 10.9

Build a control word for ADCTL register to start a single scan of channels 1–4.

PROBLEM 10.10

The content of the ADCTL register is equal to $00. How will it affect the A/D function?

11

Input Capture

"Lost time is never found again."

—Benjamin Franklin (one of the
Founding Fathers of the United States)

OUTLINE

OBJECTIVES

Upon completion of this chapter, you should be able to

1. Distinguish between input and output events.
2. Understand the uniqueness of the IC4 function.
3. Identify the three basic modules of the input-capture system.
4. Configure and use basic timing registers such as TCNT, TCTL2, TMSK1/2, TFLG1/2, etc.
5. Explain the input edge detection logic and interrupt generation logic.
6. Understand the input-capture function of HC11.
7. Use the input-capture registers TICx.
8. Use the HC11 input-capture technique in real-time problem solving

Key Terms: Embedded System; Event; Input Compare; Input-Compare Registers; Interrupt; Interrupt Service Routine (ISR); Interrupt Vector; Output Compare; Output-Compare Registers; TCNT; TCTL1; TCTL2; TMSK1; TMSK2; Transition.

11.1 Introduction

Recording time is extremely critical in sprint races such as the 100-meter dash. These races can last for as short time as 10 seconds. How is the time when the runner crosses the finish line recorded in the Olympics? One method to record the finishing time is to project a laser from one end of the finish line to the other. A light sensor, also known as a photoelectric cell or electric eye, receives this beam at one end. As a runner crosses the line, the beam is blocked, and the electric eye sends a signal to the timing console or the microcontroller to record the runner's time. Another method of recording is performed by using a high-speed digital video camera aligned with the finish line that is capable of scanning an image through a thin slit up to 2,000 times a second. When the leading edge of each runner's torso crosses the line, the camera sends an electric signal to the timing console or the microcontroller to record the time. The times that we see in our digital scoreboard are received from the timing console or the microcontroller.

The crossing of the finish line by the athlete is an event. There are thousands of applications that sense a wide range of events. An event may be defined as a change in behavior such as a signal transition. The change in state of a signal from high to low, or low to high is called a signal transition. Events are categorized into internal and external events. If an event is created from within HC11 and sent to some external device, it is called an output event. We dealt with output events in Chapter 9. Additionally, if an external source sends a signal that is sensed by the HC11, it is called an input event.

If an edge transition occurs in the input signal, the system can detect it as an input event. For example, an airplane control system is required to keep track of what time the airplane reaches a certain pressure at high altitudes. The embedded system employed for such monitoring would examine the voltage at the input signal port. The input signal port would probably be connected to a pressure sensor that would provide this voltage. When the pressure has not reached the specified value, the voltage would be low. However, as soon as the pressure reaches the specified value, the voltage at the input signal port would change its state from low to high. This change of state, or transition, would be detected by the system. It will be considered an input event. Input capture is a mechanism designed to capture the time at which an input event takes place. Since the timing of input events cannot be generally controlled by microcontrollers, the recording of the timing for such events is more important than anything else. Input-capture primarily handles this task of saving the timings of input events. Other derivatives of this functionality are to determine the frequency of input events for a given time range, find the width of the input pulse, or compute the period of a signal by subtracting the time between two consecutive rising edges. The primary goal of this chapter is to become familiar with the input-capture functionality of HC11. The project covered at the end of this chapter will use the input-capture functionality of HC11 to determine the period of an input signal by subtracting the time between two consecutive falling edges.

Helpful Hint: Embedded systems using input capture will store a time stamp in memory when an input event occurs.

11.2 Basic Modules of Input Capture

Input capture is a mechanism to store a time stamp in memory when an input event occurs. The HC11 input-capture functions are listed in Table 9.2. These are abbreviated as IC1, IC2, IC3, and IC4. The Timer Input-Capture registers (TICx and TI4/O5) latch the time stamp from the TCNT register. Recall from Chapter 9 that TCNT register is a 16-bit free-running register. The TCNT acts as a reference of time. The Input-Capture registers (TICx and TI4/O5) are referred to as input-capture latches due to their property of latching on the time stamp of the input event. Recall from Chapter 9 how the OMx and OLx bits in TCTL1 register could produce four different output events, for example, no effect, toggle the OCx pin, clear the OCx pin, and set the OCx pin. In the same way, in the case of input-capture, the programmer can program the input-capture functions such that it could detect four combinations of edge behavior, for example, nothing, rising edges only, falling edges only, or any edge. By any edge, we mean any of the rising or falling edges. We will study details of these edge detection settings in subsequent sections. The input-capture functions are realized by three basic modules. These modules are shown in Figure 11.1 and are listed as follows:

- An input-capture register
- Input edge detection logic
- Interrupt generation logic

In the subsequent sections, we will take a closer look at each of these modules. Common hardware is allotted to the OC5 and IC4 functions. This sets IC4 apart from the rest of the ICx functions. The Pulse Accumulator Control (PACTL) register shown in Figure 9.18 hosts the special control bits for the proper operation of IC4. The IC4 function shares a 16-bit register with the OC5 called TI4/O5. DDRA3 is the data direction control bit for bit 3 of the Port A register (PA3). For the IC4 function to work properly, DDRA3 should be cleared. The bit I4/O5 located at b_2 in the PACTL register determines which functions out of OC5 and IC4 are active. When the I4/O5 bit is set, the IC4 function is activated, and the OC5 function is disabled. While reading these sections, you will find many similarities between output-compare and input-capture. However, keep a closer eye on the registers and flags because

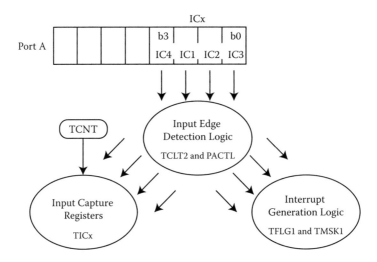

FIGURE 11.1
Basic modules of input capture.

Helpful Hint: The IC4 function does not have a dedicated TICx register.

one can easily get confused between input-capture and output-compare functions.

Section 11.2 Review Quiz

List the basic modules of the input-capture system.

11.3 Input-Capture Registers

The registers used in the input-capture function are listed in Table 11.1. Out of these registers, let us focus on the TIx (i.e., TIC1, TIC2, and TIC3) and TI4/O5. The basic structure of these registers is shown in Figure 11.2. These registers are independent of each other and can operate simultaneously. TIC1, TIC2, and TIC3 are 16-bit read-only registers. Here, read-only means that they cannot be written by software. They are not affected by a reset. The TI4/O5 register is reset to $FFFF. When configured for the IC4 function, it becomes a read-only register. All the 16-bit timer input-capture registers capture the 16-bit value from the TCNT register in response to the occurrence of an input event. This time stamp value can be interpreted as a relative time at which the input event took place. Each of the TICx registers can be accessed by software as a pair

Helpful Hint: TICx registers are capable of operating simultaneously.

Best Practice: Programmers always prefer to use a double-byte load instruction such as load D (LDD) to access data from the TICx registers.

TABLE 11.1

Input-Capture Registers

Name	Address	Brief Description
TICx	$1010 to $1015	Timer input-capture register
T14/05	$101E to $101F	Shared timer input-capture register
TCTL2	$1021	Timer control register 2
TMSK1	$1022	Timer interrupt mask register 1
TFLG1	$1023	Timer interrupt flag register 1
PACTL	$1026	Pulse accumulator control register

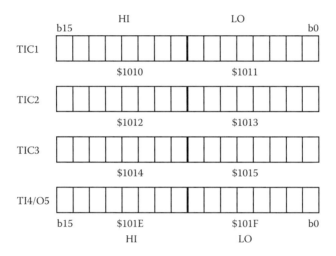

FIGURE 11.2
Structure of TIC registers.

of 8-bit registers. However, the time stamp transfer from the TCNT register to TICx is performed as a single 16-bit parallel word.

Section 11.3 Review Quiz

The TICx registers are affected by a reset. (True/False).

11.4 Input Edge Detection Logic

As mentioned earlier, the four event configurations possible with input-capture functions are: no detection, rising edge only detection, falling edge only detection, and any edge detection. Each of the input-capture events can

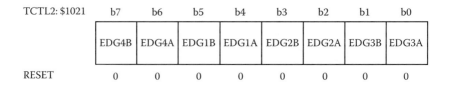

TCTL2: $1021	b7	b6	b5	b4	b3	b2	b1	b0
	EDG4B	EDG4A	EDG1B	EDG1A	EDG2B	EDG2A	EDG3B	EDG3A
RESET	0	0	0	0	0	0	0	0

FIGURE 11.3
Structure of the TCTL2 register.

TABLE 11.2

Basic Rules For TCTL2 Configuration

EDGxB	EDGxA	Event Configuration
0	0	Input-capture function is disabled (no detection possible)
0	1	Capture on rising edges only
1	0	Capture on falling edges only
1	1	Capture on any edges (both rising or falling)

be programmed independently by a programmer. It is up to the programmer to look at the design and interface and decide what configuration he or she would have to set in order to capture the correct input from an input event. The event configuration is determined by the Timer Control Register 2 (TCTL2). The layout for the TCTL2 register is given in Figure 11.3.

As shown in the figure, the bits in the TCTL2 register have the logical names: EDGxB and EDGxA. Again, here x takes values 1, 2, 3, or 4. Table 11.2 itemizes the effect that the bits in TCTL2 have on the event configuration. At reset, all the bits in the TCTL2 register are cleared. This is the default state. This means that with EDGxB and EDGxA all set to zero, ICx is disabled. This is shown in Table 11.2, where EDGxB and EDGxA are equal to zero. Let us look at the settings a programmer will have to make in order to activate the IC1 function to capture on rising or falling edges. As per Table 11.2, each

Helpful Hint: A control word can be built so that the TCTL2 register can be configured with a single write.

EDG1B (b_5) and EDG1A (b_4) must be set to 1. Similarly, for the IC1 function to capture on rising edge only, EDG1B (b_5) and EDG1A (b_4) must be set to 0 and 1, respectively.

Example 11.1

Use Table 11.2 to create a control word to activate IC1 function in order to capture "falling edge only." Write a code fragment to achieve this configuration.

SOLUTION

Accumulator A will be used to load a control word for the TCTL2 register. Since we are dealing with IC1, and the falling edge only, EDG1B (b_5) and EDG1A (b_4) must be set to 1 and 0, respectively. The rest of the bits should be zero. Therefore, we get the control word 0010 0000$_2$. In hex, it is equivalent to $20. To save the control word, we use the load and store method as follows:

Best Practice: Programmers often use the equate directive as much as possible. It improves the readability.

```
LDAA  #%00100000  ; IC1 capture on falling edge only
STAA  $1021       ; Activate ICI
```

Section 11.4 Review Quiz

OMx and OLx are the logical names for the bits in the TCTL2 register. (True/ False)

11.5 Interrupt Generation Logic

Recall Figure 9.14, where we discussed the timer system flags and interrupt enable bits. The TFLG1 register contains the input-capture status flags (ICxF, I4/O5F). If an input-capture function is active, the hardware automatically sets the corresponding flags whenever a selected edge is detected at the corresponding input-capture pin. The status flags are cleared by writing a 1 to the corresponding bit position. Likewise, the TMSK1 register contains the input-capture local interrupt enable bits (ICxI, I4/O5I). Input-capture interrupts are locally controlled by these local interrupt enable bits. Whenever the status flag in TFLG1 is set, its corresponding interrupt enable bit (if set), will request that interrupt. Figure 9.14 is repeated as Figure 11.4 for improved readability.

Common Practice: Good programmers make sure that before leaving the interrupt service routine, their program clears the ICxF bit by writing to the TFLG1 register.

TMSK1: $ 1022	b7	b6	b5	b4	b3	b2	b1	b0
	OC1I	OC2I	OC3I	OC4I	I4/O5I	IC1I	IC2I	IC3I
RESET	0	0	0	0	0	0	0	0

TFLG1: $ 1023	b7	b6	b5	b4	b3	b2	b1	b0
	OC1F	OC2F	OC3F	OC4F	I4/O5F	IC1F	IC2F	IC3F
RESET	0	0	0	0	0	0	0	0

FIGURE 11.4
Timer system flags and interrupt enable bits.

Section 11.5 Review Quiz

With IC1I set in TMSK1, if an input-capture event occurs on pin PA2, the TCNT register's current value will be latched into register TIC1. (True/ False)

11.6 A Project with Input Capture

In general, the period is the time for one complete cycle of an oscillation of a wave. The frequency is the number of periods per unit time (per second) and is typically measured in hertz. We will write a program that will measure the period of an input waveform. The period of an input waveform is measured by capturing two successive edges of the same polarity. By the same polarity, we mean that either both of them should be rising edges, or both of them should be falling edges. The period measured by the counts in the TCNT register will be equal to the difference in time of the two events. If a user wants to measure the width of a pulse, the same method can be used but instead of the same polarity, the opposite polarity will be set. Figure 11.5 illustrates the relationship between the pulse width and period of an input waveform.

First, we will use the equate directive for timer registers TIC1, TCTL2, and TFLG1. Note that the REGBASE is the register base address. This is all done in the top portion of the program.

```
MAIN       EQU    $0120      ;equating main
DATA       EQU    $0000      ;equating data
REGBASE    EQU    $1000      ;register base address
TIC1       EQU    $10        ;Timer input capture register TIC1
TCTL2      EQU    $21        ;Timer Control Register 2
TFLG1      EQU    $23        ;Timer Interrupt Flag Register 1
```

Next, we reserve 2 bytes for storage space and call them EVENT1 and PERIOD. EVENT1 will be used as a temporary time storage for the first falling edge. PERIOD will be used to store the period, which is calculated based on the difference between the times of the two falling edges.

```
EVENT1   RMB   2            ;event1 temporary time storage
PERIOD   RMB   2            ;computed period storage
```

We will be using IC1; therefore, the Timer Input-Capture register (TIC1) will latch the time stamp from the TCNT register. Recall that the TCNT register is a 16-bit free-running register and acts as a reference of time. The TIC1 register is 16-bit read-only register. It captures the 16-bit value from the TCNT register in response to an input event. Refer to Example 11.1 to review the

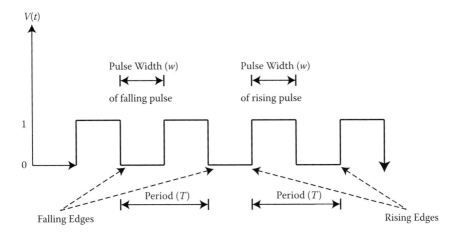

FIGURE 11.5
Elements of an input waveform for input-capture calculations.

details of activating the IC1 function to capture the falling edge only. The example contains a code fragment also.

```
LDY   #REGBASE   ;Load register base address in Y
LDAA  #%00100000 ;build IC1 falling edge control word
STAA  TCTL2,Y    ;set IC1 to detect falling edge
```

The TFLG1 register contains the input-capture status flag (IC1F). The status flag is cleared next.

```
LOOP    BCLR  TFLG1,Y $FB ;clear IC1 flag
```

The next step is for the program to wait for a falling edge. The program waits for the IC1F to be set. Since the input-capture IC1 function is active, the hardware will automatically set the corresponding flags (IC1F) whenever a falling edge is detected at the corresponding input-capture pin. This will be the first falling edge detected by the system. The time stamp transfer from the TCNT register to the TIC1 register will be performed as a single 16-bit parallel word. We will use a double-byte load instruction load D (LDD) to access data from the TIC1 registers. The time stamp is stored in EVENT1.

```
W1  BRCLR TFLG1,Y $04 W1   ;wait for IC1 to set by input
    LDD   TIC1,Y           ;get time of input event
    STD   EVENT1           ;store input event time in EVENT1
```

The status flag IC1F is again cleared, and the program again waits for the IC1F to be set. This time it is waiting for the second falling edge to be detected.

```
    BCLR  TFLG1,Y $FB      ;clear IC1 flag
W2  BRCLR TFLG1,Y $04 W2   ;wait for IC1 to set by input
```

As soon as the hardware automatically sets the corresponding flags (IC1F) in response to a falling edge, the time stamp transfer from the TCNT register to TIC1 will be performed as a single 16-bit parallel word. We will again use LDD to access data from the TIC1 registers. The time stored in EVENT1 is subtracted from this time stamp, and stored in PERIOD.

```
LDD    TIC1,Y              ;get time of second input event
SUBD   EVENT1              ;find EVENT2 - EVENT1
STD    PERIOD              ;store period in PERIOD area
```

The program can be made to keep repeating this edge detection and period calculation with a BRA instruction.

```
BRA    LOOP                ;keep looping
```

The program flow is shown in Figure 11.6. The complete program is as follows:

Helpful Hint: The difference calculated through input-capture function is the period of the input waveform measured in counts of the TCNT register.

```
* * * * * * * * * * * * * * * * * * * * * * * * * * * * * * * * * * * * * * * * * * * * * * * * * *
* Name: InCap.asm
* Objective: Determine the period of input signal
* Usage: Call 120
* Output: Saved in memory at PERIOD
* * * * * * * * * * * * * * * * * * * * * * * * * * * * * * * * * * * * * * * * * * * * * * * * * *
MAIN         EQU     $0120            ;equating main
DATA         EQU     $0000            ;equating data
REGBASE      EQU     $1000            ;register base address
TIC1         EQU     $10              ;Timer input capture register
TIC1
TCTL2        EQU     $21              ;Timer Control Register 2
TFLG1        EQU     $23              ;Timer Interrupt Flag Register 1

             ORG     DATA             ;standard ORG statement
EVENT1       RMB     2                ;event1 temporary time storage
PERIOD       RMB     2                ;computed period storage

             ORG     DATA             ;standard ORG statement
             LDY     #REGBASE         ;Load register base address
             LDAA    #%00100000       ;build IC1 falling edge
             STAA    TCTL2,Y          ;set IC1 to detect falling edge
LOOP         BCLR    TFLG1,Y $FB      ;clear IC1 flag
W1           BRCLR   TFLG1,Y $04  W1  ;wait for input
             LDD     TIC1,Y           ;get time of input event
             STD     EVENT1           ;store input event time
             BCLR    TFLG1,Y $FB      ;clear IC1 flag
W2           BRCLR   TFLG1,Y $04  W2  ;wait for input
             LDD     TIC1,Y           ;get time of second input event
             SUBD    EVENT1           ;find EVENT2 - EVENT1
             STD     PERIOD           ;store period in PERIOD area
             BRA     LOOP             ;keep looping
```

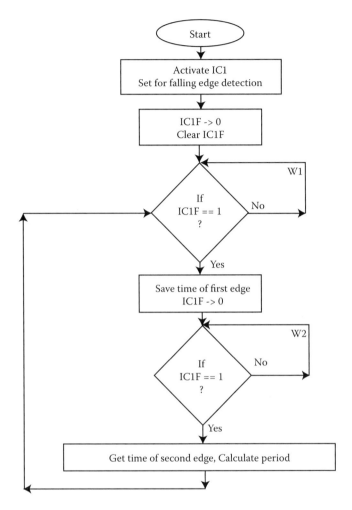

FIGURE 11.6
Program flow for measuring the period of an input waveform.

Section 11.6 Review Quiz

Name three HC11 timer-system-related registers that were used in the project.

11.7 Summary

1. The input-capture feature of the HC11 monitors three input pins of Port A. These pins are PA2 (IC1), PA1 (IC2), and PA0 (IC3).
2. Each input-capture function comprises a 16-bit latch, input edge detection logic, and interrupt generation logic.

3. Edges are of two types: rising or falling.

4. Control bits are included in the edge detection logic in order to select the type of the edge to be detected.

5. A pair of control bits (EDGxB, EDGxA) in the TCTL2 register is used to select the type of edges detected by each input-capture function.

6. The interrupt generation logic is comprised of a status flag and a local interrupt enable bit.

7. The status flag of interrupt generation logic indicates that an edge has been detected.

8. The local interrupt enable bit of the interrupt generation logic is responsible for generating a hardware interrupt request.

9. If a selected edge is detected on one of the input-capture pins, the timer "captures" the event.

10. "Capture" means that the 16-bit latch records the current timer value from the TCNT register into a register (TIC1, TIC2, or TIC3) and sets a flag that means that the input event occurred (IC1F, IC2F, or IC3F in TFLG1).

11. If the ICxI is equal to zero, the corresponding input-capture interrupt is inhibited. This turns the input capture to operate in polled mode. In polled mode, the ICxF bit must be read by user software to find when an edge has been detected.

12. If the ICxI control bit is equal to one, and if the corresponding ICxF bit is set to one, a hardware interrupt request will be generated.

13. Input-capture events can be used to measure the width of a pulse.

Glossary

Embedded System: An embedded system is some combination of computer hardware and software, either fixed in capability or programmable, that is specifically designed for a particular function.

Event: A change in behavior such as a signal transition.

Input Compare: A mechanism to capture the time at which an input event takes place.

Input-Compare Registers: Four 16-bit built-in input-capture registers designed to capture 16-bit values from the TCNT register when an input-capture event occurs. They are TIC1, TIC2, TIC3, and TI4/O5.

Input Event: An external source sending a signal to be sensed by the HC11.

Interrupt: An asynchronous signal indicating the need for attention or a synchronous event in software indicating the need for a change in execution.

Interrupt Service Routine (ISR): A function called when a particular interrupt occurs. The instructions executed during an interrupt constitute an interrupt service routine.

Interrupt Vector: The address of the first instruction of the ISR.

Output Compare: Mechanism that causes output events such as pin actions and signal transition to occur on the OCx output pins at specified times.

Output-Compare Registers: Five 16-bit built-in output compare counters to control output compare functions. They are TOC1, TOC2, TOC3, TOC4, and TI4/O5.

TCNT: A 16-bit built-in register that is driven by the main clock and is read only for the user.

TCTL1 (Timer Control Register 1): An 8-bit built-in register used to control the action of each of the output-compare functions of the HC11.

TCTL2 (Timer Control Register 2): An 8-bit built-in register used to program the edges that the input-capture-functions of HC11 will react to.

TFLG1 (Timer Flag Register 1): An 8-bit built-in register that contains five Output Compare Flags (OC1F to OC5F) and three Input-Capture Flags (IC1 to IC3F).

TMSK1 (Timer Interrupt Mask Register 1): An 8-bit built-in register that contains, among other control bits, the three Input-Capture Interrupt Enable control bits (IC1I to IC3I) primarily used by the input-capture function of the HC11.

TMSK2 (Timer Interrupt Mask Register 2): An 8-bit built-in register located at $1024. The bits it contains are TOI, RTII, PAOVI, PAII, PR1, and PR0.

Transition: The change in state of a signal from high to low, or low to high.

Answers to Section Review Quiz

11.2 An input-capture register, input edge detection logic, and interrupt generation logic

11.3 False

11.4 False

11.5 True

11.6 TIC1, TCTL2, and TFLG1

True/False Quiz

1. In HC11, an 8-bit input latch captures the current value of the free-running counter (TCNT).

2. The TICx registers are not affected by reset. They cannot be written by software.

3. Each of the three input-capture functions can be independently configured for edge detection.

4. The ICxF status bit is always set to 1 each time a falling edge is detected at the corresponding input-compare pin.

5. The ICxF status bit is never set to 1 each time a rising edge is detected at the corresponding output-capture pin.

6. The ICxF status bit is automatically set to 1 each time a selected edge is detected at the corresponding input-capture pin.

7. A programmer can program each input-capture function to detect a particular edge polarity on the corresponding timer input pin.

8. The input-capture functionality of HC11 can be used to find the period of a signal.

9. The input-capture functionality of HC11 can be used to find the width of a pulse.

10. With TOF set, the value of TCNT is interpreted as the number of counts since the latest TCNT overflow.

11. A programmer can never change the E clock rate.

12. The two Timer Interrupt Flag registers (TFLG1 and TFLG2) contain 12 status flag bits.

13. The period of a signal can be measured by computing the time between two rising edges.

14. The period of a signal can be measured by computing the time between two falling edges.

15. It is extremely important that the software clear the ICxF bit by writing to the TFLG1 register before leaving the ISR.

Questions

QUESTION 11.1

How can we measure a time greater than the range of the 16-bit main timer counter with an input capture?

QUESTION 11.2

What are some uses of the input-capture functionality of HC11?

QUESTION 11.3

An unfiltered bouncing switch contact is a common example where there could be a number of closely spaced edges. How are these undesirable extra captures avoided?

QUESTION 11.4

What is the role of the TCNT register in the input-capture functionality of HC11?

QUESTION 11.5

How does the input-capture system behave if the interrupt request is inhibited?

Problems

PROBLEM 11.1

Use Table 11.2 to create a control word to activate IC2 function in order to capture any edge. Write a code fragment to achieve this configuration.

PROBLEM 11.2

Show how you will clear the status flag IC2F.

PROBLEM 11.3

What might be the problems in the following equate direction?

```
REGBASE    EQU    $1000    ;register base address
TIC1       EQU    $10      ;Timer input capture register TIC1
TCTL2      EQU    $23      ;Timer Control Register 2
TFLG1      EQU    $21      ;Timer Interrupt Flag Register 1
```

12

Higher-Level Programming

"When you take something incredibly complex and try to wrap it in something simpler, you often just shroud the complexity."

—Anders Hejlsberg (lead architect of the C# programming language)

OUTLINE

OBJECTIVES

Upon completion of this chapter you should be able to

1. Understand the levels in programming languages
2. Understand the C program syntax
3. List the fundamental data types in C
4. Use the arithmetic operators in C
5. Understand memory allocation for integer and floating point data types
6. Solve moderately difficult problems using pointers and arrays

Key Terms: Arithmetic, Array, Assembly Language, Assembler, Compiler, Code Block, CPU, Data Type, Declaration, Escape Sequence, Function, Higher-Level Languages, Loops, Lower-Level Languages, Machine Language, Operators, Pointer, Subroutine, Variable.

12.1 Introduction

From organizing a closet to deciding to buy a house, people regularly solve problems on a day-to-day basis. Problems can range from small to very large. An algorithm may be loosely defined as a set of instructions for solving a problem. An algorithm is an effective method for solving a problem expressed as a finite sequence of steps. We know by now that proficiency with algorithms is of strategic value in using the computer as a problem-solving tool, since a computer can solve a problem only after it has been told how to solve it. Our strategy in solving problems in the projects covered in previous chapters was to first perform requirements analysis, then develop a design, followed by implementation and testing. In all these projects, our effort was required to develop a detailed solution, which subsequently was communicated to the computer. The coding performed in those projects used an assembly programming language suitable for HC11 microcontroller. A hierarchy diagram of computer programming languages relative to the computer hardware was shown in Figure 4.1. We repeat the same figure here as Figure 12.1. As shown in the figure, the level above assembly language is higher-level language, which is closer to human language and further from machine language. A compiler is used to translate higher-level language into machine language. In this chapter, we will learn the basics of C programming in order for us to be able to later write simple to moderately difficult programs in C programming language. We will cover numerous examples in order to illustrate basic C programming for firm foundation building. This chapter will serve as a pointer to those students who want to move from assembly programming to higher-level programming for microcontroller or embedded systems. A project in C will be covered at the end of this chapter.

12.2 Levels in Programming Languages

FORTRAN or formula translation was the first practical higher-level programming language invented by John Backus for IBM in 1954. It was later released for commercial use in 1957. Fortran is still used today in computationally intensive areas such as numerical weather prediction, finite element analysis, computational fluid dynamics, computational physics, and computational chemistry. John Backus had a vision of creating a programming language that was closer in appearance to human language, which is the definition of a higher-level language. A higher-level programming language hides from view the details of CPU operations such as memory access models and management of scope. Additionally, as mentioned earlier,

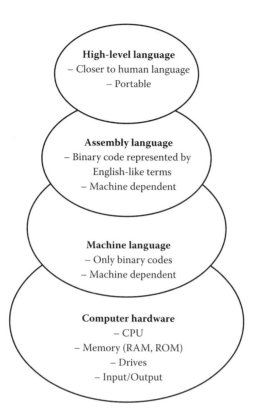

FIGURE 12.1
Hierarchy of programming languages relative to computer hardware.

such languages are mainly considered higher-level because they are closer to human languages and farther from machine languages. Therefore, the main advantage of high-level languages over lower-level languages is that they are easier to read, write, and maintain. Some examples of higher-language programming languages are C, C++, C#, JAVA, etc. Rather than dealing with registers, memory addresses and call stacks, higher-level languages deal with variables, arrays, objects, complex arithmetic, threads, locks, and other abstract computer science concepts, with a focus on usability over optimal program efficiency. Another advantage of higher-level language over assembly language is that it is portable, which means that a program can run on a variety of computers.

A lower-level programming language on the other hand provides little or no

Common Misconception: The term "higher-level language" does not mean that the language is superior to lower-level programming languages.

Helpful Hint: A higher-level language isolates the execution semantics of computer architecture from the specification of the program. This simplifies the program development when compared to a lower-level language.

abstraction from a computer's instruction set architecture. The word "lower" refers to the small or nonexistent amount of abstraction between the language and machine language. This is the reason why lower-level languages are often described as being "close to the hardware." Almost all the assembly languages belong to the set of lower-level languages. The most basic difference is that the statements in a lower-level language can be directly correlated to processor instructions, while a single statement in a higher-level language may execute dozens of processor instructions. A program created in a lower-level language can be made to run faster, and with a smaller memory footprint as compared to its equivalent program created with higher-level language.

Helpful Hint: Lower-level languages are simple, but are considered difficult to use, due to numerous technical details which must be remembered by the programmer.

Example 12.1

Which of the following belong to higher-level programming languages?

(a) C#
(b) Delphi
(c) Fortran
(d) Java
(e) Perl
(f) All of the above

Helpful Hint: Lower-level languages have the advantage that the programmer is able to tune the code to be smaller or more efficient, and that more system-dependent features are sometimes available.

SOLUTION

(f) All of the above.

Example 12.2

Dataflow languages rely on a (usually visual) representation of the flow of data to specify the program. They are frequently used for reacting to discrete events or for processing streams of data. Give some examples of dataflow languages.

SOLUTION

Some dataflow languages are Hartmann pipelines, LabVIEW, Prograph, Max, Pure data, gAlan, Beast/BSE etc.

Section 12.2 Review Quiz

In computer programming, a ____-level programming language is a programming language that provides little or no abstraction from a computer's instruction set architecture. (lower/ higher)

12.3 C Programming

C is a programming language originated for developing the Unix operating system. It is a higher-level and an imperative (procedural) systems implementation language. C is one of the most popular programming languages of all time, and there are very few computer architectures for which a C compiler does not exist. C is widely used for developing portable application software and system-level software. C has some very attractive characteristics that make it one of the best candidates for systems programming. Some of these characteristics are code portability and efficiency, ability to access specific hardware addresses, and low runtime demand on system resources. Therefore, it is often used for "system programming," including implementing operating systems and embedded system applications. We will learn about the fundamentals of C programming in this section.

12.3.1 Getting Started with C

As seen in Chapter 7, probably the best way to start learning a programming language is by writing a program. Let us look at a "Hello World!" program written in C. Many programmers run this program in order to ensure that a language's compiler or assembler, development environment, and runtime environment are correctly installed. We already know what output will be created by this program. Yes, it is the string "Hello World!"

A "Hello World!" in C looks like the following:

```
/*
Hello World! program in C
This program outputs a string
*/
#include <stdio.h>
  int main()
  {
        printf("Hello World!"); //output the string
        return 0;
  }
```

You can save the code in a file with the name and extension as "hello.c". This program can be compiled by typing:

```
>gcc hello.c
```

This will create an executable file a.out, which is then executed simply by typing its name. The result is that the characters "Hello World!" are printed out.

C supports two comment formats. The first allows a programmer to write comments over multiple lines.

```
/*
Hello World! program in C
This program outputs a string
*/
```

The second comment type was added in the C99 standard. By using a double slash, a programmer indicates that anything from that point until the end of the line is a comment.

```
printf("Hello World!"); //output the string
```

The first statement "#include < stdio.h>" includes a specification of the C input/output library. All variables in C must be explicitly defined before use. The directive "#include" tells the C compiler to insert the contents of the specified file at that point in the code. The "<...>" notation instructs the compiler to look for the file in certain "standard" system directories.

Helpful Hint: The ".h" files are by convention "header files" which contain definitions of variables and functions necessary for the execution of a program.

Functions in C are similar to subroutines. Functions specify the tasks to be performed by the program. The statement "int main()" tells the compiler that there is a function named "main," and that the function returns an integer, hence int. The "curly braces" ({ and }) signal the beginning and end of functions and other code blocks. A *code block* is a section of code that is grouped together. Blocks consist of one or more declarations and statements. Code blocks are often delimited by curly braces. A program must contain one and only one main (). Recall Figure 7.1 where we discussed the fixed field format for .ASM source code. Unlike most of the lower-level programming languages, all C statements are defined in free format (i.e., with no specified layout or column assignment). Whitespace (tabs or spaces) is never significant, except inside quotes as part of a character string.

C programs are built by a series of program statements. These statements are terminated by a semicolon. The semicolon is part of the syntax of C. It tells the compiler that you are at the end of a command. You can see these semicolons in the program just shown. A clear understanding of statements and their appropriate use is critical in order to implement a solution correctly.

12.3.2 Data Types

When writing a C program, a programmer has to tell the system beforehand about what type of numbers or characters will be used in the program. These

TABLE 12.1

Data Types in C

Data Type	Keyword	Bytes Required
Character	char	1
Unsigned character	unsigned char	1
Integer	int	2
Short Integer	short int	2
Long Integer	long int	4
Unsigned Integer	unsigned int	2
Unsigned Short Integer	unsigned short int	2
Unsigned Long Integer	unsigned long int	4
Float	float	4
Double	double	8
Long Double	long double	10

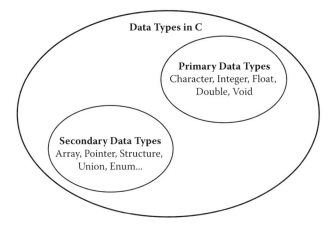

FIGURE 12.2
Categories of data types in C.

are data types. A data type is a set of data with values having predefined characteristics. Examples of data types are integer, floating point unit number, character, string, etc. There are many data types in C language. A programmer has to use appropriate data type as per his or her requirement. Some data types used in C are shown in Table 12.1, and their categorization is shown in Figure 12.2.

Integers are whole numbers. The range of values for the integers is machine dependent. C has 3 classes of integer storage namely short int, int, and long int. Figure 12.3 illustrates an example of memory allocation for these classes of integers. You can see that a short int requires half the space than normal

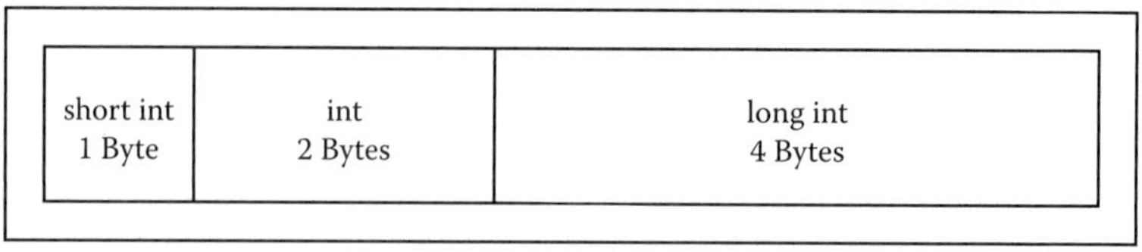

short int	int	long int
1 Byte	2 Bytes	4 Bytes

FIGURE 12.3
Memory allocation for integer data types.

integer values. Integers can be typed as signed and unsigned. Unsigned numbers are always positive. The use of int is shown as follows:

```
{
    int count;
    count = 15;
}
```

In the above example, the code block consists of four lines. Two of them are occupied by curly braces. The symbolic name "count" is called a variable. Variables are memory locations that are given names and can be assigned values. We use variables to store data in memory for later use. A declaration statement provides a data type and a variable name. The third line in the code block above is a declaration statement with an equal sign (=). This is a binary operator called the *assignment operation*. We will look into operators in the next subsection.

Helpful Hint: Unless a programmer specifies otherwise, all forms of integers are signed by default.

Helpful Hint: The long and unsigned integers are used to declare a longer range of values.

In computing, floating point describes a system for representing real numbers that supports a wide range of values. In C, a floating point number represents a real number with six-digits precision. As shown in Table 12.1, floating point numbers are denoted by the keyword *float*. If more accuracy of the floating point number is required, the type double can be used to define the number. Figure 12.4 illustrates an example of memory allocation for floating point data type. You can see that a float requires half the space than normal double values. Data types float and double are used in the code blocks shown as follows:

```
{
    float yards;
    yards = 5.5;
}
{
    double cells;
    cells = 2700000;
}
```

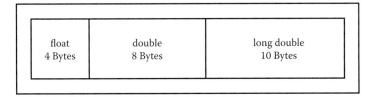

FIGURE 12.4
Memory allocation for floating point data types.

Void is also a data type. A void type has no values therefore we cannot declare it as variable as we did in the case of integer and float. A void data type is normally used with function to specify its type. Our "Hello World!" program is an example where we declared "main()" as void type because it does not return any value. A single character can be defined as a character type of data. In other words, character type variable can hold a single character. Characters are usually stored in 1 byte of internal storage. The qualifier signed or unsigned can be explicitly applied to char. The following code block shows the use of character data type.

Helpful Hint: With object-oriented programming, a programmer can create new data types to meet application needs. Languages that leave little room for programmers to define their own data types are said to be strongly typed languages.

Self-Learning: Do some research from the Internet and books on C programming and find out the standard range for various fundamental data types used in C.

```
{
    char Letter;
    Letter = 'z';
}
```

12.3.3 Operators

An operator is a symbol that helps the user to command the computer to do a certain mathematical or logical manipulations. An operand is something that an operator acts on. We have seen the assignment operator before. Operators are used to operate on data and variables. C has many operators which can be classified as

1. Mathematical operators
2. Relational operators
3. Logical operators
4. Assignment operators

Arithmetic operators fall under mathematical operators. The C programming language supports the most common arithmetic operator. All these operators are used in most of the other languages in the same way. Table 12.2 lists some of the arithmetic operators. Modulus operator (%) returns the remainder of

TABLE 12.2

Arithmetic Operators

Operator	Operator
Addition	+
Subtraction	-
Multiplication	*
Division	/
Modulus operator	%

integer division calculation. The operators have precedence rules which are the same rule in mathematics. Some examples of arithmetic operators are

```
x + y
x - y
-x + y
a * b + c
-a * b
```

Helpful Hint: The modulus operator evaluates the remainder of the operands after division.

Note that here a, b, c, x, and y are known as operands.

Example 12.3

Integer arithmetic is an operation when an arithmetic operation is performed on two whole numbers or integers. The result is always an integer. Let a = 27 and b = 5 be two integer numbers. What will be the result of the following integer operations?

    (a) a + b
    (b) a − b
    (c) a * b
    (d) a % b
    (e) a / b

SOLUTION

Following are the results of the integer operations:

    (a) a + b = 32
    (b) a − b = 22
    (c) a * b = 115
    (d) a % b = 2
    (e) a / b = 5

Helpful Hint: In integer division, the fractional part is truncated.

Example 12.4

Floating point arithmetic is an operation when an arithmetic operation is preformed on two real numbers or fraction numbers. Let a = 14.0 and b = 4.0. What will be the result of the following floating point operations?

   (a) a + b
   (b) a − b
   (c) a * b
   (d) a / b

SOLUTION

Following are the results of the integer operations:

   (a) a + b = 18.0
   (b) a − b = 10.0
   (c) a * b = 56.0
   (d) a / b = 3.50

Example 12.5

The following C program demonstrates some arithmetic operators. What will be the output of this program?

```c
#include <stdio.h>

void main(){
        int x = 10, y = 20;

        printf("x = %d\n",x);
        printf("y = %d\n",y);

        /* demonstrate = operator + */
        y = y + x;
        printf("y = y + x; y = %d\n",y);

        /* demonstrate - operator */
        y = y - 2;
        printf("y = y - 2; y = %d\n",y);

        /* keep console screen until a key stroke */
        char key;
        scanf(&key);
}
```

SOLUTION

The output on the screen is achieved using printf that writes to the standard output a sequence of data formatted as the format argument specifies. The first printf in the code above has %d as a *specifier* and x as an *argument*. Also, \n is an escape sequence. An escape sequence is a series of characters used to change the state of computers and their attached peripheral devices. The C programming language provides many escape sequences. The escape sequences for newline and carriage return are '\n' and '\r', respectively. A specifier defines

TABLE 12.3

C Operators in Order
of Precedence

Operator	Associatively
Unary -	Right to left
(,)	Left to right
*, /, %	Left to right
+, -	Left to right

the type and the interpretation of the value of the corresponding argument. The specifier %c, %d or %i, %f, and %s output character, signed decimal integer, decimal floating point, and string of characters, respectively. The output of the program is the following:

```
x = 10
y = 20
y = y + x; y = 30
y = y - 2; y = 28
```

Each operator in C has a relative order of precedence. If a statement contains two or more operators, the order of precedence provides a nonambiguous answer. However, the precedence can be altered by the use of parenthesis (). The first evaluation is always performed for operators in the parenthesis. Then, the operators outside the parenthesis are evaluated. If more than one parenthesis are present, then left-to-right direction of evaluation is followed. Table 12.3 lists order of precedence for some operators.

Helpful Hint: Operators on the same line in the Order of Operator Precedence have the same precedence, and the "Associativity" gives their evaluation order.

12.3.4 Conditional Flow and Program Loops

Recall from Section 6.4, the use of different combinations of conditional branch instructions to obtain control over the program execution flow. We have seen that the IF-THEN-ELSE structure employs conditional branch instructions to establish whether a condition is true. The general format for if-else statement in C is shown here:

```
if( condition 1 )
    statement1;
else if( condition 2 )
    statement2;
else if( condition 3 )
    statement3;
else
    statement4;
```

In the above if-else statement, the *else* clause allows action to be taken where the condition evaluates as false (zero).

Example 12.6

The following C program uses an if-else statement to validate the user input. Describe the validation.

```c
#include <stdio.h>

main()
{
        int number;
        int valid = 0;

        while( valid == 0 ) {
                printf("Enter a number: ");
                scanf("%d", &number);
                if( number < 0 ) {
                        printf("Number invalid. Please
re-enter\n");
                        valid = 0;
                }
                else if( number > 180 ) {
                        printf("Number invalid. Please
re-enter\n");
                        valid = 0;
                }
                else
                        valid = 1;
        }
        printf("The number is %d\n", number );
}
```

SOLUTION

Looking at the statements like "if(number < 0)" and "else if(number > 180)", it can be easily concluded that the program uses an if-else statement to validate the users input to be in the range 0–180.
A sample program output can be the following:

```
Enter a number: 1112
Number is above 180. Please reenter
Enter a number: 5
The number is 5
```

Common Misconception: Student often make mistakes in using the equality test. Don't forget that the equality test is '=='; a single '=' causes an assignment, not a test, and always leads to disaster!

In Chapter 6 we studied that the WHILE-DO repetition structure continues executing the body of the loop as long as the comparison test is true.

On the other hand, the DO-UNTIL repetition structure executes the loop as long as the comparison test is false. The structures for both these types were shown in a flowchart in Figure 6.11. WHILE-DO and DO-WHILE are two of the many flavors of repetitive structures used in C programming. A DO-WHILE loop statement allows a user to execute code block in loop body at least one. Here is a DO-WHILE loop syntax:

```
do {
  // statements
} while (expression);
```

Example 12.7

Using a DO-WHILE repetition structure, write a C program that prints exactly 10 times as indicated:

```
1
2
3
4
5
6
7
8
9
10
```

SOLUTION

We will use a DO-WHILE repetition structure with two integer variables, one to specify the start and the other to end. Since the output has to start from 1, and end at 10, we will set the variables i and x to values 0 and 10, respectively. The only thing we need to make sure is that we keep incrementing the variable i (our counter) by one, and compare it against the variable x (our final value) in every loop. The C program for the above mentioned output is given as follows. Note how the new line is implemented using the escape sequence "\n" in the printf statement.

```
#include <stdio.h>
void main(){
    int x = 10;
    int i = 0;
    // using do-while loop statement
    do{
        i++;
        printf("%d\n",i);
    }while(i < x);
}
```

12.3.5 Subroutines

C uses the term functions for subroutines. We have discussed subroutine in Section 6.6, where we saw that as we develop bigger programs, we quickly find that there are program sections that are so useful that they can be used in different places. A subroutine is a program section structured in such a way that it can be called from anywhere in the program. Once it has been executed, the program continues to execute from wherever it was before. We have also seen that there are many advantages of breaking a program into subroutines. On a large programming task, a team of programmers can divide various parts of the programs into subroutines. This division of complex and large problem solution helps in the better understanding of the code as each subroutine becomes smaller in size as compared to the full program.

Helpful Hint: Functions result in elimination of duplicate code and enable the reuse of code across multiple programs.

Example 12.8

Complete and explain the following C program that is in need of a function that multiplies two numbers given in the main program. The main program is shown as follows, without the required libraries.

```
void main()
{
      int input1, input2, result;

      input1 = 25;
      input2 = 10;

      result = multiply(input1,input2); //subroutine call

      printf("Multiplication of %d and %d is %d",input1,input2,
result);
}
```

SOLUTION

When executing the statement "result = multiply(input1,input2);", the program execution jumps to the function named multiply. The "multiply" function is given as follows:

```
/* function definition block */
int multiply(int x, int y)
{
      return x*y;
}
```

Therefore, multiply() returns its answer, which is saved into the variable 'result'. After the value is returned, execution returns back to main and proceeds to the next statement.

12.3.6 Pointers and Arrays

A pointer is a memory address. Let us declare a variable with the name count.

```
int count;
```

The variable count occupies some memory. Depending upon the system, it may occupy 2 bytes of memory. This is because in the system an int will be 2 bytes wide. Let us declare another variable as shown here:

```
int *count_ptr = &count;
```

count_ptr is declared as a pointer to int. In the above declaration, we have initialized the pointer to point to count. We know that the variable count occupies some memory. Its location in memory is called its address. The expression &count is the address of count. That is the reason we call "&" as "address-of-operator." The "*" informs the compiler that the declaration is about a pointer variable. This is needed by the compiler to set aside however many bytes is required to store an address in memory. To understand it pictorially, let us study the Figure 12.5. The boxes in the figure are variables. Our variable count is also a box with size equal to 2 bytes. The size in bytes can be determined by sizeof(int). The address of this box is its location in memory, which is &count. When we access the address, we actually access the contents of the box it points to. Since the pointer count_ptr points to count, its contents are equal to the address of the variable count. Therefore, the pointer to count is the content of count_ptr.

To clarify, we summarize the about declaration as:

- count is an integer variable
- count_ptr is a pointer to int
- &count is the address of count
- The content of count_ptr is equal to &count due to the initialization of count_ptr.

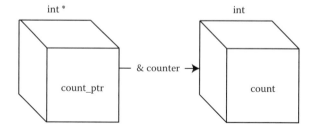

FIGURE 12.5
Basic concept of pointer.

Using a pointer we can directly access the value stored in the variable which it points to. We simply have to precede the pointer's identifier with an asterisk (*), which acts as a dereference operator. It is commonly spoken of as "value pointed by." We will use dereference operator in Example 12.11.

Array by definition is a variable that holds multiple elements that have the same data type. The C language provides a capability that enables a programmer to define an array. Let us take an example of a course consisting of a number of students. In order to find the class average on the grades, we will have to read the grades into the computer. This could be easily done by using an array for storing and computing purposes. We can define a variable as a grade that represents not a single value of grade but an entire set of grades. Each element of the set can then be referenced by means of a number called an *index number* or *subscript*. The declaration would look like:

```
float grade[50];
```

The floating point data type is the type of element that will be contained in the array grade, and 50 is the size that indicates the maximum number of elements that can be stored inside this array. The array grade is shown in Figure 12.6. To access individual array elements (for example, to store its first element in a variable called first), we just have to write:

```
first = grade[0];
```

We will take a look at some examples that use the concept of pointer and array in the next section.

Helpful Hint: Working with pointers brings a programmer in closer contact with the machine's low-level workings.

Helpful Hint: Pointers can be confusing for beginners, so take your time and follow the explanations and examples closely.

Helpful Hint: A pointer is a memory address.

Helpful Hint: A single declaration can declare multiple variables of the same type by simply providing a comma-separated list.

Helpful Hint: Array elements may have some random value until a programmer formally initializes them.

Section 12.3 Review Quiz

The length of a source program can be reduced by using _____ at appropriate places. (functions/variables/operators)

12.4 Examples

Here we look at some more examples to explore some more features of C programming. We will use a few basic concepts of arrays and pointers in these examples.

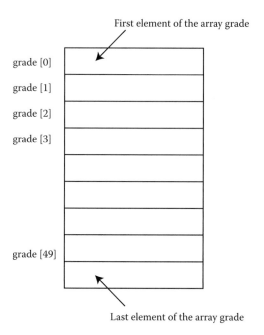

First element of the array grade

grade [0]

grade [1]

grade [2]

grade [3]

grade [49]

Last element of the array grade

FIGURE 12.6
An example of grade array.

Example 12.9

Write a simple function to add up all of the integers in a single dimensioned array. Do not comment your code; rather name the variables such that it makes the code self-explanatory.

SOLUTION

The function addArray given as follows provides the addition of all the elements of an array. The name of the array is passed as a function parameter. The name contains the base address. Base address is defined as the address of the 0th element. The size of the array has been passed as an additional argument (parameter) to the function.

```
int addArray(int array[], int size)
{        int count;
         int total = 0;

         for(count = 0; count < size; count ++)
                total += array[count];

         return(total);
}
```

Self-Learning: Search from the Internet t concept of passing parameters to a fu tion. Search with keywords like "call value" and "call by reference."

Example 12.10

Consider the following code fragment given here:

```
Bob = 55;
Alice = Bob;
Fred = &Bob;
```

Show using a diagram the relationship between Bob, Alice, and Fred after the execution of this code fragment. Make appropriate assumptions.

SOLUTION

The relationship is shown in Figure 12.7. First, we assign the value 55 to Bob. We assume that the address in memory for Bob is 6935. The second statement would copy the content of variable Bob to Alice. This is a standard assignment operation. Therefore, Alice takes the value 55. Finally, the third statement copies to Fred not the value contained in Bob but a reference to it. In other words, Bob's address which is 6935 is assigned to Fred. The reference operator (&) indicates that we are no longer referring to the value of Bob but to its reference, i.e., its address in memory. The variable that stores the reference to another variable (like Fred in this example) is what we call a pointer.

Example 12.11

After the execution of the code fragment from Example 12.10, we execute the following statement.

```
Mike = *Fred;
```

Show using a diagram the relationship and values between Bob, Fred, and Mike after the execution of this statement. Make appropriate assumptions.

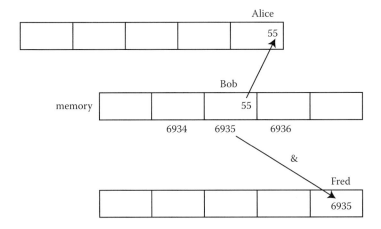

FIGURE 12.7
Relationship for Example 12.10.

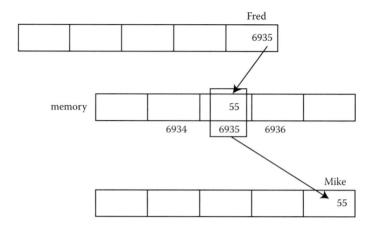

FIGURE 12.8
Relationship for Example 12.11.

SOLUTION

We read the statement as "Mike equal to value pointed by Fred". Therefore, Mike would take the value 55, since Fred is 6935, and the value pointed by 6935 is 55. Figure 12.8 shows this relationship.

Helpful Hint: Notice the difference between the reference and dereference operators. The symbol & is the reference operator and should be read as "address of." One the other hand, the symbol * is the dereference operator and should be read as "value pointed by."

Example 12.12

Study the code given here and rewrite it such that it performs the same task.

```
#include < stdio.h>

void main()
{
        int i, counter;
        counter = 0;
        i = getchar();
        while (i != EOF) {
                counter = counter + 1;
                i = getchar();
        }
        printf("Number of characters in file = %d\n",
counter);
}
```

SOLUTION

In the code above, EOF is a special return value which is defined in stdio.h. When an end-of-file marker is encountered by a getchar() function, EOF is returned. The code above counts the number of characters in the input stream. This is done by reading characters from standard input and adding them to the "counter," until it encounters the EOF. The output is performed by printf statement where the number of characters is displayed to the user. The above program can be rewritten as shown here:

```
#include < stdio.h>

void main()
{
    int c, counter = 0;
    while ( (c = getchar()) != EOF ) counter++;
    printf("Number of characters in file = %d\n", counter);
}
```

Helpful Hint: Although C allows immense brevity of expression, the code readability suffers for the programmers who are new to computer programming.

Section 12.4 Review Quiz

Which of the following statements are true?

(a) An array is a collection of variables of the same type. (b) Individual array elements are identified a double index. (c) In C the index begins at one and is never written inside square brackets.

12.5 A Project with C

A prime number is a natural number that has exactly two distinct natural number divisors: 1 and itself. In this project, we will write a program that can calculate and display the first 500 prime numbers. The program will start from the prime number 2 and end at the 500th prime number (i.e., 3571).

The algorithm to find out if a natural number n is a prime number is simple. We can use the definition of prime number (i.e., a prime number is a natural number that has exactly two distinct natural number divisors: 1 and itself). To test if n is prime, we can loop through 2 to n–1 and use the modulus operator to find if n is properly divisible by these numbers. Recall that the modulus operator evaluates the remainder of the operands after division. We can use the modulus operator to test whether n provides a zero remainder when evaluated with numbers from 2 to n–1. If any of them results in zero remainder, then n cannot be prime. However, since by another definition, any number that is not prime can be properly divided by at least one other prime number, an elegant and much more efficient way to perform this algorithm is to test only with the prime numbers less than n. In order to do this, we will maintain a list of prime numbers already found, and use that to test n. The flow chart for this program is shown in the Figure 12.9.

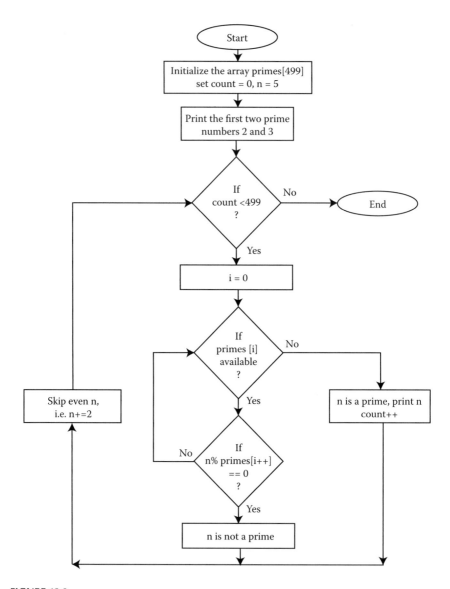

FIGURE 12.9
Flowchart for the program to print first 500 prime numbers.

As shown in the Figure 12.9, the program first initializes an array of size 499 to values 3 and 0. The size is not 500 because we will print the first two primes, that is, 2 and 3, directly from our program. In short, we have hard-coded these two primes in the beginning of the program. We set the integer variables count and n to 0 and 5, respectively. The variable n is set to 5 because we will print 2 and 3 first, and 4 we know is not a prime number, therefore, it is obvious to start our test for the prime number from the digit 5.

The variable count is used to move one by one to all the first 500 primes. Since we already have 2 and 3, we will check so that n does not exceed beyond 499 in the "while" condition. As shown in the flow chart, integer variable i is set to zero. This variable lets us move in the array named "primes" so that we can use the prime numbers that we have already found. We next check in a while condition if a previously found prime is available or not. If available, we perform our test of n%primes[i++], and check if the outcome is zero or not. A zero from this modulus operation will indicate that n is not a prime. Otherwise, we will keep moving in our primes list and perform the same test. If none of the primes are able to yield a zero on the modulus test, then it will indicate that n is a prime. We will add it to our primes list and increment the counter by one. We skip the even numbers because even numbers except 2 are not prime numbers. We keep looping in our while condition of checking if the variable count is less than 499, and follow the same pattern to come up with a list of first 500 primes. The program with this method is shown here:

```c
/*
Program Name: prime.c
Objective: Find first 500 prime numbers
Output: On screen
*/
#include <stdio.h>
#include <stdlib.h>
int main(void) {
        int primes[499] ={3, 0};
        int n = 5, i;
        int count = 0;
        int isPrime;
        // print first 2 primes
        printf("%8s%8d%8d", "Primes", 2, 3);

    /*
    We already have 2 and 3
    now find the next 498
    */
    while ( count < 499 ) {
            i = 0;
            isPrime = 1;
        // test if number divides by any previous primes
        while ( primes[i] ) {
                if ( (n % primes[i++]) == 0 ) {
                        isPrime = 0;     // this means that it is
not prime
                        break;                // come out of the loop
                } //end of if
        } // end of while loop
```

```
    if ( isPrime ) {                    // if a prime
            printf("%8d", n);           // print it to the screen
            primes[i] = n;              // add it to the list
            ++count;
    // start a new line for every 10 primes
    if ( ((count + 3) % 10) == 0 )
            putchar('\n');
    }// end of if
    // skip even numbers as they are not prime
    n += 2;
    } // end of while loop
    putchar('\n');
    getchar();
    return 0;
}
```

A sample program output for this project is shown in Figure 12.10. Note that only the first 20 rows are shown in the output.

Common Practice: One of the debugging techniques that beginners use is to prin out the values of variables as the application execute. That will provide a peek into what's happening behind the scenes.

Team Discussion: How would you imple ment a program that would factor a given number?

Section 12.5 Review Quiz

(a) Each array element occupies consecutive memory locations. (True/False)
(b) Array name is a pointer that points to the first element. (True/False)

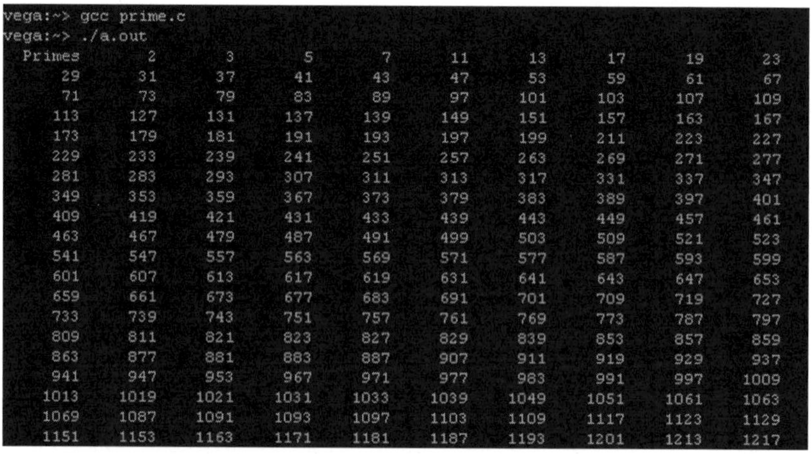

```
vega:~> gcc prime.c
vega:~> ./a.out
  Primes      2      3      5      7     11     13     17     19     23
      29     31     37     41     43     47     53     59     61     67
      71     73     79     83     89     97    101    103    107    109
     113    127    131    137    139    149    151    157    163    167
     173    179    181    191    193    197    199    211    223    227
     229    233    239    241    251    257    263    269    271    277
     281    283    293    307    311    313    317    331    337    347
     349    353    359    367    373    379    383    389    397    401
     409    419    421    431    433    439    443    449    457    461
     463    467    479    487    491    499    503    509    521    523
     541    547    557    563    569    571    577    587    593    599
     601    607    613    617    619    631    641    643    647    653
     659    661    673    677    683    691    701    709    719    727
     733    739    743    751    757    761    769    773    787    797
     809    811    821    823    827    829    839    853    857    859
     863    877    881    883    887    907    911    919    929    937
     941    947    953    967    971    977    983    991    997   1009
    1013   1019   1021   1031   1033   1039   1049   1051   1061   1063
    1069   1087   1091   1093   1097   1103   1109   1117   1123   1129
    1151   1153   1163   1171   1181   1187   1193   1201   1213   1217
```

FIGURE 12.10
Sample output for the program to print first 500 prime numbers.

12.6 Summary

1. A higher-level programming language is a programming language with strong abstraction from the details of the computer.

2. A lower-level programming language, on the other hand, provides little or no abstraction from a computer's instruction set architecture.

3. A program created in a lower-level language can be made to run very fast, and with a very small memory footprint, as compared to its equivalent program created with higher-level language.

4. C is a higher-level programming language and an imperative (procedural) systems implementation language.

5. The statement "int main()" tells the compiler that there is a function named "main," and that the function returns an integer, hence int.

6. The "curly braces" ({ and }) signal the beginning and end of functions and other code blocks.

7. A program must contain one and only one main ().

8. Unlike the lower-level programming language, all C statements are defined in free format, i.e., with no specified layout or column assignment.

9. A data type is a set of data with values having predefined characteristics. Examples of data types are: integer, floating point unit number, character, string, etc.

10. An operator is a symbol which helps the user to command the computer to do a certain mathematical or logical manipulations. C has many operators which can be classified as: mathematical operators, relational operators, logical operators, and assignment operators.

11. C uses the term functions for subroutines. Judicious use of subroutines often significantly reduces the cost of developing and maintaining a large program. It also increases its quality and reliability.

12. Subroutines are an important mechanism for sharing and trading software.

13. The variable that stores the reference to another variable is called a pointer.

14. Pointers are a very powerful feature of the C language that has many uses in advanced programming.

15. The C language provides a capability that enables the user to define a set of ordered data items known as an array.

Glossary

Arithmetic: Traditional operations such as addition, subtraction, multiplication, and division.

Array: A variable that holds multiple elements which has the same data type.

Assembler: A software package that is used to convert assembly language into machine language.

Assembly Language: A low-level programming language unique to each microcontroller. It is converted, or assembled, into machine code before it can be executed.

Code Block: A block is a section of code which is grouped together.

Compiler: A software package that converts a higher-level language program into machine language code.

CPU: Central Processing Unit.

Data Type: A classification identifying one of various types of data, such as floating point, integer, or Boolean.

Declaration: A declaration specifies the identifier, type, and other aspects of language elements such as variables and functions.

Escape Sequence: A series of characters used to change the state of computers and their attached peripheral devices.

Function: See subroutines.

Higher-Level language: A type of computer language closest to human language that is a level above assembly language.

Loops: Repeat a statement a certain number of times or while a condition is fulfilled.

Lower-Level Language: A lower-level programming language provides little or no abstraction from a computer's instruction set architecture. The word "lower" refers to the small or nonexistent amount of abstraction between the language and machine language.

Machine Language: Computer instruction written in binary code that is understood by a computer; the lowest level of programming language.

Operators: An operator is a symbol that helps the user to command the computer to do a certain mathematical or logical manipulations.

Pointer: The variable that stores the reference to another variable is called a *pointer.*

Subroutine: A program stored in higher memory that can be used repeatedly as part of a main program.

Variable: Variables in C are memory locations that are given names and can be assigned values.

Answers to Section Review Quiz

12.2 lower

12.3 functions

12.4 (a)

12.5 (a) True, (b) True

True/False Quiz

1. An advantage of higher-level language over assembly language is that it is portable.

2. Some examples of lower-language programming languages are C, C++, C#, JAVA, etc.

3. A lower-level language can be directly correlated to processor instructions.

4. The ".h" files are by convention "header files" that can never contain definitions of variables and functions necessary for the execution of a program.

5. In C programming, whitespace (tabs or spaces) is never significant, except inside quotes as part of a character string.

6. Statements in C programs are terminated by a colon.

7. The range of values for the integers in C programming is machine dependent.

8. C uses the long and unsigned integers to declare a longer range of values.

9. An operator is a symbol that helps the user to command the computer to do certain mathematical or logical manipulations.

10. Modulus operator (%) returns the remainder of integer division calculation.

11. C uses the term utility for subroutines.

12. Judicious use of subroutines (functions) often increases the cost of developing and maintaining a large program.

13. Judicious use of subroutines (functions) decrease code quality and reliability.

14. Subroutines, often collected into libraries, are an important mechanism for sharing and trading software.

15. C# is a programming language originally developed for developing the Unix operating system.

Questions

QUESTION 12.1

Which of the following statements are true with respect to functions in C?

(a) Functions can have no arguments and no return values.
(b) Functions can have arguments and no return values.
(c) Functions can have arguments and return values.
(d) Functions can return multiple values.
(e) Functions can have no arguments and return values.

QUESTION 12.2

What are some advantages of using functions in C?

QUESTION 12.3

Explain the concept of end-of-file (EOF) with respect to C.

QUESTION 12.4

The code following uses while loop. Rewrite it in an easier syntax by using a loop.

```
i = startValue;
    while (i <= lastValue)
    {
            ...block of statements...
            i = i + incrementValue;
    }
```

QUESTION 12.5

List a few of the arithmetic operators that are used in C programming.

QUESTION 12.6

How is an array declared in a C program? Give an example.

Problems

PROBLEM 12.1

The following program illustrates a nested if-else structure and multiple parameters to the scanf function. Comment on the program and describe how the output of this program would look like.

```c
#include <stdio.h>
main()
{
      int invalidOperator = 0;
      char operator;
      float N1, N2, result;

      printf("Enter two numbers and an operator\n");
      printf("Format: N1 operator N2\n");
      scanf("%f %c %f", &N1, &operator, &N2);

      if(operator == '*')
        result = N1 * N2;
      else if(operator == '/')
        result = N1 / N2;
      else if(operator == '+')
        result = N1 + N2;
      else if(operator == '-')
        result = N1 - N2;
      else
        invalidOperator = 1;

      if( invalidOperator != 1 )
        printf("%f %c %f is %f\n", N1, operator, N2, result );
      else
        printf("Invalid operator.\n");
}
```

PROBLEM 12.2

Write a C program to determine if a given year is a leap year or not.

PROBLEM 12.3

In Example 7.11, we wrote an assembly program to generate a Fibonacci series. Write a C program to print a Fibonacci series.

PROBLEM 12.4

What is the output of this program? What does the program compute for a given VALUE (6 here)?

```c
#include <stdio.h>
```

```
#define VALUE 6
        int i,j;
        void main()
        {
                j=1;
                for (i=1; i<=VALUE; i++)
                j=j*i;

                printf("The computation on %d is %d\n",VALUE,j);
        }
```

PROBLEM 12.5

Find the error in the following program and correct it.

```
#include <stdio.h>
#define LAST 10

int main()
{
        int i, sum == 0;

        for ( i = 1 i <= LAST; i++ ) {
                sum += i;
        } / end of for loop /

        printf("sum = %d\n, sum);

        return 0
}
```

PROBLEM 12.6

Write a C program to change the value of a variable as per the following description:

Let myNumber be an integer and myPointer by a pointer to integer. Initialize myNumber to be equal to 15. Set myPointer to point to myNumber. Print out the value of myNumber. The value should print as 15. Next, change myNumber from 15 to 25 through myPointer. To show that myNumber has changed, print its value again to the screen. The value should print as 25.

Appendix 1—Supplemental Web Site

The Web site *Microcontroller Guide* (www.microcontrollerguide.com) is maintained by Syed R. Rizvi, the author of this book. There are several case studies and interesting facts about microcontroller projects for both students and hobbyists. The Web site contains video tutorials in order to demonstrate best practices in building robust microcontroller-based projects. There are discussions on issues and challenges in building modern embedded systems. Several examples are available on a wide range of electronics topics. Many of these examples include animated illustrations and video demos that may help students or hobbyists understand how each circuit functions.

There is a section dedicated to best practices for high-level programming. For example, HC11 can be programmed with the high-level programming language C. Although the language has been around for close to 30 years, its appeal has not yet worn off. It continues to attract a large number of people who must develop new skills for writing new applications, or for porting or maintaining existing applications. The Web site provides useful tips on the best practices for programming in C. Additionally, program samples, articles, tutorials, and tips for relatively modern programming languages like C# are also available for knowledge enrichment.

The Web site also contains information about retail stores and online distributors. It is constantly updated to unleash the cutting edge technologies and explore state-of-the-art microcontroller hardware and software.

Appendix 2—States and Resolution for Binary Numbers

TABLE A2.1

States and Resolution for Binary Numbers

Word Length in Bits n	Maximum Number of Combinations 2^n	Resolution of a Binary Ladder ppm
1	2	500,000
2	4	250,000
3	8	125,000
4	16	62,500
5	32	31,250
6	64	15,625
7	128	7,812.5
8	256	3,906.25
9	512	1,953.13
10	1,024	976.56
11	2,048	488.28
12	4,096	244.14
13	8,192	122.07
14	16,384	61.04
15	32,768	30.52
16	65,536	15.26
17	131,072	7.63
18	262,144	3.81
19	524,288	1.91
20	1,048,576	0.95
21	2,097,152	0.48
22	4,194,304	0.24
23	8,388,608	0.12
24	16,777,216	0.06

Appendix 3—Basic Boolean Theorems and Identities

1. Identity elements

$1 . A = A$

$0 + A = A$

2. Inverse elements

$\overline{A} . A = 0$

$\overline{A} + A = 1$

3. Idempotent laws

$A + A = A$

$A . A = A$

4. Boundess laws

$A + 1 = 1$

$A . 0 = 0$

5. Distributive laws

$A . (B + C) = A.B + A.C$

$A + (B . C) = (A + B) . (A + C)$

6. Order exchange laws

$A . B = B . A$

$A + B = B + A$

7. Absorption laws

$A + (A . B) = A$

$A . (A + B) = A$

8. Associative laws

$A + (B + C) = (A + B) + C$

$A . (B . C) = (A . B) . C$

9. Elimination laws

$A + (\overline{A} . B) = A + B$

$A . (\overline{A} + B) = A . B$

10. De Morgan theorem

$\overline{(A + B)} = \overline{A} . \overline{B}$

$\overline{(A . B)} = \overline{A} + \overline{B}$

Appendix 4—The Resistor Color Code

TABLE A4.1

The Resistor Color Code

Color	Value	Multiplier	Tolerance
Black	0	0	NA
Brown	1	10	NA
Red	2	100	2%
Orange	3	1,000	NA
Yellow	4	10,000	NA
Green	5	100,000	NA
Blue	6	1,000,000	NA
Violet	7	10,000,000	NA
Gray	8	100,000,000	NA
White	9	1,000,000,000	NA
Gold	NA	0.1	5%
Silver	NA	0.01	10%

An Example of Resistor Color Code

Assume that in a resistor the first, second, third, and fourth bands are red, violet, green, and gold, respectively. The following is a step-by-step method of determining the value of the resistance:

1st band is red. Therefore, the first digit is 2.

2nd band is violet. Therefore, the second digit is 7.

3rd band is green. Therefore, the multiplier is 100,000.

4th band is gold. Therefore, the tolerance is 5%.

Combining them, we get $27 \times 100000 \pm 5\%$ or $2.7 \text{ M}\Omega \pm 5\%$.

Helpful Hint: The color code of resistor can be easily remembered by a statement such as "BB Roy Great Britain Very Good Wife." The letters in uppercase represent the first letter of the color. Remember that the colors become lighter as their value increases.

Appendix 5—Waterfall Software Development Lifecycle Model

The waterfall model is the classic model of software engineering. It is a sequential design process, in which progress is seen as flowing constantly downward through the phases of Analysis, Design, Implantation, Testing, and Maintenance. Due to its characteristic feature of detailed initial planning in the early stages, it catches design flaws before they develop. For projects where quality control is a major concern, this model works due to detailed documentation and planning activities. In Royce's original waterfall model, the following phases are in order:

1. Requirements Analysis
2. Design Development
3. Implementation or Coding
4. Product Integration
5. Verification
6. Release/Deployment
7. Maintenance

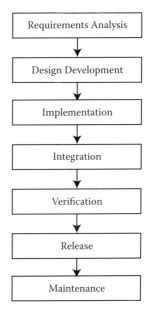

The name *waterfall* comes from the fact that this model is based on the principle that one should move to a phase only when its preceding phase is completed. However, in the real world, it is difficult to achieve perfect completion of any phase as a prerequisite for the start of the subsequent phase. Therefore, there are various modified waterfall models that may include slight or major variations upon this process (see Figure A5.1).

FIGURE A5.1
Phases in waterfall software development lifecycle model.

Appendix 6—Loading Your Program into the EEPROM

This is a step-by-step procedure to load your program into the EEPROM if using BUFFALO 3.4 and PROCOMM. Other versions of BUFFALO and terminal emulators will have a slightly modified procedure.

1. Set the appropriate program start address: Your program should be assembled in the EEPROM range ($B600–B7FF). This restricts you to only 512 bytes of working space for your code. Use ORG $B600 in your program to start your first instruction from the location $B600. Several lines of code must be copied directly from the Buffalo Monitor program into the EEPROM program to perform the minimum initialization of the EVBU board in order to get the EVBU to boot directly to the program in EEPROM. These lines of codes are called the initialization sequence. For example, you can load them at the memory location $B600 by performing the MOVE from block $E000 to $E030 to the block $B600 to $B630.

2. Erase data from EEPROM: Use the BULK command on the BUFFALO prompt in order to bulk-erase the contents of EEPROM.

3. Set the EVBU baud rate to 300: Use the MM command on the BUFFALO prompt to change the contents of the BAUD rate register (location $102B) from 30 (default baud rate of 9600) to 35 (baud rate of 300). Make sure you do not hit the reset at this point of time.

4. Set the PROCOMM baud rate to 300: Enter the baud rate screen by pressing ALT-P and select 8 (300, N, 8, 1). Press the ESC key to quit the screen.

5. Verify 300 baud: By pressing the ENTER one or more times, verify that PROCOMM and EVB communicate with the 300 baud rate.

6. Load your program: Load and verify your program using the normal process.

7. Reset the PROCOMM baud rate to 9600. The 9600 rate can be reset by selecting 11 (9600, N, 8, 1) followed by pressing ESC to quit the screen.

8. Reset the EVBI baud rate to 9600: Reset EVBU back to the 9600 baud rate. Now, your program can run from EEPROM. Just use the Go command and the appropriate start address, for example, $B600.

   Helpful Hint: To load the initialization code, use the BUFFALO command MM and enter the code. Another way of doing it is to make it a part of your downloadable code, or create a separate program and download it followed by a download of your program starting from where the first one ended.

9. Set the program boot from EEPROM: Here, we configure EVBU so that the program is booted from EEPROM. Move the Jumper (J2) on the EVBU from its default position (pins 2–3) to pins (1–2). This way, the BUFFALO monitor program will know to boot the program loaded into the EEPROM.

Best Practice: When writing code to be placed in EEPROM, experienced programmers never declare constants with FCB if in the RAM area. They use RMB for variables and their program creates them. Only read-only variables should be put in the EEPROM area.

Appendix 7—Pulse-Width Modulation

Pulse-width modulation (PWM) is a technique frequently used by modern electronic power switches to control power input to inertial electrical devices such as DC motors, etc. Other applications include digital-to-analog conversion and communication, to cause lights to change their intensities, etc. By turning the switch between supply and load ON and OFF at high speed, the average value of voltage or current is input to the load. The longer the switch is in the ON state compared to the OFF state, the higher the power supplied to the load is.

In Chapter 11, Figure 11.5 shows a waveform where the period of the wave and pulse width are denoted by T and w, respectively. It is assumed that pulse width for falling and rising edges are the same. The duty cycle is a measure of the average value of the waveform. Some important formulas related to PWM are

- % Duty cycle = $(w/T) * 100\%$
- $V_{eff} = (1/T) \int v(t)\, dt$
- $V_{average} = V_{eff} = (w/T) * V_m$

In a square wave, if $w = T/2$, the duty cycle will be 50%. The V_{eff} will be equal to $V_m * (\frac{1}{2}) = V_m/2$. If the time base is fixed, that is, if T is fixed, a programmer can change the average voltage of the PWM waveform by just changing the pulse width (w). Various interrupts can be employed together to achieve a variety of PWM in microcontrollers.

Appendix 8—HC11 Instruction Set

TABLE A8.1

CCR Legend

Symbol	Description
—	Not changed
x	Updated according to data
?	Undefined
0	Reset to 0
1	Set to 1
3	c | =(msn>9)
4	Most significant bit of b
5	Set when interrupt occurs; if previously set, a nonmaskable interrupt is required to exit the wait state

TABLE A8.2

HC11 Instruction Set

Instruction	IMM	DIR	INDX	INDY	EXT	INH	Function	H	I	N	Z	V	C
aba	—	—	—	—	—	1b 2 1	$(A) + (b) \to A$	×	—	×	×	×	×
abx	—	—	—	—	—	3a 3 1	$(x) + \$00{:}(B) \to X$	—	—	—	—	—	—
aby	—	—	—	—	—	183a 4 2	$(y) + \$00{:}(B) \to y$	—	—	—	—	—	—
adca	89 2 2	99 3 2	a9 4 2	18a9 5 3	a9 4 3	—	$a{+}{=}m + c$	×	—	×	×	×	×
adcb	c9 2 2	d9 3 2	e9 4 2	18e9 5 3	f9 4 3	—	$b{+}{=}m + c$	×	—	×	×	×	×
adda	8b 2 2	9b 3 2	ab 4 2	18ab 5 3	bb 4 3	—	$(A) + (M) \to A$	×	—	×	×	×	×
addb	cb 2 2	db 3 2	eb 4 2	18eb 5 3	fb 4 3	—	$(b) + (M) \to b$	×	—	×	×	×	×
addd	c3 4 3	d3 5 2	e3 6 2	18e3 7 3	f3 6 3	—	$(d) + (M){:}(M+1) \to d$	—	—	×	×	×	×
anda	84 2 2	94 3 2	a4 4 2	18a4 5 3	b4 4 3	—	$(a) \cdot (m) \to a$	—	—	×	×	0	—
andb	c4 2 2	d4 3 2	e4 4 2	18e4 5 3	f4 4 3	—	$(b) \cdot (m) \to b$	—	—	×	×	0	—
asl	—	—	68 6 2	1868 7 3	78 6 3	—	$m{<}{<}{=}1$	—	—	×	×	×	×
asla	—	—	—	—	—	48 2 1 a	$<{<}{=}1$	—	—	×	×	×	×
aslb	—	—	—	—	—	58 2 1	$b{<}{<}{=}1$	—	—	×	×	×	×
asld	—	—	—	—	—	05 3 1	$d{<}{<}{=}1$	—	—	×	×	×	×
asr	—	—	67 6 2	1867 7 3	77 6 3	—	$[i](m){>}{>}{=}1$	—	—	×	×	×	×
asra	—	—	—	—	—	47 2 1	$[i](a){>}{>}{=}1$	—	—	×	×	×	×
asrb	—	—	—	—	—	57 2 1	$[i](b){>}{>}{=}1$	—	—	×	×	×	×
bcc	24 3 2	—	—	—	—	—	bra(cc)	—	—	—	—	—	—
bclr	—	15 6 3	1d 7 3	181d 8 4	—	—	m&=im	—	—	×	×	0	—
bcs	25 3 2	—	—	—	—	—	bra(cs)	—	—	—	—	—	—
beq	27 3 2	—	—	—	—	—	bra(eq)	—	—	—	—	—	—
bge	2c 3 2	—	—	—	—	—	bra(ge)	—	—	—	—	—	—
bgt	2e 3 2	—	—	—	—	—	bra(gt)	—	—	—	—	—	—
bhi	22 3 2	—	—	—	—	—	bra(hi)	—	—	—	—	—	—

Mnemonic	IMM/REL	DIR	IND,X	IND,Y	EXT	Operation	S	X	H	I	N	Z	V	C
bhs	24 3 2	—	—	—	—	bra(hs)	—	—	—	—	—	—	—	—
bita	85 2 2	95 3 2	a5 4 2	18a5 5 3	b5 4 3	a·(m)	—	—	—	—	x	x	0	—
bitb	c5 2 2	d5 3 2	e5 4 2	18e5 5 3	f5 4 3	b·(m)	—	—	—	—	x	x	0	—
ble	2f 3 2	—	—	—	—	bra(le)	—	—	—	—	—	—	—	—
blo	25 3 2	—	—	—	—	bra(lo)	—	—	—	—	—	—	—	—
bls	23 3 2	—	—	—	—	bra(ls)	—	—	—	—	—	—	—	—
blt	2d 3 2	—	—	—	—	bra(lt)	—	—	—	—	—	—	—	—
bmi	2b 3 2	—	—	—	—	bra(mi)	—	—	—	—	—	—	—	—
bne	26 3 2	—	—	—	—	bra(ne)	—	—	—	—	—	—	—	—
bpl	2a 3 2	—	—	—	—	bra(pl)	—	—	—	—	—	—	—	—
bra	20 3 2	—	—	—	—	bra	—	—	—	—	—	—	—	—
brclr	—	13 6 4	1f 7 4	181f 8 5	—	IF (M) · mm = 0 THEN P + ssrr = P	—	—	—	—	—	—	—	—
brn	21 3 2	—	—	—	—	bra(0)	—	—	—	—	—	—	—	—
brset	—	12 6 4	1e 7 4	181e 8 5	—	If (M') · mm = 0 then P + ssrr = P	—	—	—	—	—	—	—	—
bset	—	14 6 3	1c 7 3	181c 8 4	—	(M)+mm → M	—	—	—	—	x	x	0	—
bsr	8d 6 2	—	—	—	—	bsr	—	—	—	—	—	—	—	—
bvc	28 3 2	—	—	—	—	bra(vc)	—	—	—	—	—	—	—	—
bvs	29 3 2	—	—	—	—	bra(vs)	—	—	—	—	—	—	—	—
cba	11 2 1	—	—	—	—	(a) − (b)	—	—	—	—	x	x	x	x
clc	0c 2 1	—	—	—	—	0 → c	—	—	—	—	—	—	—	0
cli	0e 2 1	—	—	—	—	0 → I	—	—	—	0	—	—	—	—
clr	—	—	6f 6 2	186f 7 3	7f 6 3	$00 → M	—	—	—	—	0	1	0	0
clra	4f 2 1	—	—	—	—	$00 → A	—	—	—	—	0	1	0	0
clrb	5f 2 1	—	—	—	—	$00 → B	—	—	—	—	0	1	0	0
clv	0a 2 1	—	—	—	—	0 → v	—	—	—	—	—	—	0	—

Continued

TABLE A8.2 (continued)

HC11 Instruction Set

Instruction	IMM	DIR	INDX	INDY	EXT	INH	Function	H	I	N	Z	V	C
cmpa	81 2 2	91 3 2	a1 4 2	18a1 5 3	b1 4 3	—	(a) − (m)	—	—	x	x	x	x
cmpb	c1 2 2	d1 3 2	e1 4 2	18e1 5 3	f1 4 3	—	(b) − (m)	—	—	x	x	x	x
com	—	—	63 6 2	1863 7 3	73 6 3	—	!(m) → m	—	—	x	x	0	1
coma	—	—	—	—	—	43 2 1	!(a) → a	—	—	x	x	0	1
comb	—	—	—	—	—	53 2 1	!(b) → b	—	—	x	x	0	1
cpd	1a83 5 4	1a93 6 3	1aa3 7 3	cda3 7 3	1ab3 7 4	—	(d) − (m):(M+1)	—	—	x	x	x	x
cpx	8c 4 3	9c 5 2	ac 6 2	cdac 7 3	bc 6 3	—	(x) − (m):(M+1)	—	—	x	x	x	x
cpy	188c 5 4	189c 6 3	1aac 7 3	18ac 7 3	18bc 7 4	—	(y) − (m):(M+1)	—	—	x	x	?	3
daa	—	—	—	—	—	19 2 1	A=da(a)	—	—	x	x	?	3
dec	—	—	6a 6 2	186a 7 3	7a 6 3	—	(m) − 1 → m	—	—	x	x	x	—
deca	—	—	—	—	—	4a 2 1	(A) − 1 → a	—	—	x	x	x	—
decb	—	—	—	—	—	5a 2 1	(b) − 1 → b	—	—	x	x	x	—
des	—	—	—	—	—	34 3 1	(s) − 1 → s	—	—	—	—	—	—
dex	—	—	—	—	—	09 3 1	(x) − 1 → x	—	—	—	x	—	—
dey	—	—	—	—	—	1809 4 2	(y) − 1 → y	—	—	—	x	—	—
eora	88 2 2	98 3 2	a8 4 2	18a8 5 3	b8 4 3	—	(a) ⊕ (m) → a	—	—	x	x	0	—
eorb	c8 2 2	d8 3 2	e8 4 2	18e8 5 3	f8 4 3	—	(b) ⊕ (m) → b	—	—	x	x	0	—
fdiv	—	—	—	—	—	03 41 1	d / x → x; (d) % (x) → d	—	—	—	x	x	x
idiv	—	—	—	—	—	02 41 1	(d) / (x) → x; (d) % (x) → d	—	—	—	x	0	x
inc	—	—	6c 6 2	186c 7 3	7c 6 3	—	(m) + 1 → m	—	—	x	x	x	—
inca	—	—	—	—	—	4c 2 1	(A) + 1 → a	—	—	x	x	x	—
incb	—	—	—	—	—	5c 2 1	(b) + 1 → b	—	—	x	x	x	—

Mnemonic	IMM	DIR	IND,X	IND,Y	EXT	INH	Operation	S	X	H	I	N	Z	V	C
ins	—	—	—	—	—	31 3 1	(s) + 1 → s	—	—	—	—	—	—	—	—
inx	—	—	—	—	—	08 3 1	(x) + 1 → x	—	—	—	—	—	×	—	—
iny	—	—	—	—	—	1808 4 2	(y) + 1 → y	—	—	—	—	—	×	—	—
jmp	—	—	6e 3 2	186e 4 3	7e 3 3	—	jmp	—	—	—	—	—	—	—	—
jsr	—	9d 5 2	ad 6 2	18ad 7 3	bd 7 3	—	jsr	—	—	—	—	—	—	—	—
ldaa	86 2 2	96 3 2	a6 4 2	18a6 5 3	b6 4 3	—	(M) → A	—	—	—	—	×	×	0	—
ldab	c6 2 2	d6 3 2	e6 4 2	18e6 5 3	f6 4 3	—	(M) → B	—	—	—	—	×	×	0	—
ldd	cc 3 3	dc 4 2	ec 5 2	18ec 6 3	fc 5 3	—	(M) → A; (M) → B	—	—	—	—	×	×	0	—
lds	8e 3 3	9e 4 2	ae 5 2	18ae 6 3	be 5 3	—	(M):(M+1) → s	—	—	—	—	×	×	0	—
ldx	ce 3 3	de 4 2	ee 5 2	cdee 6 3	fe 5 3	—	(M):(M+1) → X	—	—	—	—	×	×	0	—
ldy	18ce 4 4	18de 5 4	1aee 6 3	18ee 6 3	18fe 6 4	—	(M):(M+1) → Y	—	—	—	—	×	×	0	—
lsl	—	—	68 6 2	1868 7 3	78 6 3	—	{u|(m)<<=1	—	—	—	—	×	×	×	×
lsla	—	—	—	—	—	48 2 1	{u|(a)<<=1	—	—	—	—	×	×	×	×
lslb	—	—	—	—	—	58 2 1	{u|(b)<<=1	—	—	—	—	×	×	×	×
lsld	—	—	—	—	—	05 3 1	{u|(d)<<=1	—	—	—	—	×	×	×	×
lsr	—	—	64 6 2	1864 7 3	74 6 3	—	{u|(m)>>=1	—	—	—	—	0	×	×	×
lsra	—	—	—	—	—	44 2 1	{u|(a)>>=1	—	—	—	—	0	×	×	×
lsrb	—	—	—	—	—	54 2 1	{u|(b)>>=1	—	—	—	—	0	×	×	×
lsrd	—	—	—	—	—	04 3 1	{u|(d)>>=1	—	—	—	—	0	×	×	×
mul	—	—	—	—	—	3d 10 1	(a) · (b) → d	—	—	—	—	—	—	—	4
neg	—	—	60 6 2	1860 7 3	70 6 3	—	$00 – (m) → m	—	—	—	—	×	×	×	×
nega	—	—	—	—	—	40 2 1	$00 – (a) → a	—	—	—	—	×	×	×	×
negb	—	—	—	—	—	50 2 1	$00 – (b) → b	—	—	—	—	×	×	×	×
nop	—	—	—	—	—	01 2 1	nop	—	—	—	—	—	—	—	—
oraa	8a 2 2	9a 3 2	aa 4 2	18aa 5 3	ba 4 3	—	(a) + (m) → a	—	—	—	—	×	×	0	—

Continued

TABLE A8.2 (continued)

HC11 Instruction Set

Instruction	IMM	DIR	INDX	INDY	EXT	INH	Function	H	I	N	Z	V	C
orab	ca 2 2	da 3 2	ea 4 2	18ea 5 3	fa 4 3	—	$(b) + (m) \to b$	—	—	×	×	0	—
psha	—	—	—	—	—	36 3 1	$(A) \to m_S$ $(S) - 1 \to S$	—	—	—	—	—	—
pshb	—	—	—	—	—	37 3 1	$(B) \to m_S$ $(S) - 1 \to S$	—	—	—	—	—	—
pshx	—	—	—	—	—	3c 4 1	$(X_L) \to m_S$ $(S) - 1 \to S$ $(X_H) \to m_S$ $(S) - 1 \to S$	—	—	—	—	—	—
pshy	—	—	—	—	—	183c 5 2	$(Y_L) \to m_S$ $(S) - 1 \to S$ $(Y_H) \to m_S$ $(S) - 1 \to S$	—	—	—	—	—	—
pula	—	—	—	—	—	32 4 1	$(S) + 1 \to S$ $m_S \to (A)$	—	—	—	—	—	—
pulb	—	—	—	—	—	33 4 1	$(S) + 1 \to S$ $m_S \to (B)$	—	—	—	—	—	—
pulx	—	—	—	—	—	38 5 1	$(S) + 1 \to S$ $m_S \to X_H$ $(S) + 1 \to S$ $m_S \to X_L$	—	—	—	—	—	—
puly	—	—	—	—	—	1838 6 2	$(S) + 1 \to S$ $m_S \to Y_H$ $(S) + 1 \to S$ $m_S \to Y_L$	—	—	—	—	—	—
rol	—	—	69 6 2	1869 7 3	79 6 3	—	$rol(m) \to m$	—	—	×	×	×	×
rola	—	—	—	—	—	49 2 1	$rol(a) \to a$	—	—	×	×	×	×

Mnemonic	IMM	DIR	IND,X	IND,Y	EXT	INH	Operation	S	X	H	I	N	Z	V	C
rolb	—	—	—	—	—	59 2 1	rol(b) → b	—	—	—	—	x	x	x	x
ror	—	—	66 6 2	1866 7 3	76 6 3	—	ror(m) → m	—	—	—	—	x	x	x	x
rora	—	—	—	—	—	46 2 1	ror(a) → a	—	—	—	—	x	x	x	x
rorb	—	—	—	—	—	56 2 1	ror(b) → b	—	—	—	—	x	x	x	x
rti	—	—	—	—	—	3b 12 1	rti	x	x	x	x	x	x	x	x
rts	—	—	—	—	—	39 5 1	rts	—	—	—	—	—	—	—	—
sba	—	—	—	—	—	10 2 1	(A) − (b) → a	—	—	—	—	x	x	x	x
sbca	82 2 2	92 3 2	a2 4 2	18a2 5 3	b2 4 3	—	a=m+c	—	—	—	—	x	x	x	x
sbcb	c2 2 2	d2 3 2	e2 4 2	18e2 5 3	f2 4 3	—	b=m+c	—	—	—	—	x	x	x	x
sec	—	—	—	—	—	0d 2 1	1 → c	—	—	—	—	—	—	—	1
sei	—	—	—	—	—	0f 2 1	1 → I	—	—	—	1	—	—	—	—
sev	—	—	—	—	—	0b 2 1	1 → v	—	—	—	—	—	—	1	—
staa	—	97 3 2	a7 4 2	18a7 5 3	b7 4 3	—	(A) → M	—	—	—	—	x	x	0	—
stab	—	d7 3 2	e7 4 2	18e7 5 3	f7 4 3	—	(B) → M	—	—	—	—	x	x	0	—
std	—	dd 4 2	ed 5 2	18ed 6 3	fd 5 3	—	(A) → M, (B) → M+1	—	—	—	—	x	x	0	—
stop	—	—	—	—	—	cf 2 1	stop	—	—	—	—	—	—	—	—
sts	—	9f 4 2	af 5 2	18af 6 3	bf 5 3	—	(S) → M:M+1	—	—	—	—	x	x	0	—
stx	—	df 4 2	ef 5 2	cdef 6 3	ff 5 3	—	(X) → M:M+1	—	—	—	—	x	x	0	—
sty	—	18df 5 3	1aef 6 3	18ef 6 3	18ff 6 4	—	(Y) → M:M+1	—	—	—	—	x	x	0	—
suba	80 2 2	90 3 2	a0 4 2	18a0 5 3	b0 4 3	—	(A) − (M) → a	—	—	—	—	x	x	x	x
subb	c0 2 2	d0 3 2	e0 4 2	18e0 5 3	f0 4 3	—	(b) − (M) → b	—	—	—	—	x	x	x	x
subd	83 3 4	93 5 2	a3 6 2	18a3 7 3	b3 6 3	—	d − (M):(m + 1) → d	—	—	—	—	x	x	x	x
swi	—	—	—	—	—	3f 14 1	swi	—	—	—	1	—	—	—	—
tab	—	—	—	—	—	16 2 1	(A) → B	—	—	—	—	x	x	0	—
tap	—	—	—	—	—	06 2 1	(A) → CCR	x	x	x	x	x	x	x	x
tba	—	—	—	—	—	17 2 1	(B) → A	—	—	—	—	x	x	0	—

Continued

TABLE A8.2 (continued)

HC11 Instruction Set

Instruction	IMM	DIR	INDX	INDY	EXT	INH	Function	H	I	N	Z	V	C
tpa	—	—	—	—	—	07 2 1	$(CCR) \rightarrow A$	—	—	—	—	—	—
tst	—	—	6d 6 2	186d 7 3	7d 6 3	—	$(m) - \$00$	—	—	x	x	0	0
tsta	—	—	—	—	—	4d 2 1	$(a) - \$00$	—	—	x	x	0	0
tstb	—	—	—	—	—	5d 2 1	$(b) - \$00$	—	—	x	x	0	0
tsx	—	—	—	—	—	30 3 1	$(S) + 1 \rightarrow X$	—	—	—	—	—	—
tsy	—	—	—	—	—	1830 4 2	$(S) + 1 \rightarrow Y$	—	—	—	—	—	—
txs	—	—	—	—	—	35 2 1	$(X) - 1 \rightarrow S$	—	—	—	—	—	—
tys	—	—	—	—	—	1835 4 2	$(y) + 1 \rightarrow s$	—	—	—	—	—	—
wai	—	—	—	—	—	3e 14+1	wait	—	5	—	—	—	—
xgdx	—	—	—	—	—	8f 3 1	swap(d,x)	—	—	—	—	—	—
xgdy	—	—	—	—	—	188f 4 2	swap(d,y)	—	—	—	—	—	—

Note: Accumulator A and B are 8 bits. Register D (a = MSB; b = LSB), X, and Y are 16 bits.

Appendix 9—Comprehensive Glossary

Absolute Addressing: Addressing in which the instruction contains the address of the data to be operated on.

Accumulator: A parallel register in a microcontroller that is the focal point of all arithmetic and logic operations.

ADC: An analog-to-digital converter is a device that converts a continuous quantity to a discrete digital number.

Address: An address is an identifier for a memory location, at which a computer program or a hardware device can store data and later retrieve it.

Address Bus: An address bus is a unidirectional, 16-bit bus that carries the 16-bit address code from the processor to the memory unit to select the memory location that the processor is accessing for a READ or WRITE operation.

Alphanumeric: Characters that contain letters as well as numbers and symbols.

Analog: A system that deals with continuously varying physical quantities such as voltage, temperature, pressure, or velocity. Most quantities in nature occur in analog form, yielding an infinite number of different levels.

Analog-To-Digital (A/D): See ADC.

And Gate: A logic gate that produces a high output only when all the inputs are high.

And Masking: A type of masking that uses the AND operation to clear specific bits of a data word while not affecting the others.

Arithmetic: Traditional operations such as addition, subtraction, multiplication, and division.

Arithmetic Logic Unit (ALU): An ALU is a digital circuit that performs arithmetic and logical operations such as addition and subtraction, logical AND, OR, and NOT operations and data shifting.

Arithmetic Shift: Moves all the bits in an operand to the left or to the right by one bit position. When shifting to the right, the MSB is shifted and the original sign bit remains in the MSB position. When shifting to the left, the LSB is shifted into the C flag of the CCR.

Array: A variable that holds multiple elements that has the same data type.

ASCII Code: American Standard Code For Information Interchange. ASCII is a 7-bit code used in digital systems to represent all letters, symbols, and numbers to be input or output to the outside world.

Assemble: The processes of creating object code done by an assembler by translating assembly instruction mnemonics into opcodes, and by resolving symbolic names for memory locations and other entities.

Assembler: A software package that is used to convert assembly language into machine language.

Assembler Directive: An assembler directive is a message to the assembler that gives it the information it requires to execute the assembly process.

Assembly Language: A low-level programming language unique to each microcontroller. It is converted, or assembled, into machine code before it can be executed.

Baud Rate: The measurement of the number of times per second a signal in a communications channel changes.

BCD: Binary-coded decimal. A 4-bit code used to represent the 10 decimal digits 0 to 9.

Bias: The voltage necessary to cause a semiconductor device to conduct or cut off current flow. A device can be forward biased or reverse biased, depending on what action is desired.

Bidirectional: Data flow in both directions.

Binary: The base 2 numbering system. Binary numbers are made up of 1's and 0's, each position being equal to a different power of 2.

Bipolar: A class of integrate logic circuits implemented with bipolar transistors; also known as TTL.

Bit: A single binary digit. The binary number 1101 is a 4-bit number.

Boolean Algebra: The mathematics of logic circuits.

Boolean Equation: An algebraic expression that illustrates the functional operation of a logic gate or combination of logic gates.

Branch: A jump that can be conditional or unconditional.

Breakpoint: A place in a program where the processor will suspend execution and branch to a special routine that allows the operation to examine the processor registers.

Buffalo: BUFFALO stands for Bit User Fast Friendly Aid to Logical Operations. It is a monitor program that provides a controlled environment for the HC11 chip.

Bus: A bus consists of wires that are used to transfer data either in serial or parallel transmission.

Byte: A group of 8 bits.

Carry Flag (C): A bit in the CCR that indicates that the result of an operation produced an answer greater than the number of available bits.

Central Processing Unit (CPU): CPU is used to refer to the part of a computing system that carries out the instructions of a microcontroller program, and is the primary element carrying out the microcontroller's functions.

Channel: With reference to ADC, it is a path for analog data to be converted to digital codes.

Checksum: A checksum is a fixed-size datum computed from an arbitrary block of digital data for the purpose of detecting accidental errors that may have been introduced during its transmission or storage.

Chip: The term given to an IC.

Clear: Equal to zero.

CMOS: Complementary metal-oxide semiconductor; a class of integrated logic circuits that is implemented with a type of field-effect transistor.

Code Block: A block is a section of code that is grouped together.

Comment: A comment is a programming language construct used to embed programmer-readable annotations in the source code of a computer program.

Compiler: A software package that converts a higher-level language program into machine language code.

Complement: The inverse or opposite of a number. Low is the complement of high, and 0 is the complement of 1.

Condition Code Register (CCR): Referred to as "C," "CCR," or "Status" register; it is an 8-bit status and control register. Out of the eight bits in CCR, three bits are S, X, and I. These are called *control bits*.

Conditional Branch: A jump that occurs only if certain conditions are satisfied.

Control Bits: Out of the eight bits in CCR, five bits are H, N, Z, V, and C. These are called *status flags*. The remaining three bits are S, X, and I. These are called *control bits*.

Control Bus: A control bus is a grouping of all the timing and control signals needed to synchronize the operation of the processor with the other units of the microcontroller.

Control Statement: The ability to change the computer's control from automatically reading the next line of code to reading a different one.

Control Unit: A control unit is primarily responsible for reading the instruction from memory and ensuring the execution of the instructions.

Controller Inverter: A logic circuit that produces the 1's complement of the input word.

Cutoff: A term used in transistor switching that signifies that the collector-to-emitter junction is turned off, or is not allowing current flow.

DAC: A digital-to-analog converter is a device that converts a digital code to an analog signal.

Data Bus: A data bus is a bidirectional, 8-bit bus over which 8-bit words can be sent from the processor to memory (that is, WRITE operation) or from the memory to the processor (a READ operation).

Data Direction Control Bit: A bit that determines the direction of the corresponding bit in the data register.

Data Direction Register: A register whose bits determine the direction of the corresponding data register.

Data Movement: Moving data from one location to another. The source and destination locations are determined by the addressing modes, and can be registers or memory.

Data Type: A classification identifying one of various types of data, such as floating-point, integer, or Boolean.

Debugger: A debugger provides an iterative method of executing and debugging the user's software, one or a few instructions at a time, allowing the user to see the effects of small pieces of the program, and thereby isolate programming errors.

Debugging: Procedures for checking a program for logic and syntax errors.

Decimal: The base 10 numbering system. The 10 decimal digits are 0 to 9. Each decimal position is a different power of 10.

Declaration: A declaration specifies the identifier, type, and other aspects of language elements such as variables and functions.

Digital: A system that deals with discrete digits or quantities. Digital electronics deals exclusively with 1's and 0's, or ONs and OFFs. Digital codes (such as ASCII) are then used to convert the 1's and 0's to a meaningful number, letter, or symbol for some output display.

Digital-To-Analog: See DAC.

Direct Addressing (DIR): Two bytes and typically takes only three clock cycles to execute.

Disabled: To disallow or deactivate a function or circuit.

Dos: Disk operating system.

Dual-In-Line Packages (DIP): The most common pin layout for integrated circuits. The pins are aligned in two straight lines, one on each side of the IC.

EEPROM: A type of nonvolatile programmable link based on electrically erasable programmable read-only memory cells and can be turned on or off repeatedly by programming.

Embedded Systems: A computer system designed to perform one or a few dedicated functions, often with real-time computing constraints.

Enable: To activate or put into an operational mode an input on a logic circuit that enables its operation.

Enabled: To allow or activate a function or circuit.

EPROM: A type of nonvolatile programmable link based on electrically programmable read-only memory cells and can be turned either on or off once with programming.

Escape Sequence: A series of characters used to change the state of computers and their attached peripheral devices.

EVBU: The EVBU is the Motorola M68HC11 Universal Evaluation Board, a development tool for HC11 microcontroller-based designs.

Even Parity: An even number of 1's in a binary word.

Event: Change in behavior such as signal transition.

Exclusive-Nor Gate: A logic gate that produces a logical 0 when the two inputs are at opposite levels.

Exclusive-Or (XOR) Gate: A logic gate that produces a logical 1 when the two inputs are at opposite levels.

Executable Code: Software (in machine language) in a form that can be run in the computer.

Execute: To run a program that causes the computer to carry out its instructions.

Extended Addressing (EXT): Two-byte operand address following the opcode.

External World: See peripherals.

Flowchart: A diagram used by the programmer to map out the looping and conditional branching that a program must make. It becomes the blueprint for the program.

Full-Input Handshaking Mode: Handshaking when the HNDS bit of PIOC is equal to 1 and the OIN bit is equal to 0.

Full-Output Handshaking Mode: Handshaking when the HNDS bit of PIOC is equal to 1 and the OIN bit is equal to 1.

Function: See subroutines.

Gate: A logic circuit with one or more input signals but only one output signal.

Global Control Bit: The entire class of interrupts is controlled with a single bit.

Graphical User Interface (GUI): Often pronounced "gooey," it is a type of user interface that allows users to interact with programs in more ways than typing such as computers.

Half-Carry Flag (H): A bit in the CCR that indicates whether or not a carry is produced from bit position 3 to bit position 4 during the addition of two 8-bit binary numbers.

Hand Assembly: The act of converting assembly language instructions into machine language codes by hand, using a reference chart.

Hands-On: Involving practical experience of equipment, etc.

Handshaking: A method of data transfer that uses control signals between the computer and the peripheral device.

Hard Real-Time System: Tasks must be completed within specified periods.

Hardware: The integrated circuits and electronic devices that make up a computer system.

Hello World Program: A "Hello World" program is a computer program that prints out "Hello World!" on a display device.

Hexadecimal: The base 16 numbering system. The 16 hexadecimal digits are 0 to 9 and A to F. Each hexadecimal position represents a different power of 16.

Higher-Level Language: A type of computer language closest to human language that is a level above assembly language.

Immediate Addressing (IMM): The byte following the opcode is the operand.

Implementation: Implementation is the realization of an application, or execution of a plan, idea, model, design, specification, standard, algorithm, or policy.

Indexed Addressing (INDX, INDY): The operand address is obtained by adding the offset byte that follows the opcode to the contents of an index register.

Inherent Addressing (INH): Single-byte instructions that do not require an operand address.

Input/Output (I/O): The communication between a processor and the outside world.

Input Compare: Mechanism to capture the time at which an input event takes place.

Input-Compare Registers: Four 16-bit built-in input capture registers designed to capture 16-bit values from the TCNT register when an input-capture event occurs. They are TIC1, TIC2, TIC3, and TI4/O5.

Input Event: External source sending a signal to be sensed by the HC11.

Instruction: Commands that control the actions of the processor.

Integrated Circuit (IC): The fabrication of several semiconductor and electronic devices such as transistors, diodes, and resistors onto a single piece of silicon crystal.

Integrated Development Environment (IDE): An IDE also known as an integrated design environment or integrated debugging environment is a software application that provides comprehensive facilities to computer programmers for software development.

Interrupt: An asynchronous signal indicating the need for attention or a synchronous event in software indicating the need for a change in execution.

Interrupt Mask Flag (I): A bit in the CCR that allows interrupts to be enabled or disabled by respectively setting or clearing this flag.

Interrupt Request (IRQ): The external stimulus is called an interrupt request.

Interrupt Service Routine (ISR): A function called when a particular interrupt occurs. The instructions executed during an interrupt constitute an interrupt service routine.

Interrupt Vector: The address of the first instruction of the ISR.

Inverter: Logic circuit that inverts or complements its inputs.

Jump: The process of changing the execution sequence of the processor.

Jump Table: An efficient method of transferring program control (branching) to another part of a program using a table of branch instructions.

Keypad: A keypad is a set of buttons arranged in a block or "pad" that usually bear digits and other symbols and usually a complete set of alphabetical letters.

Label: A name given to an instruction in an assembly-language program. To jump to this instruction, you can use the label rather than the address. The assembler will work out the correct address of the label and will use it in the machine-language program.

Least Significant Bit (LSB): The bit having the least significance in a binary string. The LSB will be in the position of the lowest power of 2 within the binary number.

Least Significant Digit (LSD): The digit having the least significance in a digital string.

LIFO: Last in first out. In a LIFO structured linear list, elements can be added or taken off from only one end. Therefore, the last element to be placed on the open end is also the first to be taken off the open end.

Light-Emitting Diode (LED): A semiconductor light source used as indicator lamp in many devices.

Linker: A utility program that links a compiled or assembled program to a particular environment.

Liquid Crystal Display (LCD): An LCD is a thin, flat electronic visual display that uses the light modulating properties of liquid crystals (LCs).

List File: A file that typically shows the relationship between source and output.

Local Control Bit: Provides individual control of each interrupt within the class.

Logic: Study of arguments.

Logic Circuit: A circuit whose input and output signals are 2-state, either low or high voltages. The basic logic circuits are OR, AND, and NOT gates.

Logical Shifting: Causes "0" to be shifted into one end of the data word, pushing all of the data bits over one position in the data word.

Loops: Repeat a statement a certain number of times or while a condition is fulfilled.

Lower-Level Language: A lower-level programming language that provides little or no abstraction from a computer's instruction set architecture. The word *lower* refers to the small or nonexistent amount of abstraction between the language and machine language.

Machine Language: Computer instruction written in binary code that is understood by a computer; the lowest level of programming language.

Machine Code: The binary codes that make up a microcontroller's program instructions.

Make: MAKE is a utility that automatically builds executable programs and libraries from source code by reading files called *Make-files*. These Make-files specify how to derive the target program.

Maskable Interrupt: Type of interrupt procedure that allows the interrupt request to be ignored if the interrupt disable flag is set.

Maximum Range: With reference to ADC, the maximum voltage range is defined as a difference between the maximum and minimum reference voltage.

Microcontroller: A small computer on a single IC containing a processor, memory, and I/O peripherals.

Mission Critical: Any factor whose failure or disruption is essential to the core function of an organization, and will result in the failure of business operations.

Mnemonics: The abbreviated spellings of instructions used in assembly language.

Monitor Program: The computer software program initiated at power-up that supervises system operating tasks, such as reading the keyboard and driving the computer monitor.

MOSFET: Metal oxide semiconductor field-effect transistor.

Most Significant Bit (MSB): The bit having the most significance in a binary string. The MSB will be in the position of the highest power of 2 within the binary number.

Most Significant Digit (MSD): The digit having the most significance in a digital string.

Nand Gate: A logic gate that produces a low output when all the inputs are high.

Negative Flag (N): A bit in the CCR that indicates that the result of a mathematical operation is negative.

Nested Subroutine: The main program jumps to a subroutine that contains a jump to a subroutine, and so on.

Nibble: A group of four bits.

Non-Real-Time System: Systems not subject to operational deadlines from event to system response.

Nonmaskable Interrupt: Type of interrupt procedure that forces the processor to respond to the interrupt request.

Nonmaskable Interrupt Flag (X): A bit in the CCR that masks the XIRQ request when set. It is set by the hardware and cleared by the software as well as set by unmaskable XIRQ.

Nor Gate: A logic gate that produces a low output when one or more of the inputs are high.

Octal: The base 8 numbering system. The eight octal numbers are 0 to 7. Each octal position represents a different power of 8.

Octet: A group of eight bits.

Odd Parity: An odd number of 1's in a binary word.

Offset: A byte that follows the opcode for a conditional branch instruction such as BEQ. The offset is added to the PC to determine the address to which the processor will branch.

Opcode: Operation code. It is the unique multibit code given to identify each instruction to the microcontroller.

Operand: The parameters that follow the assembly language mnemonic to complete the specification of the instruction.

Operand Address: Address in memory where an operand is currently stored or is to be stored.

Operating System: See Monitor program.

Operators: An operator is a symbol that helps the user to command the computer to perform certain mathematical or logical manipulations.

Or Gate: A logic gate that produces a high output when one or more of its inputs are high.

Or Masking: A type of masking where the OR operation is used to set specific bits of a data word while not affecting the others.

Output Compare: Mechanism that causes output events such as pin actions and signal transitions to occur on the OCx output pins at specified times.

Output-Compare Registers: Five 16-bit built-in output compare counters to control output compare functions. They are TOC1, TOC2, TOC3, TOC4, and TI4/O5.

Outside World: See peripherals.

Overflow Flag (V): A bit in the CCR that indicates that the result of an operation has overflowed according to the processor's word representation, similar to the carry flag but for signed operations.

Parity: In relation to binary codes, the condition of evenness or oddness of the number of 1's in a code group.

Parity Generator: A circuit that produces either an odd- or even-parity bit to go along with the data.

Peripheral: A device such as a printer or model that provides communication with a computer.

Pointer: The variable that stores the reference to another variable is called a *pointer.*

Polling: Actively sampling the status of an external device by a client program as a synchronous activity.

Port Number: An 8-bit number used to select a particular I/O port.

Ports: An interface between the processor and the outside world. Ports support a variety of input/output functions.

Prebyte: First byte of a 2-byte opcode.

Processor Register (or General-Purpose Register): A processor register is a small amount of storage available on the CPU whose contents can be accessed more quickly than storage available elsewhere.

Program Counter (PC): A 16-bit internal register that contains the address of the next program instruction to be executed.

Program: Complete sequence of instructions that direct a computer to perform a specific task or solve a problem.

Programmable Logic Controller (PLC): A programmable logic controller is a digital computer used for automation of electromechanical processes, such as control of machinery on factory assembly lines, amusement rides, or lighting fixtures.

Programmer: One who designs or writes a program.

Protocol: A technical specification consisting of a set of rules that govern the information exchanged between devices.

Pull: To read data from the stack.

Push: To save data in the stack.

Range: With reference to ADC, the voltage range is defined as the difference between the high and low reference voltage.

Reactive System: Maintain an ongoing interaction with their environment rather than produce some final value upon termination.

Real-Time Constraint: Operational deadline.

Real-Time System: System subject to operational deadlines from event to system response.

Relative Addressing (REL): Used in conditional branch instructions to determine the branching address by adding the offset to the PC.

Requirements Analysis: Requirements analysis encompasses those tasks that go into determining the needs or conditions to meet for a new or altered product.

Reset: Sets initial conditions within the system and begin executing instructions from a predetermined address.

Resolution: With reference to ADC, resolution is the relationship of each step to the range.

Rotate Operation: Shift operation where the bit that gets shifted out of the high or low bit position gets placed in the bit position vacated on the other side of the byte.

S-Record File: Hex format file that can be programmed into memory.

Saturation: A term used in transistor switching that signifies that the collector-to-emitter junction is turned on, or conducting current heavily.

Sensor: A device that measures a physical quantity and converts it into a signal that can be read by an observer or by an instrument.

Sequential Execution: When the instructions in a program are executed one after the other in the order in which they were written.

Set: Equal to one.

Seven-Segment Display: A seven-segment display is a form of electronic display device for displaying decimal numerals that is an alternative to the more complex dot-matrix displays.

Simple Strobed Mode: Simple handshaking in HC11.

Single-Byte Instruction: The single-byte instruction contains only an 8-bit opcode. There is no operand address specified.

Single Stepping: A debugging procedure whereby the processor executes an instruction and halts until it receives a command to continue to the next instruction. In this way, the register and memory contents can be examined by the operator after each execution.

Soft Real-Time System: Tasks may not be completed within specified periods.

Software: Computer program statements that give step-by-step instructions to a computer to solve a problem.

Software Design: Software design is a process of problem-solving and planning for a software solution.

Software Development Life Cycle Model: Description of phases of the software cycle and the order in which those phases are executed.

Software Testing: Software testing is an investigation conducted to provide stakeholders with information about the quality of the product or service under test.

Source Program: In the context of HC11, a program written in assembly language.

Stack: A portion of RAM reserved for the temporary storage and retrieval of information, typically the contents of the processor's internal registers.

Stack Pointer: A 16-bit internal register that contains the address of the last entry on the RAM stack.

State Diagram: A state diagram is a type of diagram used to describe the behavior of systems.

Statement Label: A meaningful name given to certain assembly language program lines so that they can be referred to from different parts of the program.

Status Flags: Out of the eight bits in CCR, five bits are H, N, Z, V, and C. These are called *status flags*.

Step Boundaries: Upper and lower limits to each step are called *step boundaries*.

Steps: A step is a small portion of the range that has a name or a digital code associated with it.

Stop Disable Flag (S): A bit in the CCR that when set to 1 will prevent the STOP instruction from being executed.

Subroutine: A program stored in higher memory that can be used repeatedly as part of a main program.

Surface-Mounted Device (SMD): The newest style of integrated circuits, soldered directly to the surface of a printed circuit board. They are much smaller and lighter than the equivalent logic constructed in the DIP through-hole style logic.

TCNT: A 16-bit built-in register that is driven by the main clock and is read-only for the user.

TCTL1 (Timer Control Register 1): An 8-bit built-in register used to control the action of each of the output-compare-functions of the HC11.

TCTL2 (Timer Control Register 2): An 8-bit built-in register used to program the edges that the input-capture-functions of HC11 will react to.

TFLG1 (Timer Flag Register 1): An 8-bit built-in register that contains five output compare flags (OC1F to OC5F) and three input capture flags (IC1 to IC3F).

Three-Byte Instruction: In a 3-byte instruction, the first byte is the opcode, second, and third bytes for a 16-bit operand address.

Timing Diagram: A timing diagram is a graphical method of showing the exact output behavior of a logic circuit for every possible set of input conditions.

TMSK1 (Timer Interrupt Mask Register 1): An 8-bit built-in register that contains among other control bits, the three input capture interrupt enable control bits (IC1I to IC3I) primarily used by the input capture function of the HC11.

TMSK2 (Timer Interrupt Mask Register 2): An 8-bit built-in register located at $1024. It contains the bit TOI, RTII, PAOVI, PAII, PR1, and PR0.

Transducer: A transducer is a device that converts one type of energy to another.

Transistor-Transistor Logic (TTL): The most common integrated circuit used in the digital world today.

Transition: The change in state of a signal from high to low or low to high.

Truth Table: Truth table is a means for describing how a logic circuit's output depends on the logic levels present at the circuit's input.

Two-Byte Instruction: In a two-byte instruction, the first byte of the two-byte instruction is an opcode, and the second byte is an 8-bit address code that specifies the operand address. These two bytes are always stored in memory in this order.

Unconditional Branch: Instructions that do not test any condition. They either always cause a branch (e.g., BRA) or never cause a branch (e.g., BRN).

Unidirectional: Data flow in only one direction.

User Interface (UI): A user interface is the space where interaction between humans and machines occurs.

Utility Subroutine: A predefined subroutine available in BUFFALO monitor.

Validation: Validation refers to an activity to ensure that the software that is being created is as per the requirements agreed upon at the analysis phase and to ensure the product's quality.

Variable: Variables in C are memory locations that are given names and can be assigned values.

Vector Address: A specific address in memory where the interrupt vector is stored.

Verification: Verification refers to an activity to ensure that specific functions are correctly implemented.

Very Large-Scale Integration (VLSI): The process of creating an IC by combining thousands of transistors into a single chip.

Voltage Range: See Range.

Word: A string of bits that represents a coded instruction or data.

Zero Filling: The process of filling all bit positions with zeros that are not occupied by data.

Zero Flag (Z): A bit in the CCR that indicates that the result of a mathematical or logical operation was zero.

Index